Plant Development
and
Biotechnology

Plant Development and Biotechnology

EDITED BY

Robert N. Trigiano Ph.D.
Dennis J. Gray Ph.D.

CRC PRESS

Boca Raton   London   New York   Washington, D.C.

Back cover: Digital micrograph of transgenic grape somatic embryos showing expression of GFP (appears on p. 276).

Library of Congress Cataloging-in-Publication Data

Plant development and biotechnology / edited by Robert N. Trigiano and Dennis J. Gray.
    p. cm.
Includes bibliographical references and index.
ISBN 0-8493-1614-6 (alk. paper)
    1. Plant tissue culture. 2. Plants—Development. 3. Plant biotechnology. I. Trigiano, R. N. (Robert Nicholas), 1953- II. Gray, Dennis J. (Dennis John), 1953-

QK725.P562 2004
571.5'382—dc22

    2004045725

Visit the CRC Press Web site at www.crcpress.com

Acknowledgments

We would like to thank the contributing authors for their outstanding efforts and patience throughout this lengthy process; the University of Tennessee's Institute of Agriculture and the University of Florida's Institute for Food and Agricultural Sciences for providing financial support for this project; Kay Trigiano for her superb editorial assistance, proofreading, and encouragement; John Sulzycki, senior editor at CRC Press, for his understanding and compassion; and Pat Roberson and Suzanne Lassandro at CRC Press, whose tireless efforts contributed greatly to this project. Finally, as with our other CRC projects, we are most indebted to Andrew N. Trigiano and Bob D. Gray, who provided constant inspiration and a welcome source of diversions during the completion of this book.

Editors

Robert N. Trigiano, Ph.D., is professor of ornamental plant biotechnology and plant pathology in the Department of Entomology and Plant Pathology at the University of Tennessee Agricultural Experiment Station at Knoxville.

Dr. Trigiano received his B.S. degree with an emphasis in biology and chemistry from Juniata College, Huntingdon, Pennsylvania in 1975, and an M.S. in biology (mycology) from the Pennsylvania State University in 1977. He was an associate research agronomist, specializing in mushroom culture and plant pathology, for Green Giant Co. in Le Sueur, Minnesota until 1979, and then a mushroom grower for Rol-Land Farms, Ltd., Blenheim, Ontario, Canada in 1979 and 1980. Dr. Trigiano completed a Ph.D. in botany and plant pathology at North Carolina State University at Raleigh in 1983. After concluding postdoctoral work in the Plant and Soil Science Department at the University of Tennessee, he became an assistant professor in the Department of Ornamental Horticulture and Landscape Design at the same university in 1987, and was promoted to associate professor in 1991 and to professor in 1997. He joined the Department of Entomology and Plant Pathology in 2002.

Dr. Trigiano is a member of the American Society for Horticultural Science (ASHS) and the Mycological Society of America, and the honorary societies of Gamma Sigma Delta, Sigma Xi, and Phi Kappa Phi. He has been an associate editor for the *Journal of the American Society for Horticultural Science* and for *Plant Cell, Tissue and Organ Culture*, and an editor for *Plant Cell Reports*. Dr. Trigiano is coeditor of *Critical Reviews in Plant Sciences* and the popular textbooks *Plant Tissue Culture Concepts and Laboratory Exercises* (2nd edition) and *Plant Pathology Concepts and Laboratory Exercises*, all published by CRC Press. He received the T. J. Whatley Distinguished Young Scientist Award (The University of Tennessee, Institute of Agriculture, 1991) and the Gamma Sigma Delta Research Award of Merit (The University of Tennessee, 1991). In 1998, he received the ASHS Publication Award for the most outstanding educational paper and the Southern region ASHS L. M. Ware Distinguished Research Award.

Dr. Trigiano has been the recipient of several research grants from the United States Department of Agriculture (USDA), the Horticultural Research Institute, and private industries and foundations. He has published more than 120 research papers, book chapters, and popular press articles. He teaches undergraduate and graduate courses in plant tissue culture, DNA analysis, protein gel electrophoresis, and plant microtechnique. Current research interests include somatic embryogenesis and micropropagation of ornamental species, fungal physiology, population analysis, DNA profiling of fungi and plants, and gene discovery.

Dennis J. Gray, Ph.D., is a member of the University of Florida/IFAS Horticulture Department. He directs the plant biotechnology program at the Mid-Florida Research and Education Center.

Dr. Gray graduated with a B.A. degree in biology from California State College, Stanislaus in 1976 and received an M.S. degree in mycology, with a minor in botany, from Auburn University in 1979. He earned a Ph.D. degree in botany, with a minor in plant pathology, from North Carolina State University in 1982. After a postdoctoral fellowship at the University of Tennessee, he joined the faculty of the University of Florida in 1984, reaching the rank of professor in 1993.

Dr. Gray has been a member of the American Association for the Advancement of Science, the American Institute of Biological Sciences, the American Society for Horticultural Science, the Botanical Society of America, the Council for Agricultural Science and Technology, the Society for *In Vitro* Biology, the International Association for Plant Tissue Culture and Biotechnology, the

International Horticultural Society, and Sigma Xi. He was associate editor, then managing editor, of the internationally recognized, refereed journal *Plant Cell, Tissue and Organ Culture* from 1988 through 1994. He currently is coeditor-in-chief of *Critical Reviews in Plant Sciences.*

Dr. Gray has received research grants and support from the Binational Research and Development Fund, the Florida Department of Agriculture and Consumer Affairs, the Florida High Technology and Industry Council, the United States Department of Agriculture, and private industry. He has been an author or coauthor of more than 200 publications and holds several patents. He was awarded the rank of University of Florida Research Foundation Professorship from 1998 to 2001. He led a research team that received the USDA's Secretary's Award for research in grapevine biotechnology in 2002. His current interests include the developmental biology of regenerative plant cells and the integration of contemporary and newly emerging technologies for crop improvement.

Contributors

David W. Altman
BioCon Associates
Franklin, Tennessee

Robert M. Beaty
Department of Botany
The University of Tennessee
Knoxville, Tennessee

Caula A. Beyl
Department of Plant and Soil Science
Alabama A&M University
Normal, Alabama

James D. Caponetti
Department of Botany
The University of Tennessee
Knoxville, Tennessee

Alan C. Cassells
Department of Zoology, Applied Ecology, and
  Plant Science
National University of Ireland
Cork, Ireland

Erika Charbit
Division of Biology
Kansas State University
Manhattan, Kansas

Zong-Ming Cheng
Department of Plant Sciences
The University of Tennessee
Knoxville, Tennessee

Michael E. Compton
School of Agriculture
University of Wisconsin – Platteville
Platteville, Wisconsin

Sacco C. de Vries
Department of Agrotechnology and Food
  Sciences
Wageningen University
Wageningen, the Netherlands

Barbara M. Doyle
Department of Zoology, Applied Ecology, and
  Plant Science
National University of Ireland
Cork, Ireland

Victor P. Gaba
Department of Virology
Agricultural Research Organization,
  The Volcani Center
Bet Dagan, Israel

Effin T. Graham
Department of Plant Sciences
The University of Tennessee
Knoxville, Tennessee

Dennis J. Gray
Mid-Florida Research and Education Center
University of Florida/IFAS
Apopka, Florida

Matthew D. Halfhill
Department of Plant Sciences
The University of Tennessee
Knoxville, Tennessee

Ernest Hiebert
Department of Plant Pathology
University of Florida
Gainesville, Florida

Laura C. Hudson
Department of Plant Sciences
The University of Tennessee
Knoxville, Tennessee

Subramanian Jayasankar
Department of Plant Agriculture – Vineland
  Campus
University of Guelph
Vineland Station, Ontario, Canada

Michael E. Kane
Department of Environmental Horticulture
University of Florida
Gainesville, Florida

Peggy G. Lemaux
Department of Plant and Microbial Biology
University of California – Berkeley
Berkeley, California

Zhijian T. Li
Mid-Florida Research and Education Center
University of Florida/IFAS
Apopka, Florida

Mary Ann Lila
Department of Natural Resources and
  Environmental Sciences
University of Illinois
Urbana, Illinois

Chia-Min Lin
Department of Plant Pathology
University of Florida
Gainesville, Florida

Kathleen R. Malueg
Department of Plant Sciences
The University of Tennessee
Knoxville, Tennessee

Andreas P. Mordhorst
Heythuysen
The Netherlands

Kimberly A. Pickens
Department of Plant Sciences
The University of Tennessee
Knoxville, Tennessee

Sandra M. Reed
Floral and Nursery Plant Research Unit
U.S. National Arboretum
Agricultural Research Service
U.S. Department of Agriculture
McMinnville, Tennessee

Harry A. Richards
Food Safety Center
The University of Tennessee
Knoxville, Tennessee

James A. Saunders
U.S. Department of Agriculture – Agricultural
  Research Service
Beltsville, Maryland

Otto J. Schwarz
Department of Botany
The University of Tennessee
Knoxville, Tennessee

Anjuna R. Sharma
Department of Botany
The University of Tennessee
Knoxville, Tennessee

C. Neal Stewart, Jr.
Department of Plant Sciences
The University of Tennessee
Knoxville, Tennessee

Gayle R. L. Suttle
Microplant Nurseries, Inc.
Gervais, Oregon

Leigh E. Towill
U.S. Department of Agriculture – Agricultural
  Research Service
National Center for Genetic Resources
  Preservation
Fort Collins, Colorado

Robert N. Trigiano
Department of Entomology and Plant
  Pathology
Institute of Agriculture
The University of Tennessee
Knoxville, Tennessee

Richard E. Veilleux
Department of Horticulture
Virginia Polytechnic Institute and State
  University
Blacksburg, Virginia

Albrecht G. von Arnim
Department of Botany
The University of Tennessee
Knoxville, Tennessee

Shibo Zhang
Department of Plant and Microbial Biology
University of California – Berkeley
Berkeley, California

Table of Contents

Section I

Introduction

1 Introduction*

Dennis J. Gray and Robert N. Trigiano

All life on earth depends on the continuous acquisition of energy. The sun provides nearly all earth's renewable energy in the form of radiation. Radiant energy heats the earth, causing convection and contributing in great measure to weather patterns, which transfer water through the atmosphere from oceans and lakes to land and back; this creates the basic environment needed to nurture life. However, life itself depends on the availability of chemical energy, the vast majority of which also is captured from sunlight. Plants with chlorophyll perform the function of energy capture, utilizing photons to drive the cleavage of water into oxygen and hydrogen. In the reaction termed photosynthesis, carbon dioxide (some of which is created by the respiration of plants, animals, fungi, and bacteria) is ultimately recycled by its conversion into oxygen, water, and sugar. The chemical energy of sugar is then used to drive all of the other metabolic reactions needed to perpetuate life. The carbon in sugar is utilized to produce all of the more complex carbohydrates, proteins, and other structural molecules that make up plant cells, tissues, and organs. In turn, the chemical energy and nutritive substances contained in plants are used as the primary energy source for animals.

The evolutionary forces that drive the diversity of plants and animals are well known. Humankind altered evolutionary natural selection by recognizing some sources of food to be preferable to others, and hence selecting those deemed to be superior in one way or another. Selection of improved "varieties" probably began in an unintentional manner, but became more organized over the millennia. Eventually, the possibility of selective breeding of plants (and animals) was recognized and became the cornerstone of modern agricultural development. In turn, modern agriculture became the foundation of modern civilization, allowing humankind to switch from a lifestyle of bare subsistence to one that promoted pursuit of activities never before imagined. These unimaginable activities included the conception, development, and pursuit of plant science.

The development of plants into a diversity of useful items and products is made possible only by human endeavor. The development of plants using previously unimagined ideas continues today. Breeding programs are being revolutionized with biotechnology, leading to plants that have properties we never anticipated. Along with such rapid change comes controversy: new discussions, fear of the unknown, optimism, and opportunities for commerce. This book is based upon our previous textbook, *Plant Tissue Culture Concepts and Laboratory Exercises* (2nd edition), but is designed to address the thoughts set forth above. We have removed most of the practical laboratories and substituted new chapters that cover both basic and highly evolved aspects of modern plant science. While in no way intended to be a comprehensive treatment, this book does address many major aspects of plant development and biotechnology.

The purpose of this chapter is to provide an initial focus for *Plant Development and Biotechnology* and to begin the process of defining terms and ideas unique to this subject area. Keeping in mind that the mission of the entire book is to introduce, define, and provide training, we will use this introductory chapter primarily to orient the reader to the book's structure and to highlight information that is discussed in depth in subsequent chapters. The book is intentionally written to be rather informal — it generally provides the reader with a minimum number of references, but does not sacrifice essential information or accuracy. Chapters on broad topics are written by

---

* Florida Agricultural Journal Series No. R-10163

0-8493-1614-6/05/$0.00+$1.50

specialists with considerable experience in the field. The chapters necessarily assume that the student has some basic understanding of botany and botanical terminology. As such, it is recommended that companion plant anatomy, genetics, morphology, and physiology textbooks be available as needed in order to provide access to basic botanical knowledge.

In addition to this introductory section, the textbook is divided into the following five primary sections: "History of Plant Tissue Culture"; "Supporting Methodologies"; "Propagation and Development Concepts"; "Crop Improvement Techniques"; and "Special Topics." Each section combines related facets of plant development and biotechnology.

Section II, "History of Plant Tissue Culture," is an abbreviated account of how the field developed from an early theoretical base to the highly technical discipline that exists today. This treatment is somewhat unique because it documents research progress as modulated by significant world events. Beyond recording the people, places, and dates for pertinent discoveries, we feel it is interesting for students to see the challenges encountered by researchers that result in the often uneven pace of research. We end Chapter 2 on a contemporary "hot" topic by discussing progress being made with genetically engineered plants in light of society's reaction, much of which is information gleaned from news reports. In addition to a past historical account, we hope that Chapter 2 will provide a snapshot of the controversial present.

Section III, "Supporting Methodologies," begins the process of teaching by discussing various methods for preparing media, providing complete and logical examples for how to make solutions and dilutions, and demonstrating how to accomplish sterile culture work. An experienced teacher with laboratory resources and methodologies already in place may choose to pay less attention to Chapter 3, whereas the instructor of a newly established course will find it indispensable. This is one of the few chapters that contain the protocols and procedure boxes that were presented in our previous textbooks. From the many comments that we have had on this chapter over the past several years, we believe that many students and researchers will applaud its inclusion as a very valuable and an excellent reference. Section III continues with chapters designed to emphasize common methods for visualizing and documenting studies (histology and microscopy/photography, respectively) of plant development and other plant processes, and for quantifying responses (using statistical analysis) of tissue culture in research. Again, we have retained Chapter 4 because of the overwhelming response of readers of the previous edition. Chapters 7 and 8 introduce students to the concepts of plant anatomy and plant growth regulators, respectively. Section III ends with two very informative chapters: a chapter on software and databases for analyzing molecular and protein data (Chapter 9) and a chapter concerned with molecular approaches to studying plant development (Chapter 10). These chapters were included in the present edition as a response to users' needs and requests.

Section IV, "Propagation Techniques," encompasses the essential foundation of plant tissue culture. In this section the three types of commonly used culture regeneration systems are introduced and discussed from traditional viewpoints and in relation to molecular developmental aspects.

We begin Section IV by discussing propagation from preexisting meristems (Chapter 11, "Shoot Culture Procedures"), a process that is more commonly termed "micropropagation." Micropropagation is the simplest and most commercially useful tissue culture method. The tissue culture industry uses micropropagation almost exclusively for ornamental plant production.

The second propagation system discussed in Section IV is "Propagation from Nonmeristematic Tissues: Organogenesis" (Chapter 12). Organogenesis is the de novo development of organs, typically shoots and/or roots, from cells and tissues that would not normally form them. The term "adventitious" has also been used to describe the plant parts formed by the process of organogenesis. Shoot organogenesis is another means of propagating plants. While not used much in commercial production, organogenesis is used extensively in genetic engineering as a means to produce plants from genetically altered cells. In Chapter 12, both the theory and the developmental sequences of how cells are induced to follow such a developmental pathway are discussed. Chapter 13 discusses some of the more recently discovered molecular aspects of meristem formation, development, and control. These are presented in depth and illustrate the significance of our burgeoning molecular biological knowledge to practical usage.

The third propagation system discussed in Section IV is "Propagation from Nonmeristematic Tissues: Nonzygotic Embryogenesis" (Chapter 14). Nonzygotic embryogenesis is a broadly defined term meant to cover all instances where embryogenesis occurs outside of the normal developmental pathway found in the seed. One type of nonzygotic embryogenesis is termed somatic embryogenesis. This is a unique phenomenon exhibited by vascular plants, in which somatic (nonsexual) cells are induced to behave like zygotes. Such induced cells begin a complex, genetically programmed series of divisions and eventual differentiation to form an embryo that is more or less identical to a zygotic embryo. This type of propagation system is important because the embryos develop from single cells, which can be genetically engineered and are complete individuals that are capable of germinating directly into plants. Thus, somatic embryogenesis also represents a potentially efficient propagation system. Chapter 14 discusses the developmental processes and significance of nonzygotic embryogenesis. In addition, the process is very efficient, such that there has been interest in developing nonzygotic embryogenesis into a propagation system. Chapter 15 summarizes some of the newer molecular studies concerned with initiation, development, and control of the nonzygotic embryogenic process.

In Section V, "Crop Improvement Techniques," the aforementioned propagation techniques are integrated with other methodologies in order to modify and manipulate germplasm. Chapter 16 discusses the use of plant protoplasts. Protoplasts are plant cells from which the cell wall has been enzymatically removed, making the cells amenable to cell fusion and other methods of germplasm manipulation. Chapter 17 details the use of haploid cultures in plant improvement. Haploid cultures usually are derived from microspore mother cells and result in cells, tissues, and plants with half the normal somatic cell chromosome number. Such plants are of great use in genetic studies and breeding, because all of the recessive genes are expressed; by doubling the haploid plants back to the diploid ploidy level, dihaploid plants are produced, which are completely homozygous. True homozygous plants are time consuming and often impossible to produce by conventional breeding. Embryo rescue was conspicuously lacking in our previous editions. Chapter 18 details the technique of embryo rescue and discusses its usage as a commonly applied methodology.

Chapter 19 discusses genetic engineering technologies, a current hot topic in agriculture. Transformation, wherein genes from unrelated organisms can be integrated into plants without sexual reproduction, resulting in "transgenic plants," is the most significant application for plant tissue culture when considering its impact on humankind. Chapter 20 includes a discussion on genetically modified organisms and the controversy that surrounds them, especially as it impacts food supplies, commerce, and natural ecosystems. Many favorable comments convinced us to retain the next chapter in this edition. It is very practical and is the only chapter to contain a bona fide laboratory exercise. Chapter 21 describes the construction of a device for particle bombardment of plants, an alternate method to Agrobacterim mediated for transformation. In this method, DNA is coated onto small tungsten or gold particles and literally shot into the plant cells. The last of the chapters to deal with transformation is a guide to building a very simple illumination system for visualizing green fluorescent protein (GFP) (Chapter 22). The gene that codes for GFP was isolated from a jellyfish. GFP is used as a very convenient tool to study genetic transformation in both plants and animals because expression of the highly fluorescent protein is very easy to visualize and allows cells, tissues, and entire organisms to be nondestructively evaluated for transformation.

Chapter 23 describes the use of cryopreservation of germplasm. Cryopreservation is an efficient means of safeguarding valuable plant germplasm by freezing all metabolic activity. Cryopreserved cells and tissues can be kept for extended periods of time without mutations or any physiological decline. Chapter 24 describes the production of secondary products by plant cells in a culture. This subject is somewhat unique in the context of previous topics because the end product is not a regenerated plant, but rather a chemical. The use of plants as biofactories to produce complex pharmaceuticals is of great interest, particularly due to its potential application in health care.

SectionVI, "Special Topics," is a bit of a catch-all section where we have placed topics that we considered important enough to warrant inclusion in the book, but which did not quite fit in the other sections. Section VI begins with a chapter concerned with *in vitro* plant pathology (Chapter 25). This topic is introduced to demonstrate the use of *in vitro* systems to mimic whole plants in the development of disease symptoms — overall, a convenient means of studying host–pathogen interactions. Chapter 26 discusses reasons that genetic and phenotypic variations occur as a result of culture and presents up-to-date research findings. Chapter 27 is a look into the mechanics of commercial plant production, which is the primary method for producing many house and fruit plants and is an important industry worldwide. Chapter 28 describes the importance of clean cultures and the problems associated with maintaining *in vitro* cultures. This chapter includes both traditional methods and newer, more sophisticated molecular and serological techniques for indexing plants for pathogens. The last chapter (Chapter 29) is a look at entrepreneurship. This chapter outlines the dos (and some of the don'ts) when considering launching a new enterprise.

Based upon the many telephone calls, letters and e-mails that we have received, the first and second editions of *Plant Tissue Culture Concepts and Laboratory Exercises* successfully facilitated the training of students in the current principles and methodologies of our rapidly evolving field. We asked for constructive criticisms from users of the book and employed them in order to make a number of improvements in *Plant Development and Biotechnology*. Although users primarily interested in tissue cultures found the previous editions useful, a significant number of readers and instructors wanted a book composed primarily of lecture topics, one that included some of the molecular aspects of plant development. We believe that we and our contributing authors have addressed these needs. Hopefully, *Plant Development and Biotechnology* will enjoy at least the same level of success as our previous book! As always, we welcome comments from colleagues and students as they put the textbook to use.

Section II

History of Plant Tissue Culture

2 History of Plant Tissue and Cell Culture

James D. Caponetti, Dennis J. Gray, and Robert N. Trigiano

CHAPTER 2 CONCEPTS

- The first successful plant tissue and cell culture was developed by Gottlieb Haberlandt near the turn of the 20th century.

- Plant tissue culture techniques progressed rapidly during the 1930s due to the discovery that B vitamins and natural auxin were necessary for the growth of isolated tissues.

- By the 1970s, the role of the five major plant growth regulator classes in tissue culture had been recognized and investigated. The classes of growth regulators are auxins, cytokinins, gibberellins, ethylene, and abscisic acid.

- By the mid-1950s, plant tissue culture methods had progressed to the point of making a major impact on research in plant developmental biology. For example, somatic cells of a carrot would differentiate into embryos when cultured under the proper conditions. The ability to regenerate plants from single somatic cells through such a "normal" developmental pathway was envisioned to have great applications in propagation and genetic engineering.

- The recovery of plants from haploid cells began to appear in the 1960s. This discovery received significant attention from plant breeders, since plants recovered from doubled haploid cells are homozygous and express all recessive genes, making them ideal pure breeding lines.

- The use of plant tissue culture technology to enable the development of genetically modified (GM) crops illustrates its key, but often overlooked, importance in our lives. Plant tissue culture is also an increasingly important tool for advanced studies of plant development, especially with the integration of molecular biological techniques.

The field of plant tissue culture is based on the premise that plants can be separated into their component parts (organs, tissues, or cells), which can be manipulated *in vitro* and then grown back into complete plants. This idea of handling higher plants with the ease and convenience of microorganisms has conjured up many wonderful possibilities for study and use; these possibilities, in turn, have been the longstanding stimuli driving research and development in the field. Plant tissue culture, along with molecular genetics, is a core technology for genetic engineering. After many years of overly optimistic promise, genetically engineered plants reached the marketplace in the 1990s. At the end of this chapter, we discuss how the greatly anticipated societal benefits of this technology have been complicated by a good deal of concern regarding its safety.

The first successful plant tissue and cell culture was created by Gottlieb Haberlandt near the turn of the 20th century, when he reported the culture of leaf mesophyll tissue and hair cells (see Steward, 1968; Krikorian and Berquam, 1969). This was a remarkable accomplishment, considering that little was known about plant physiology at the time. In retrospect, however, Haberlandt must have drawn on a body of previous knowledge in plant biology. We must assume that he was familiar with the writings of early philosophers such as Aristotle, Theophrastus, Pliny the Elder, Dioscorides, Avicenna, Magnus, Angelicus, and Goethe, and that his studies surely must have included the anatomical observations of Hooke, Malpighi, Grew, Nageli, and Hanstein. His own research must have led him to the investigations by early plant physiologists such as van Helmont, Mariotte, Hales, Priestley, Ingenhousz, and Senebier, and he must have had access to the morphological and physiological investigations of 19th century botanical researchers such as Schleiden, Schimper, Pringsheim, Unger, Hedwig, Hofmeister, Vochting, Sachs, Goebel, Bower, and Farlow. The information available from these sources, coupled with improved light microscopes, must have given Haberlandt the insight necessary to culture plant cells and to predict that they could not only grow, but divide and develop into embryos and then into whole plants, a process referred to as totipotency by Steward (1968).

Unfortunately, the cells that Haberlandt cultured did not divide, and thus his ideas of plant development and totipotency did not come to fruition. The probable reason for the failure was that plant growth regulators (PGRs—Chapter 8) needed for cell division, proliferation, and embryo induction were not present in the culture medium. Indeed, PGRs had not yet been discovered. Apparently, Haberlandt became discouraged and pursued other physiological investigations. However, his ideas of embryo induction in culture did not go unnoticed in the scientific world. For example, Hannig (1904) cultured nearly mature embryos excised from seeds of several species of crucifers. Moreover, Haberlandt's lack of success was not in vain because one of his students, Kotte (1922), reported the growth of isolated root tips on a medium consisting primarily of inorganic salts. At the same time, and quite independently, Robbins (1922) reported a similar success with root and stem tips, and White (1934) reported that, not only could cultured tomato root tips grow, but they could be repeatedly subcultured to a fresh medium of inorganic salts supplemented with yeast extract, which was, unbeknownst to him, a good source of B vitamins.

Innovative plant tissue culture techniques progressed rapidly during the 1930s, due to the discovery that B vitamins and natural auxin were necessary for the growth of isolated tissues containing meristems. The growth-promoting effects of thiamine on isolated tomato root tips was reported by White (1937). A series of ingenious experiments with oat seedlings by Fritz Went in the 1920s, plus the work of other plant physiologists, including Kenneth Thimann, led to the discovery of the first PGR, indoleacetic acid (IAA). IAA is a naturally occurring member of a class of PGRs termed "auxins." The events leading to the discovery of IAA are well documented in a report published in 1937. Duhamet (1939) reported the stimulation of growth of excised roots by IAA.

The avenue was now open for rapid progress in the successful culture of plant tissues during the 1930s. The culture of meristems other than root and shoot tips was explored. With improved culture media, La Rue (1936) achieved better success at culturing plant embryos compared to the efforts of Hannig 32 years earlier. Gautheret (1934) reported successful culture of the cambium of several species of trees to produce callus on media containing B vitamins and IAA. Further research with other meristematic tissues led Nobecourt (1937) and Gautheret (1939) to obtain callus from carrot root cambium, and White (1939) to obtain tobacco tumor (crown gall) tissue. The rapid progress of plant tissue cultures was abruptly curtailed, however, by the start of World War II in 1939.

One theater of World War II was, of course, concentrated in Europe. Further progress in plant tissue culture by the European originators was almost impossible in the midst of disruption, destruction, and shortages of laboratory supplies and equipment. Scientists in other locations also suffered shortages of laboratory supplies, equipment, and personnel. Nevertheless, during the war years (1939–1945) some progress was made in plant tissue culture. Johannes Van Overbeek and his associates (1942) reported they were able to obtain seedlings from heart-shaped embryos by enriching culture media with coconut milk, in addition to the usual salts, vitamins, and other nutrients. Also, by this time Panchanan Maheshwari, along with his associates and students in India, was very active in angiosperm embryology research that began before the war and progressed into the 1940s, as described in his book (1950). Armin Braun reported on tumor induction related to crown gall disease. Experiments on tobacco tissues by Folke Skoog demonstrated organ formation in cultured tissues and organs. In Western Europe, some progress occurred; Guy Camus was the first to report grafting experiments in tissue cultures, and Georges Morel reported on developing techniques to culture parasitized plant tissues.

After the war's end in the spring of 1945, a resumption of prewar tissue culture activity could not occur immediately, but nonetheless some progress in plant tissue culture technology occurred. Ernest Ball reported on greatly improving the potential for culturing shoot tips, beginning with those of nasturtium and lupine. Albert Hildebrandt and his coworkers made improvements to the medium for the growth of tobacco and sunflower tissues, and Morel was well along in his research on applying tissue culture techniques to the study of parasites associated with plant tissues. Herbert Street and his colleagues began a series of extensive studies on the nutrition of excised tomato root tips.

After 1950, rapid progress was made in plant tissue culture techniques. Also, many advancements were accomplished in the knowledge of plant development, especially in the area of the effects of PGRs. While intensive studies on the *in vivo* and *in vitro* effects of auxins continued, other classes of PGRs were now recognized. One such class, the cytokinins, emerged from the investigations of Skoog and his associates on the nutritional requirements of tobacco callus, which extended to include study of the induction of bud formation on tobacco stem segments by adenine sulfate (Skoog and Tsui, 1951). These initial investigations led to the discovery of kinetin, a cytokinin that acts as a cell division promoter, by Skoog and Miller (1957). This research subsequently led to the discovery of other cytokinins during the latter half of the 1950s and the 1960s. Intensive research then began with adenine, kinetin, and several other newly discovered cytokinins on their *in vitro* shoot-promoting effects for the rapid propagation of plants, especially the economically important horticultural and agronomic cultivars.

Meanwhile, according to Stowe and Yamaki (1957), the Western world became familiar with a third class of PGRs, the gibberellins, through the efforts of a group at Imperial Chemical Industries in Britain and a group at the U.S. Department of Agriculture. These groups discovered research on gibberellins that had been conducted by many Japanese plant physiologists, beginning with Kurosawa, in what was then Formosa. After the Western world became aware of gibberellins in the early 1950s, research on the *in vivo* effects of gibberellins accelerated at a rapid pace worldwide. *In vitro* research with gibberellins was slow to begin in the late 1950s, but increased substantially in the 1960s and 1970s.

A similar situation occurred with the two other classes of PGRs, each of which contains a single representative, ethylene, and abscisic acid. Crocker et al. (1935) were the first to propose that ethylene is involved in fruit ripening. *In vivo* investigations in the 1960s confirmed the role of ethylene in fruit ripening, as first proposed in the 1930s, and also in early seedling growth, leaf abscission and epinasty, senescence, and other growth promoting and inhibitory effects. *In vitro* studies with ethylene did not begin until the 1970s. The *in vivo* effects of abscisic acid on plant development were not recognized until the 1960s. Studies into the 1970s were directed toward leaf and fruit abscission and bud dormancy. Other plant growth promoting and inhibiting studies soon followed. *In vitro* investigations with abscisic acid began in the 1970s with experiments involving zygotic and somatic embryogenesis.

By the mid-1950s, plant tissue culture methods had progressed to the point of making a major impact on research in plant developmental biology. Those investigators who had studied plant development over the previous 50 years, using the conventional techniques of morphology, anatomy, physiology, biochemistry, cytology, and genetics, progressively incorporated the tool of plant tissue culture into their research. Major breakthroughs of knowledge in plant development came sooner than would have been possible without the techniques of sterile culture (see Chapter 3). The progress also was hastened by innovations in laboratory equipment and supplies, and by improved methods of worldwide transportation and communication between developmental botanists around the world. In particular, the development of high-efficiency particulate air (HEPA) filters, which screen fungal spores and bacteria from air, made laminar flow transfer hoods possible. Laminar flow hoods became commonly available in the late 1960s to early 1970s and, finally, made sterile cultures routine. Some of the European and American originators of the basic techniques of plant tissue culture from the 1930s and 1940s traveled around the world after the 1950s to many research laboratories as visiting botanists or to attend meetings, and became teachers of the basic methods for numerous colleagues and students.

Concomitantly, another aspect of plant tissue culture greatly aided the increased research activity in plant development. The methods of obtaining plant callus tissue from several sources were well developed by the mid-1950s. The study of the causes of mammalian cancer (especially human cancers) became very popular during this time period. It was logical, then, that plant callus development was equated to mammalian tumor development. Thus, research on plant callus and cell suspension cultures intensified in many research laboratories because several agencies that were interested in mammalian cancer research awarded generous grants to plant developmental botanists to study "plant cancers." Among these agencies were the American Cancer Society; the National Cancer Institute; the National Institutes of Health; and the U.S. Department of Health, Education, and Welfare. A few notable pioneer investigators among many in the study of plant callus and cell suspensions from the early 1950s to the mid-1960s included Morel and Wetmore (1951), Henderson and Bonner (1952), Steward et al. (1952; 1954; 1958), Muir et al. (1954), Braun (1954), Nickell (1955), Partanen et al. (1955), Das et al. (1956), Nitsch and Nitsch (1956), Torrey (1957), Reinert (1958), Klein (1958), Bergmann (1960), and Murashige and Skoog (1962).

In the 1950s, an early prediction that somatic plant cells could undergo embryogenesis was finally validated by Steward et al. (1958) and Reinert (1958), who showed that somatic carrot cells would differentiate into embryos when cultured within a proper nutrient-PGR regime. The ability to regenerate plants from single somatic cells through such a "normal" developmental pathway was envisioned to have great applications in propagation and genetic engineering (see below). Somatic embryogenesis, a type of nonzygotic embryogenesis, has now been demonstrated in most species of higher plants that have been tested (see Chapter 14).

In the early 1960s, Murashige completed a study, while working in Skoog's laboratory, that led to a commercial application for tissue culture. A culture medium developed by Murashige was originally devised for rapid growth and bioassays with tobacco callus (Murashige and Skoog, 1962). However, research by several nurseries in California and elsewhere, through the advice of Murashige and with knowledge of the work of Morel on orchid propagation, showed that practical shoot tip propagation of several ornamental plants could be accomplished with the Murashige and Skoog medium (1962). Today, the ornamental plant industry depends heavily on tissue and organ culture micropropagation to supply high-quality, low-cost stocks of many species (see Chapters 11 and 27).

Reports concerning the recovery of plants from haploid cells began to appear in the 1960s (see Chapter 17). The first successes were obtained with *Datura* (Guha and Maheshwari, 1966) and tobacco (Bourgin and Nitsch, 1967). This discovery received significant attention from plant breeders because plants recovered from doubled haploid cells are homozygous and express all recessive genes, making them ideal pure breeding lines. Haploid-based breeding programs now are in place for several major agronomic crops.

Plant protoplasts were isolated and began to be cultured in the 1960s (Cocking, 1960). The removal of the plant cell wall allowed many novel experiments on membrane transport to be undertaken and, with the use of totipotent protoplasts, allowed the somatic hybridization of sexually incompatible species to be accomplished. Somatic hybridization has now been used successfully in a number of cultivar development programs (see Chapter 16). In addition, the uptake of DNA into the plant cell was facilitated by removal of the wall, leading to the first reports of plant genetic transformation (see Chapter 19 and below).

Perhaps the greatest stimulus and change in the direction of plant cell and tissue culture research occurred after the discovery of restriction endonuclease enzymes in the early 1970s. These enzymes cleave DNA molecules at predictable sites and allow specific genes to be removed, modified, and inserted into other DNA strands. The watershed of technological development that occurred after this breakthrough, and is ongoing today, has provided the basic tools needed for genetic engineering (Chapter 19). The term "biotechnology" was coined to denote this new research field. The availability of totipotent plant cells that could conceivably be altered by insertion of specific genes caused a revolution in plant research because the obvious implications of such genetically altered plants for agriculture were so momentous. The potential commercial value of such genetically modified (GM) plants (Chapter 20) attracted an unprecedented amount of industrial and investor interest such that, in the late 1970s and 1980s, a number of new plant-biotechnology-based companies sprang up around the world, primarily in the U.S. Some of these companies still exist today, but many either failed or were absorbed into larger companies.

The practical benefits obtained through the merging of plant tissue culture and molecular biology began to be realized in 1996 with the commercialization of GM crops. Establishment of GM crops has been very rapid; to document growth of this new sector, the following statistics were gleaned from news reports. In 1995, there were no commercial plantings of GM crops. However, out of 80 million acres of corn planted yearly in the U.S., 400,000 acres of GM corn were planted in 1996, increasing to three million acres in 1997 and 17 million acres in 1998. Similarly, of 71 million acres of soybeans planted in 1997, 20 million acres were genetically modified, and some estimates suggest that nearly 100% of the soybean crop will be composed of GM varieties within the next few years. By 1998, approximately 45% of the U.S. cotton crop was genetically modifed. As of 2002, 145 million acres of farmland across the globe were estimated to be devoted to GM crop production. Sixty-two percent of global soybean production (90 million acres) was derived from GM cultivars. Corn was second with 21% of global production (30.6 million acres). Cotton and canola were the next most commonly planted transgenic crops. In the U.S., 88 million acres of farmland were devoted to GM crops in 2001, accounting for 68% of global GM crop acreage. Argentina was the second largest producer (22%), followed by Canada (6%), and China (3%).

Herbicide resistance is the dominant trait of GM crops; *Bacillus thuringiensis* (Bt) insect resistance is second. Although much less acreage is devoted to "quality" traits, such as the production of "golden rice" with enhanced vitamin A (created to solve a major nutritional problem in developing countries), it is expected that such traits like controlled ripening and enhanced oil profiles will continue to be developed and commercialized.

When considering these GM crops, along with other additives in use today, such as enzymes from microorganisms in processed products like soft drinks, cakes, cheese, bread, fish, and meats, over 90% of common foodstuffs already may contain GM components. This rapid integration of GM crops into the food supply has caused concern in certain parts of the world (Chapter 20). The concern is based upon the perception that there has been inadequate testing of GM crops and centers on several issues, including the following: (1) the unwanted spread of transgenes into wild species by pollination, resulting in, for example, herbicide-resistant weeds; (2) the development of resistant insects, due to overuse, for example, of Bt crops; (3) the potential for increased use of herbicides in herbicide-resistant GM crop production; (4) human health concerns, such as the transfer of antibiotic resistance genes from digested GM food to bacteria in the gut, and the possibly unrecognized toxicity of transgenic proteins. While controversial, most experts consider such concerns to be overreaction that is often politically motivated because GM crops already have been subjected to rigorous testing. Several countries have considered multiyear moratoriums on the planting of GM crops. However, it is expected that the overall benign impact of GM crops on the environment, including reduction of chemical inputs, along with consumer recognition of resultant increased quality of foodstuffs, will be combined with better education to solve these concerns.

The use of plant tissue culture technology to enable the development of GM crops illustrates the key importance, often overlooked, of this technology in our lives. Plant tissue culture is also an increasingly important tool for advanced studies of plant development, especially with the integration of molecular biological techniques. As will become evident throughout this book, however, much remains to be learned in terms of the basic methods and procedures needed for efficient manipulation of plants *in vitro*.

LITERATURE CITED AND SUGGESTED READINGS

Ball, E. 1946. Development in sterile culture of stem tips and subadjacent regions of *Tropaeolum majus* and of *Lupinus albus* L. *Am. J. Bot.* 33:301–318.

Bergmann, L. 1960. Growth and division of single cells of higher plants *in vitro*. *J. Gen. Physiol.* 43:841–851.

Bhojwani, S.S., V. Dhawan, and E.C. Cocking. 1986. *Plant Tissue Culture: A Classified Bibliography*. Elsevier, New York, 789.

Bourgin, J.P. and J.P. Nitsch. 1967. Obtention de *Nicotiana* haploides a partir d'étamines cultivées *in vitro*. *Ann. Physiol. Veg.* 9:377–382.

Braun, A.C. 1954. Studies on the origin of the crown gall tumor cell. In: *Abnormal and Pathological Plant Growth*, Brookhaven National Lab., Symp. No. 6:115–127.

Camus, G. 1943. Sur le greffage des bourgeons d'endive sur des fragments de tissus cultivés *in vitro*. *C. R. Acad. Sci.* 137:184–185.

Cocking, E.C. 1960. A method for the isolation of plant protoplasts and vacuoles. *Nature* 187:962–963.

Crocker, W., A.E. Hitchcock, and P.W. Zimmerman. 1935. Similarities in the effects of ethylene and the plant auxins. *Contrib. Boyce Thompson Inst.* 7:231–248.

Das, N.K., K. Patau, and F. Skoog. 1956. Initiation of mitosis and cell division by kinetin and indoleacetic acid in excised tobacco pith tissue. *Physiol. Plant.* 9:640–651.

Dormer, K.J. and H.E. Street. 1949. The carbohydrate nutrition of tomato roots. *Ann. Bot. N.S.* 13:199–217.

Duhamet, L. 1939. Action de l'heteroauxine sur la croissance de racines isolées de *Luipinus albus*. *C. R. Acad. Sci.* 208:1838–1840.

Durosawa, E. 1926. Experimental studies on the secretions of *Fusarium heterosporum* on rice plants. *Trans. Nat. Hist. Soc. Formosa* 16:123–127.

Gautheret, R. J. 1934. Culture de tissu cambial. *C. R. Acad. Sci.* 198:2195–2196.

Gautheret, R. J. 1939. Sur la possibilité de réaliser la culture indéfinie des tissus de tubercules de carotte. *C. R. Acad. Sci.* 208:118–120.

Guha, S. and S. C. Maheshwari. 1966. Cell division and differentiation in the pollen grains of *Datura in vitro*. *Nature* 212:97–98.

Hannig, E. 1904. Zur physiologie pflanzlicher embryonen. I. Uber die kultur von cruciferen-embryonen ausserhalb des embryosacks. *Bot. Ztg.* 62:45–80.

Henderson, J.H.M. and J. Bonner. 1952. Auxin metabolism in normal and crown gall tissue of sunflower. *Am. J. Bot.* 39:444–451.

Hildebrandt, A.C., A.J. Riker, and B.M. Duggar. 1946. The influence of the composition of the medium on growth *in vitro* of excised tobacco and sunflower tissue culture. *Am. J. Bot.* 33:591–597.

Hogland, D.R. and D.I. Arnon. 1938. The water culture method for growing plants without soil. *Calif. Agric. Expt. Stat. Circ.* 347, Berkeley, CA.

Klein, R.M. 1958. Activation of metabolic systems during tumor cell formation. *Proc. Natl. Acad. Sci. U.S.A.* 44:349–354.

Kotte, W. 1922. Kulturversuche mit isolierten wurzelspitzen. *Beitr. Allg. Bot.* 2:413–434.

Krikorian, A.D. and D.L. Berquam. 1969. Plant cell and tissue culture — the role of Haberlandt. *Bot. Rev.* 35:59–88.

La Rue, C.D. 1936. The growth of plant embryos in culture. *Bull. Torrey Bot. Club* 63:365–382.

Maheshwari, P. 1950. *An Introduction to the Embryology of Angiosperms*. McGraw-Hill, New York, 453.

Morel, G. 1944. Le developpement du mildou sur des tissues de vigne cultivés *in vitro*. *C. R. Acad. Sci.* 218:50–52.

Morel, G. 1948. Recherches sur la culture associée de parasites obligatoires de tissus végétaux. *Ann. Epiphyt. N. S.* 14:123–234.

Morel, G. 1965. Clonal propagation of orchids by meristem culture. *Cymbidium Soc. News.* 20:3–11.

Morel, G. and R.H. Wetmore. 1951. Fern callus tissue cultures. *Am. J. Bot.* 39:141–143.

Muir, W.H., A.C. Hildebrandt, and A.J. Riker. 1954. Plant tissue cultures produced from single isolated cells. *Science* 119:877–878.

Murashige, T. and F. Skoog. 1962. A revised medium for rapid growth and bioassays with tobacco tissue cultures. *Physiol. Plant.* 15:473–497.

Nickell, L.G. 1955. Nutrition of pathological tissues caused by plant viruses. *Annee. Biol.* 31:107–121.

Nitsch, J.P. and C. Nitsch. 1956. Auxin dependent growth of excised *Helianthus tuberosus* tissues. *Am. J. Bot.* 43:839–851.

Nobecourt, P. 1937. Cultures en série de tissus végétaux sur milieu artificiel. *C. R. Seanc. Soc. Biol.* 205:521–523.

Partanen, C.R., I.M. Sussex, and T.A. Steeves. 1955. Nuclear behavior in relation to abnormal growth in fern prothalli. *Am. J. Bot.* 42:245–256.

Reinert, J. 1958. Morphogenese und ihre kontrolle an gewebekulturen aus carotten. *Naturwiss.* 45:344–345.

Robbins, W. J. 1922. Cultivation of excised root tips and stem tips under sterile conditions. *Bot. Gaz.* 73:376–390.

Skoog, F. 1944. Growth and organ formation in tobacco tissue cultures. *Am. J. Bot.* 31:19–24.

Skoog, F. and C.O. Miller. 1957. Chemical regulation of growth and organ formation in plant tissues cultured *in vitro*. *Symp. Soc. Exp. Biol.* 11:118–131.

Skoog, F. and C. Tsui. 1951. Growth substances and the formation of buds in plant tissues. In: *Plant Growth Substances*. Ed. F. Skoog. University of Wisconsin Press, Madison, WI, 263–285.

Steward, F.C. 1968. *Growth and Organization in Plants*. Addison-Wesley, Reading, MA, 564.

Steward, F.C. and S.M. Caplin. 1954. The growth of carrot tissue explants and its relation to the growth factors present in coconut milk. I(A) The development of the quantitative method and the factors affecting the growth of carrot tissue explants. *Ann. Biol.* 30:385–394.

Steward, F.C., S.M. Caplin, and F.K. Millar. 1952. Investigations on growth and metabolism of plant cells. I. New techniques for the investigation of metabolism, nutrition and growth of undifferentiated cells. *Ann. Bot. N. S.* 16:57–77.

Steward, F.C., M.O. Mapes, and M.J. Smith. 1958. Growth and organized development of cultured cells. I. Growth and division of freely suspended cells. *Am. J. Bot.* 45:693–703.

Stowe, B.B. and T. Yamaki. 1957. The history and physiological action of the gibberellins. *Ann. Rev. Plant Physiol.* 8:181–216.

Street, H.E. and J.S. Lowe. 1950. The carbohydrate nutrition of tomato roots. II. The mechanism of sucrose absorption by excised roots. *Ann. Bot. N.S.* 14:307–329.

Torrey, J.G. 1957. Cell division in isolated single plant cells *in vitro*. *Proc. Nat. Acad. Sci. U.S.A.* 43:887–891.

Van Overbeek, J., M.E. Conklin, and A.F. Blakeslee. 1942. Cultivation *in vitro* of small *Datura* embryos. *Am. J. Bot.* 29:472–477.

Went, F.W. and K.V. Thimann. 1937. *Phytohormones*. Macmillan, New York, 294.

Wetmore, R.H. 1954. The use of *in vitro* cultures in the investigation of growth and differentiation in vascular plants. In: *Abnormal and Pathological Plant Growth*, Brookhaven National Lab., Symp. No. 6:22–40.

White, P.R. 1934. Potentially unlimited growth or excised tomato root tips in liquid medium. *Plant Physiol.* 9:585–600.

White, P.R. 1937. Vitamin B1 in the nutrition of excised tomato roots. *Plant Physiol.* 12:803–811.

White, P.R. 1939. Potentially unlimited growth of excised plant callus in an artificial medium. *Am. J. Bot.* 26:59–64.

Section III

Supporting Methodologies

3 Getting Started with Tissue Culture: Media Preparation, Sterile Technique, and Laboratory Equipment

Caula A. Beyl

CHAPTER 3 CONCEPTS

- A tissue laboratory needs adequate physical space for work and storage, and for equipment such as an autoclave, a distilled water source, balances, refrigerators, various laboratory instruments, culture vessels, and flow hoods, to name a few items.

- There are many growth media available, and the type of basal culture medium selected depends upon the species to be cultured. The growth of the plant in culture is also affected by the selection of plant growth regulators (PGRs) and environmental (cultural) conditions.

- There are about 20 different components in tissue culture medium. These include inorganic mineral elements, various organic compounds, PGRs, and support substances (e.g., agar or filter paper).

- PGRs are typically expressed in media as milligrams per liter (mg/l) or micromoles (μM). When comparing the effects of several PGRs on tissues in culture, prepare media using micromolar concentrations because an equal number of molecules of the various PGRs will be present in each of the media.

INTRODUCTION

A plant tissue culture laboratory has several functional areas, whether it is designed for teaching or research, and no matter how large or elaborate it is. A laboratory has some elements that are similar to a well-run kitchen and other elements that more closely resemble an operating room. There are areas devoted to the preliminary handling of plant tissue destined for culture, media preparation, and sterilization of media and tools; a sterile transfer hood or "clean room" for aseptic manipulations; a culture growth room; and an area devoted to washing and cleaning glassware and tools (see Chapter 27). This chapter will serve as an introduction to what goes into setting up a tissue culture laboratory. It will list what supplies and equipment are necessary, and cover some basics concerning making stock solutions, calculating molar concentrations, making tissue culture media, preparing a transfer (sterile) hood, and culturing various cells, tissues, and organs.

0-8493-1614-6/05/$0.00+$1.50

EQUIPMENT AND SUPPLIES FOR A TISSUE CULTURE TEACHING LABORATORY

Ideally, there should be sufficient bench area to allow for both preparation of media and storage space for chemicals and glassware. In addition to the usual glassware and instrumentation found in laboratories, a tissue culture laboratory needs an assortment of glassware which may include graduated measuring cylinders, wide-necked Erlenmeyer flasks, medium bottles, test tubes with caps, petri dishes, volumetric flasks, beakers, and a range of pipettes. In general, glassware should be able to withstand repeated autoclaving. Baby food jars are inexpensive alternative tissue culture containers well-suited for teaching. Ample quantities can be obtained by preceding the recycling truck on its pickup day (provided you are not embarrassed by the practice). Some tissue culture laboratories find presterilized disposable culture containers and plastic petri dishes to be convenient, but the cost may be prohibitive for others on a tight budget. There are also reusable plastic containers available, but their longevity and resistance to wear, heat, and chemicals vary considerably.

It is also good to stock metal or wooden racks to support culture tubes, both for cooling and later, during their time in the culture room, metal trays (such as cafeteria trays) and carts for transport of cultures, stoppers and various closures, nonabsorbent cotton, cheese cloth, foam plugs, metal or plastic caps, aluminum foil, Parafilm™, and plastic wrap.

To teach tissue culture effectively, some equipment is necessary, such as a pH meter, balances (one analytical to four decimal places and one to two decimal places), bunsen burners, alcohol lamps or electric sterilizing devices, several hot plates with magnetic stirrers, a microwave oven for rapid melting of large volumes of agar medium, a compound microscope and haemocytometer for cell counting, a low-speed centrifuge, stereomicroscopes (ideally with fiber optic light sources), large (10 or 25-liter) plastic carboys to store high-quality (high-purity) water, a fume hood, an autoclave (or at the very least, a pressure cooker) and a refrigerator to store media, stock solutions, plant growth regulators (PGRs), and so on. A dishwasher is useful, but a large sink with drying racks, pipette and acid baths, and a forced air oven for drying glassware will also work. Also, deionized distilled water for the final rinsing of glassware is needed. Aseptic manipulations and transfers are done in multistation laminar flow hoods (one for each pair of students).

Equipment used in the sterile transfer hood usually includes a spray bottle containing 70% ethanol, spatulas (useful for transferring callus clumps), forceps (short, long, and fine-tipped), scalpel handles (#3), disposable scalpel blades (#10 and #11), a rack for holding sterile tools, a pipette bulb or pump, bunsen burner, alcohol lamp, or other sterilizing device, and a sterile surface for cutting explants (see below). If necessary for the experimental design, uniform-sized leaf explants can be obtained using a sterile cork borer.

There are a number of options for providing a sterile surface for cutting explants. A stack of paper towels wrapped in aluminum foil is effective, and as each layer becomes messy, it can be peeled off and the next layer beneath it used (Figure 3.1). Others prefer reusable surfaces such as ceramic tiles (local tile retailers are quite generous and will donate samples), metal commercial ashtrays, or glass petri dishes (100×15 mm). Sterile plastic petri dishes also can be used, but the cost may outweigh the advantages. A container is needed to hold the alcohol used for flaming instruments, if flame sterilization is used. An ideal container for this purpose is a slide-staining Coplin jar with a small wad of cheesecloth at the bottom to prevent breakage of the glass when tools are dropped in. It has the advantage that it is made from heavier glass and, since the base is flared, it is not prone to tipping over. Other containers can also serve the same purpose, such as test tubes held in a rack or placed in a flask or beaker to prevent them from spilling. Plastic containers, which can catch fire and melt, should never be used to hold alcohol.

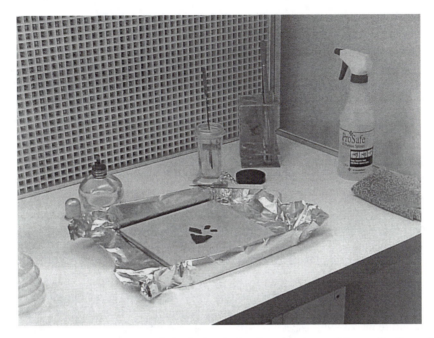

FIGURE 3.1 A typical layout of materials in the hood, showing placement of the sterile tile work surface, an alcohol lamp, a spray bottle containing 80% ethanol, a cloth for wiping down the hood, and two different kinds of tool holders — a glass-staining (Coplin) jar and a metal rack for holding test tubes.

WATER

High-quality water is a required ingredient of plant tissue culture media. Ordinary tap water contains cation, anions, particulates of various kinds, microorganisms, and gases that make it unsuitable for use in tissue culture media. Various methods are used to treat water, including filtration through activated carbon to remove organics and chlorine, deionization or demineralization by passing water through exchange resins to remove dissolved ionized impurities, and distillation, which eliminates most ionic and particulate impurities, volatile and nonvolatile chemicals, organic matter, and microorganisms. The process of reverse osmosis, which removes 99% of the dissolved ionized impurities, makes use of a semipermeable membrane through which a portion of the water is forced under pressure; the remainder, containing the concentrated impurities, is rejected. The most universally reliable method of water purification for tissue culture use is a deionization treatment followed by one or two glass distillations, although simple deionization alone is sometimes used successfully. In some cases, newer reverse osmosis purifying equipment (Milli-RO™, Millipore™, RO pure™, Barnstead™, Bion™, Pierce™), combined with cartridge ion exchange, adsorption, and membrane filtering equipment, has replaced the traditional glass distillation of water.

THE CULTURE ROOM

After the explants are plated on the tissue culture medium under the sterile transfer hood, they are moved to the culture room. This can be as simple as a room with shelves equipped with lights or as complex as a room with intricate climate control. Most culture rooms tend to be rather simple, consisting of cool white fluorescent lights mounted to illuminate each shelf. Adjustable shelves are an asset that allow for differently sized tissue culture containers and for moving the light closer to the containers to achieve higher light intensities. Putting the lights on timers allows for photoperiod manipulation. Some cultures grow equally as well in dark or light. Temperatures of 26–28°C are usually optimum. Heat buildup can be a problem if the room is small so that adequate air conditioning is required. Good air flow also helps to reduce condensation occurring inside petri dishes or other vessels. Some laboratories purchase incubators designed for plant tissue culture. If a liquid medium is used, the culture room should be equipped with a rotary or reciprocal shaker to provide sufficient oxygenation. The optimal temperature, light, and shaker conditions vary depending upon the plant species being cultured.

CHARACTERISTICS OF SOME OF THE MORE COMMON TISSUE CULTURE MEDIA

The type of tissue culture medium selected depends upon the species to be cultured. Some species are sensitive to high salts or have different requirements for PGRs. The age of the plant also has an effect. For example, juvenile tissues generally regenerate roots more readily than do adult tissues. The type of organ cultured is important; for example, roots require thiamine. Each desired cultural effect has its own unique requirements, such as auxin (see below) for induction of adventitious roots, and altering the cytokinin:auxin ratio for initiation and development of adventitious shoots.

Development of culture medium formulations was a result of systematic trial and experimentation. Table 3.1 provides a comparison of the composition of several of the most commonly used plant tissue culture media with respect to their components in milligrams per liter and molar units. Murashige and Skoog (MS) medium (1962) is the most suitable and most commonly used basic tissue culture medium for plant regeneration from tissues and callus. It was developed for tobacco, and based primarily upon the mineral analysis of tobacco tissue. This is a "high salt" medium, due to its content of K and N salts. Linsmaier and Skoog medium (1965) is basically Murashige and Skoog (1962) medium with respect to its inorganic portion, but only inositol and thiamine HCl are retained among the organic components. To counteract salt sensitivity of some woody species, Lloyd and McCown (1980) developed the woody plant medium (WPM).

Gamborg's B-5 medium (Gamborg et al., 1968) was devised for soybean callus culture, and has lesser amounts of nitrate and particularly ammonium salts than the MS medium. Although B5 was originally developed for the purpose of obtaining callus or for use with suspension culture, it also works well as a basal medium for whole plant regeneration. Schenk and Hildebrandt (1972) developed the SH medium for the callus culture of monocots and dicots. White's medium (White, 1963), which was designed for the tissue culture of tomato roots, has a lower concentration of salts than the MS medium. Nitsch's medium (Nitsch and Nitsch, 1969) was developed for anther culture and contains a salt concentration intermediate to that of MS and White's media.

Many companies sell packaged prepared mixtures of the better-known media recipes. These are easy to make because they merely involve dissolving the packaged mix in a specified volume of water. These can be purchased as the salts, the vitamins, or the entire mix with or without PGRs, agar, and sucrose. These are convenient, less prone to individual error, and make keeping stock solutions unnecessary. However, they are more expensive than making media from scratch.

COMPONENTS OF THE TISSUE CULTURE MEDIUM

Growth and development of explants *in vitro* is a product of the genetics, surrounding environment, and components of the tissue culture medium, the last of which is easiest to manipulate to our own ends. Tissue culture medium consists of 95% water, macro- and micronutrients, PGRs, vitamins, sugars (because plants *in vitro* are often not photosynthetically competent), and sometimes various other simple-to-complex organic materials. All in all, about twenty different components are usually needed.

TABLE 3.1
Composition of Five Commonly Used Tissue Culture Media in Milligrams per Liter and Molar Concentrations

Compounds	Murashige and Skoog	Gamborg B-5	WPM	Nitsch and Nitsch	Schenk and Hildebrandt	White
Macronutrients in mg/l (mM)						
NH_4NO_3	1650 (20.6)	—	400 (5.0)	—	—	—
$NH_4H_2PO_4$	—	—	—	—	300 (2.6)	—
NH_4SO_4	—	134 (1.0)	—	—	—	—
$CaCl_2 \cdot 2H_2O$	332.2 (2.3)	150 (1.0)	96 (0.7)	166 (1.1)	151 (1.0)	—
$Ca(NO_3)_2 \cdot 4H_2O$	—	—	556 (2.4)	—	—	288 (1.2)
$MgSO_4 \cdot 7H_2O$	370 (1.5)	250 (1.0)	370 (1.5)	185 (0.75)	400 (1.6)	737 (3.0)
KCl	—	—	—	—	—	65 (0.9)
KNO_3	1900 (18.8)	2500 (24.8)	—	950 (9.4)	2500 (24.8)	80 (0.8)
K_2SO_4	—	—	990	—	—	—
KH_2PO_4	170 (1.3)	—	170 (1.3)	68 (0.5)	—	—
NaH_2PO_4	—	130.5 (0.9)	—	—	—	16.5 (0.12)
Na_2SO_4	—	—	—	—	—	200 (1.4)
Micronutrients in mg/l (μM)						
H_3BO_3	6.2 (100)	3.0 (49)	6.2 (100)	10 (162)	5 (80)	1.5 (25)
$CoCl_2 \cdot 6H_2O$	0.025 (0.1)	0.025 (0.1)	—	—	0.1 (0.4)	—
$CuSO_4 \cdot 5H_2O$	0.025 (0.1)	0.025 (0.1)	0.25 (1)	0.025 (0.1)	0.2 (0.08)	0.01 (0.04)
Na_2EDTA	37.3 (100)	37.3 (100)	37.3 (100)	37.3 (100)	20.1 (54)	—
$Fe_2(SO_4)_3$	—	—	—	—	—	2.5 (6.2)
$FeSO_4 \cdot 7H_2O$	27.8 (100)	27.8 (100)	27.8 (100)	27.8 (100)	15 (54)	—
$MnSO_4 \cdot H_2O$	16.9 (100)	10.0 (59)	22.3 (132)	18.9 (112)	10.0 (59)	5.04 (30)
KI	0.83 (5)	0.75 (5)	—	—	0.1 (0.6)	0.75 (5)
$NaMoO_3$	—	—	—	—	—	0.001 (0.001)
$Na_2MoO_4 \cdot 2H_2O$	0.25 (1)	0.25 (1)	0.25 (1)	0.25 (1)	0.1 (0.4)	—
$ZnSO_4 \cdot 7H_2O$	8.6 (30)	2.0 (7.0)	8.6 (30)	10 (35)	1 (3)	2.67 (9)
Organics in mg/l (μM)						
Myo-inositol	100 (550)	100 (550)	100 (550)	100 (550)	1000 (5500)	—
Glycine	2.0 (26.6)	—	2.0 (26.6)	2.0 (26.6)	—	3.0 (40)
Nicotinic Acid	0.5 (4.1)	1.0 (8.2)	0.5 (4.1)	5 (40.6)	5.0 (41)	0.5 (4.1)
Pyridoxine HCl	0.5 (2.4)	0.1 (0.45)	0.5 (2.4)	0.5 (2.4)	0.5 (2.4)	0.1 (0.45)
Thiamine HCl	0.1 (0.3)	10.0 (30)	1.0 (3.0)	0.5 (1.5)	5.0 (14.8)	0.1 (0.3)
Biotin	—	—	—	0.2 (0.05)	—	—

Inorganic Mineral Elements

Just as a plant growing *in vivo* requires many different elements from either soil or fertilizers, a plant tissue growing *in vitro* requires a combination of macro- and micronutrients. The choice of macro- and microsalts and their concentrations is species-dependent. MS medium is very popular because most plants react to it favorably; it may not, however, necessarily result in the optimum growth and development for every species because the salt content is so high.

Macronutrients are required in millimolar (mM) quantities in most plant basal media. Nitrogen (N) is usually supplied in the form of ammonium (NH_4^+) and nitrate (NO_3^-) ions; although sometimes more complex organic sources, such as urea, amino acids like glutamine, or casein hydrolysate, which is a complex mixture of amino acids and ammonium, are used as well. Although most plants prefer NO_3^- to NH_4^+, the right balance of the two ions for optimum *in vitro* growth and development for the selected species may differ.

In addition to nitrogen, potassium, magnesium, calcium, phosphorus, and sulfur are provided in the medium as different components referred to as the macrosalts. $MgSO_4$ provides both magnesium and sulfur; $NH_4H_2PO_4$, KH_2PO_4, or NaH_2PO_4 provide phosphorus; $CaCl_2 \cdot 2H_2O$ or $Ca(NO_3)_2 \cdot 4H_2O$ provide calcium; and KCl, KNO_3, or KH_2PO_4 provide potassium. Chloride is provided by KCl and/or $CaCl_2 \cdot 2H_2O$.

Microsalts typically include boron (H_3BO_3), cobalt ($CoCl_2 \cdot 6H_2O$), iron (a complex of $FeSO_4 \cdot 7H_2O$ and Na_2EDTA, or, rarely, $Fe_2[SO_4]_3$), manganese ($MnSO_4 \cdot H_2O$), molybdenum ($NaMoO_3$), copper ($CuSO_4 \cdot 5H_2O$), and zinc ($ZnSO_4 \cdot 7H_2O$). Microsalts are needed in much lower (micromolar [μM]) concentrations than the macronutrients. Some media may contain very small amounts of iodide (KI), but sufficient quantities of many of the trace elements inadvertently may be provided because reagent grade chemicals contain inorganic contaminants.

Organic Compounds

Sugar is a very important part of any nutrient medium, and its addition is essential for *in vitro* growth and development of the culture. Most plant cultures are unable to photosynthesize effectively for a variety of reasons, including insufficiently organized cellular and tissue development, lack of chlorophyll, limited gas exchange and CO_2 in the tissue culture vessels, and less than optimum environmental conditions, such as low light. A concentration of 20–60 g/l sucrose (a disaccharide made up of glucose and fructose) is the most often used carbon or energy source because this sugar is also synthesized and transported naturally by the plant. Other mono- or disaccharides and sugar alcohols such as glucose, fructose, sorbitol, and maltose may be used. The sugar concentration chosen is dependent on the type and age of the explant in the culture. For example, very young embryos require a relatively high sugar concentration (>3%). For mulberry buds *in vitro*, fructose was found to be better than sucrose, glucose, maltose, raffinose, or lactose (Coffin et al., 1976). For apple, sorbitol and sucrose supported callus initiation and growth equally well, but sorbitol was better for peach after the fourth subculture (Oka and Ohyama, 1982).

Sugar (sucrose) that is bought from the supermarket is usually adequate, but be careful to get pure cane sugar, as corn sugar is primarily fructose. Raw cane sugar is purified and, according to the manufacturer's analysis, consists of 99.94% sucrose, 0.02% water, and 0.04% other material (inorganic elements and also raffinose, fructose, and glucose). Nutrient salts contribute approximately 20–50% to the osmotic potential of the medium, and sucrose is responsible for the remainder. The contribution of sucrose to the osmotic potential increases as it is hydrolyzed into glucose and fructose during autoclaving. This may be an important consideration when performing osmotic sensitive procedures such as protoplast isolation and culture.

Vitamins are organic substances that are parts of enzymes or cofactors for essential metabolic functions. Of the vitamins, only thiamine (vitamin B_1 at 0.1–5.0 mg/l) is essential in a culture, as it is involved in carbohydrate metabolism and the biosynthesis of some amino acids. It is usually added to tissue culture media as thiamine hydrochloride (HCl). Nicotinic acid, also known as niacin, vitamin B_3, or vitamin PP, forms part of a respiratory coenzyme and is used at concentrations between 0.1 and 5 mg/l. The MS medium contains thiamine HCl as well as two other vitamins, nicotinic acid and pyridoxine (vitamin B_6) in the HCl form. Pyridoxine is an important coenzyme in many metabolic reactions and is used in media at concentrations of 0.1–1.0 mg/l. Biotin (vitamin H) is commonly added to tissue culture media at 0.01–1.0 mg/l. Other vitamins that are sometimes used are folic acid (vitamin M; 0.1–0.5 mg/l), riboflavin (vitamin B_2; 0.1–10 mg/l), ascorbic acid (vitamin C; 1–100 mg/l), pantothenic acid (vitamin B_5; 0.5–2.5 mg/l), tocopherol (vitamin E; 1–50 mg/l) and para-aminobenzoic acid (0.5–1.0 mg/l).

Inositol is sometimes characterized as one of the B complex vitamin group, but it is really a sugar alcohol involved in the synthesis of phospholipids, cell wall pectins and membrane systems in cell cytoplasm. It is added to tissue culture media at a concentration of about 0.1–1.0 g/l and has been demonstrated to be necessary for some monocots, dicots and gymnosperms.

In addition, other amino acids are sometimes used in tissue culture media. These include L-glutamine, asparagine, serine, and proline, which are used as sources of reduced organic nitrogen, especially for inducing and maintaining somatic embryogenesis (see Chapter 14). Glycine, the simplest amino acid, is a common additive, since it is essential in purine synthesis and is a part of the porphyrin ring structure of chlorophyll.

Complex organic compounds are a group of undefined supplements such as casein hydrolysate, coconut milk (the liquid endosperm of the coconut), orange juice, tomato juice, grape juice, pineapple juice, sap from birch, banana puree, and so on. These compounds are often used when no other combination of known defined components produces the desired growth or development. However, the composition of these supplements is basically unknown and may also vary from lot to lot, causing variable responses. For example, the composition of coconut milk (used at a dilution of 50–150 ml/l), a natural source of the PGR zeatin (see below), not only differs between young and old coconuts, but also between coconuts of the same age.

Some complex organic compounds are used as organic sources of nitrogen, such as casein hydrolysate, a mixture of about 20 different amino acids and ammonium (0.1–1.0 g/l), peptone (0.25–3.0 g/l), tryptone (0.25–2.0 g/l), and malt extract (0.5–1.0 g/l). These mixtures are very complex and contain vitamins as well as amino acids. Yeast extract (0.25–2.0 g/l) is used because of the high concentration and quality of B vitamins.

Polyamines, particularly putrescine and spermidine, are sometimes beneficial for somatic embryogenesis. Polyamines are also cofactors for adventitious root formation. Putrescine is capable of synchronizing the embryogenic process of carrot.

Activated charcoal is useful for absorption of the brown or black pigments and oxidized phenolic compounds. It is incorporated into the medium at concentrations of 0.2–3.0% (w/v). It is also useful for absorbing other organic compounds, including PGRs, such as auxins and cytokinins, and other materials such as vitamins, and iron and zinc chelates (Nissen and Sutter, 1990). Carryover effects of PGRs are minimized by adding activated charcoal when transferring explants to media without PGRs. Another feature of using activated charcoal is that it changes the light environment by darkening the medium so it can help with root formation and growth. It may also promote somatic embryogenesis and enhance growth and organogenesis of woody species.

Leached pigments and oxidized polyphenolic compounds and tannins can greatly inhibit growth and development. These are formed by some explants as a result of wounding. If charcoal does not reduce the inhibitory effects of polyphenols, addition of polyvinylpyrrolidone (PVP, 250–1000 mg/l), or of antioxidants such as citric acid, ascorbic acid, or thiourea, can be tested.

PLANT GROWTH REGULATORS (PGRs)

PGRs exert dramatic effects at low concentrations (0.001–10 μM). They regulate the initiation and development of shoots and roots on explants and embryos on semisolid or in liquid medium cultures.

TABLE 3.2
Macro- and Micronutrient 100× Stock Solutions for Murashige and Skoog (MS) Medium (1962)

Stock (100×)	Component	Amount
Nitrate	NH_4NO_3	165.0 g
	KNO_3	190.0 g
Sulfate	$MgSO_4 \cdot 7H_2O$	37.0 g
	$MnSO_4 \cdot H_2O$	1.7 g
	$ZnSO_4 \cdot 7H_2O$	0.86 g
	$CuSO_4 \cdot 5H_2O$	2.50 mg[a]
Halide	$CaCl_2 \cdot 2H_2O$	44.0 g
	KI	83.0 mg
	$CoCl_2 \cdot 6H_2O$	5.0 mg[b]
PBMo	KH_2PO_4	17.0 g
	H_3BO_3	620.0 mg
	Na_2MoO_4	25.0 mg
NaFeEDTA[c]	$FeSO_4 \cdot 7H_2O$	2.78 g
	NaEDTA	3.74 g

Note: The number of grams or milligrams indicated in the amount column should be added to 1000 ml of deionized, distilled water to make 1 liter of the appropriate stock solution. For each liter of the medium made, 10 ml of each stock solution will be used. To make 100× stock solutions for any of the media listed in Table 3.1, multiply the amount of chemical listed in the table by 100 and dissolve in one liter of deionized, distilled water.

[a] Because this amount is too small to weigh conveniently, dissolve 25 mg of $CuSO_4 \cdot 5H_2O$ in 100 ml of deionized, distilled water, then add 10 ml of this solution to the sulfate stock.

[b] Because this amount is too small to weigh conveniently, dissolve 25 mg of $CoCl_2 \cdot 6H_2O$ in 100 ml of deionized, distilled water, then add 10 ml of this solution to the halide stock.

[c] Mix the $FeSO_4 \cdot 7H_2O$ and NaEDTA together and heat gently until the solution becomes orange. Store in an amber bottle or protect from light.

They also stimulate cell division and expansion. Sometimes a tissue or an explant is autotrophic and can produce its own supply of PGRs. Usually, however, PGRs must be supplied in the medium for growth and development of the culture.

The most important classes of PGRs used in tissue culture are the auxins and cytokinins. The relative effects of auxin and cytokinin ratios on the morphogenesis of cultured tissues were demonstrated by Skoog and Miller (1957) and still serve as the basis for plant tissue culture manipulations today. Some of the PGRs used are hormones (naturally synthesized by higher plants), and others are synthetic compounds. PGRs exert dramatic effects depending upon the concentration used, the target tissue, and their inherent activity, even though they are used in very low concentrations in the media ($0.1–100 \, \mu M$). The concentrations of PGRs are typically reported in milligrams per liter or in micromolar units of concentration. Comparisons of PGRs based upon their molar concentrations are more useful because the molar concentration is a reflection of the actual number of molecules of the PGR per unit volume (Table 3.2).

Auxins play a role in many developmental processes, including cell elongation and swelling of tissue, apical dominance, adventitious root formation, and somatic embryogenesis. Generally, when the concentration of auxin is low, root initiation is favored, and when the concentration is high, callus formation occurs. The most common synthetic auxins used are 1-naphthaleneacetic acid (NAA), 2,4-dichlorophenoxyacetic acid (2,4-D), and 4-amino-3,5,6-trichloro-2-pyridinecarboxylic acid (picloram). Naturally occurring indoleacetic acid (IAA) and indolebutyric acid (IBA) are also used. IBA was once considered synthetic and has been found to occur naturally in many plants including olive and tobacco (Epstein et al., 1989). Both IBA and IAA are photosensitive, so stock solutions must be stored in the dark. IAA is also easily broken down by enzymes (peroxidases and IAA oxidase). IAA is the weakest auxin and is typically used at concentrations between 0.01 and 10 mg/l. The more active auxins, such as IBA, NAA, 2,4-D, and picloram, are used at concentrations ranging from 0.001 to 10 mg/l. Picloram and 2,4-D are examples of auxins used primarily to induce and regulate somatic embryogenesis.

Cytokinins promote cell division and stimulate initiation and growth of shoots *in vitro*. The cytokinins most commonly used are zeatin, dihydrozeatin, kinetin, benzyladenine (BA), thidiazuron, and 2-isopentenyl adenine (2-iP). In higher concentrations (1–10 mg/l) they induce adventitious shoot formation, but inhibit root formation. They promote axillary shoot formation by opposing apical dominance regulated by auxins. Benzyladenine has significantly stronger cytokinin activity than the naturally occurring zeatin. However, a concentration of 0.05–0.1 μM thidiazuron, a diphenyl substituted urea, is more active than 4–10 μM BA, but thidiazuron may inhibit root formation, causing difficulties in plant regeneration. Adenine (used at concentrations of 2–120 mg/l) is occasionally added to tissue culture media and acts as a weak cytokinin by promoting shoot formation.

Gibberellins are less commonly used in plant tissue culture. Of the many gibberellins thus far described, GA_3 is the most often used, but it is very heat sensitive (after autoclaving, 90% of the biological activity is lost). Typically, it is filter sterilized and added to autoclaved medium after it has cooled. Gibberellins help stimulate elongation of internodes and have proved to be necessary for meristem growth for some species.

Abscisic acid is not normally considered an important PGR for tissue culture except for somatic embryogenesis and in the culture of some woody plants. For example, it promotes maturation and germination of somatic embryos of caraway (Ammirato, 1974) and spruce (Roberts et al., 1990).

Organ and callus cultures are able to produce the gaseous PGR ethylene. Because culture vessels are almost entirely closed, ethylene can sometimes accumulate. Many plastic containers also contribute to ethylene content in the vessel. There are contrasting reports in the literature concerning the role played by ethylene *in vitro*. It appears to influence embryogenesis and organ formation in some gymnosperms. Sometimes *in vitro* growth can be promoted by ethylene. At other times, addition of ethylene inhibitors results in better initiation or growth. For example, the ethylene inhibitors, particularly silver nitrate, are used to enhance embryogenic culture initiation in corn. High levels of 2,4-D can induce ethylene formation.

AGAR AND ALTERNATIVE CULTURE SUPPORT SYSTEMS

Agar is used to solidify tissue culture media into a gel. It enables the explant to be placed in precise contact with the medium (on the surface or embedded), but to remain aerated. Agar is a high molecular weight polysaccharide that can bind water and is derived from seaweed. It is added to the medium in concentrations ranging from 0.5 to 1.0% (w/v). High concentrations of agar result in a harder medium. If a lower concentration of agar is used (0.4%) or if the pH is low, the medium will be too soft and will not gel properly. The consistency of the agar can also influence the growth. If it is too hard, plant growth is reduced. If it is too soft, hyperhydric plants may be the result (Singha, 1982). To gel properly, a medium with 0.6% agar must have a pH above 4.8. Sometimes activated charcoal in the medium will interfere with gelling. Typical tissue culture agar melts easily at ~65°C and solidifies at ~45°C.

Agar also contains organic and inorganic contaminants, the amount of which varies between brands. Organic acids, phenolic compounds, and long chain fatty acids are common contaminants. A manufacturer's analysis shows that Difco® Bacto® agar also contains (amounts in ppm): 0.0–0.5 cadmium; 0.0–0.1 chromium; 0.5–1.5 copper; 1.5–5.0 iron; 0.0–0.5 lead; 210.0–430.0 magnesium; 0.1–0.5 manganese, and 5.0–10.0 zinc. Generally, relatively pure, plant tissue culture tested types of agar should be used. Poor quality agar can interfere or inhibit the growth of cultures.

Agarose is often used when culturing protoplasts or single cells. Agarose is a purified extract of agar that leaves behind agaropectin and sulphate groups. Since its gel strength is greater, less is used to create a suitable support or suspending medium.

Gellan gums like Gelrite™ and Phytagel™ are alternative gelling agents. They are made from a polysaccharide produced by a bacterium. Rather than being translucent (like agar), they are clear, so it is much easier to detect contamination, but they cannot be reliquified by heating and gelled again, and the concentration of divalent cations like calcium and magnesium must be within a restricted range or gelling will not occur.

Mechanical supports, such as filter paper bridges or polyethylene rafts, do not rely on a gelling agent. They can be used with liquid media, which circulate better, but keep the explant at the medium surface so that it remains oxygenated. The types of support systems that have been used are as varied as the imagination and include rock wool, cheesecloth, pieces of foam, and glass beads.

PREPARATION OF TISSUE CULTURE MEDIUM

The first step in making a tissue culture medium is to assemble the needed glassware; for example, a 1-liter beaker, a 1-liter volumetric flask, a stirring bar, a balance, pipettes, and the various stock solutions (Table 3.3). The plethora of units used to measure concentration may be confusing when first encountered, so a description of the most common units and what they mean is given below. Once familiar with these, you can confidently proceed to the section on making stock solutions.

UNITS OF CONCENTRATION CLARIFIED

Concentrations of any substance can be given in several ways. The following list gives some of the methods of indicating the concentration commonly found in literature on tissue culture.

- Percentage based upon volume (v/v). Used for coconut milk, tomato juice, and orange juice. For example, if 100 ml of 5% (v/v) coconut milk was desired, 5 ml of coconut milk would be diluted to 100 ml with water.
- Percentage based upon weight (w/v). Often used to express concentrations of agar or sugar. For example, to make a 1% (w/v) agar solution, dissolve 10 g of agar in 1 liter of nutrient medium.
- Molar solution. A mole (M) is the same number of grams as the molecular weight (Avogadro's number of molecules), so a 1-M solution represents 1 M of the substance in 1 liter of solution, and 0.01 M represents 0.01 times the molecular weight in 1 liter. A millimolar (mM) solution is 0.01 M/l, and a micromolar (mM) solution is 0.000001 M/l. Substances like plant growth regulators are active at micromolar concentrations. Molar concentration is used accurately to compare relative reactivity among different compounds. For example, a 1-μM concentration of IAA would contain the same number of molecules as a 1-μM concentration of kinetin, although the same could not be said for units based upon weight. Procedure 3.1 illustrates converting between molar concentrations and mg/ml.
- Milligrams per liter (mg/l). Although not an accurate means of comparing substances molecularly, this is simpler to calculate and use because it is a direct weight. Such direct measurement is commonly used for macronutrients and sometimes for PGRs. One milligram per liter means placing 1 mg of the desired substance in a final volume of 1 liter of solution. One milligram is 10^{-3} grams.

TABLE 3.3
Plant Growth Regulators, Their Molecular Weights, Conversions of mg/l Concentrations into µM Equivalents, and Conversions of µM Concentrations into mg/l Equivalents

Plant Growth Regulator	Abbreviation	M.W.	mg/l Equivalents for These µM Concentrations				µM Equivalents for These mg/l Concentrations			
			0.1	1.0	10.0	100.0	0.1	0.5	1.0	10.0
Abscisic acid	ABA	264.3	0.0264	0.264	2.64	26.4	0.38	1.89	3.78	37.8
Benzyladenine	BA	225.2	0.0225	0.225	2.25	22.5	0.44	2.22	4.44	44.4
Dihydrozeatin	2hZ	220.3	0.0220	0.220	2.20	22.0	0.45	2.27	4.53	45.3
Gibberellic acid	GA3	346.4	0.0346	0.346	3.46	34.6	0.29	1.44	2.89	28.9
Indoleacetic acid	IAA	175.2	0.0175	0.175	1.75	17.5	0.57	2.85	5.71	57.1
Indolebutyric acid	IBA	203.2	0.0203	0.203	2.03	20.3	0.49	2.46	4.90	49.0
Potassium salt of IBA	K-IBA	241.3	0.0241	0.241	2.41	24.1	0.41	2.07	4.14	41.4
Kinetin	Kin	215.2	0.0215	0.215	2.15	21.5	0.46	2.32	4.65	46.7
Naphthaleneacetic acid	NAA	186.2	0.0186	0.186	1.86	18.6	0.54	2.69	5.37	53.7
Picloram	Pic	241.5	0.0242	0.242	2.42	24.2	0.41	2.07	4.14	41.4
Thidiazuron	TDZ	220.3	0.0220	0.220	2.20	22.0	0.45	2.27	4.54	45.4
Zeatin	Zea	219.2	0.0219	0.219	2.19	21.9	0.46	2.28	4.56	45.6
2-Isopentenyl adenine	2-iP	203.3	0.0203	0.203	2.03	20.3	0.49	2.46	4.92	49.2
2,4-Dichlorophenoxyacetic acid	2,4-D	221.04	0.0221	0.221	2.21	22.1	0.45	2.26	4.52	45.2

Procedure 3.1
Converting Molar Solutions to mg/l and mg/l to Molar Solutions, Using Conversion Factors

---

- How to determine how many mg/l are needed for a 1.0 molar (*M*) concentration:

First, look up the molecular weight of the plant growth regulator. In this example, we will use kinetin. The molecular weight is 215.2, so a 1.0 *M* solution will consist of 215.2 g/l of solution.

By using conversion factors and crossing out terms, you cannot go wrong!

$$1\ M\ \text{solution} = \frac{\cancel{1\ \text{mole}}}{\text{liter solution}} \times \frac{215.2\ \text{g}}{\cancel{1\ \text{mole}}} = 215.2\ \text{g/l}$$

To see what a 1.0 m*M* solution of kinetin would consist of, multiply the grams necessary for a 1 *M* solution by 10^{-3}, which would give 215.2×10^{-3} g/l, or 215.2 mg/l.

$$1\ \text{m}M\ \text{solution} = \frac{\cancel{1\ \text{mmole}}}{\text{liter solution}} \times \frac{215.2\ \text{mg}}{\cancel{1\ \text{mmole}}} = 215.2\ \text{mg/l}$$

To see what a 1.0 μ*M* solution of kinetin would consist of, multiply the grams necessary for a 1 *M* solution by 10^{-6}, which would give 215.2×10^{-6} g/l, or 215.2 μg/l.

$$1\ \mu M\ \text{solution} = \frac{\cancel{1\ \mu\text{mole}}}{\text{liter solution}} \times \frac{215.2\ \mu\text{g}}{\cancel{1\ \mu\text{mole}}} = 215.2\ \mu\text{g/l or } 0.215\ \text{mg/l}$$

- How to determine the molar concentration of a solution that is in mg/l:

Let us assume you are given a 10 mg/l solution of indolebutyric acid (IBA) and you wish to know its molarity. First, look up the molecular weight of the plant growth regulator.

The molecular weight of IBA is 203.2. Again, using conversion factors and crossing out terms:

$$\frac{10\ \cancel{\text{mg}}\ \text{IBA}}{\text{liter solution}} \times \frac{\cancel{1\ \text{g}}}{1000\ \cancel{\text{mg}}} \times \frac{1\ \text{mole}}{203.2\ \cancel{\text{g}}} = \frac{0.0000492\ \text{moles}}{\text{liter of solution}} = 49.2\ \mu M\ \text{IBA}$$

Note: Remember 1 *M* = 103 m*M* = 106 μ*M*.

- Micrograms per liter (μg/l). This is used with micronutrients and also sometimes with growth regulators. It means placing 1 μg of a substance in 1 liter of a solution. One microgram = 0.001 or 10^{-3} mg = 0.000001 or 10^{-6} g.
- Parts per million (ppm). Sometimes media components are expressed in ppm, which means 1 ppm or 1 mg/l.

Instructions for making media can be found in Procedure 3.2. These instructions describe MS medium preparation, but will work just as effectively for any other media that you choose. Merely

	Procedure 3.2 **Step-by-Step Media Making**
Step	Instructions and Comments
1	If you are making 1 liter of media, put about half the volume of deionized, distilled water into a beaker and add a stirring bar. Place the beaker on a stirrer so that, when you add different components, they will be thoroughly mixed.
2	Add 10 ml of the following stock solutions (details on how to make them is provided in Table 3.3): nitrate, sulfate, halide, PBMo, and NaFeEDTA.
3	Add the appropriate amount of the vitamin stocks. Instructions for making the vitamin and PGR stocks can be found in Procedure 3.2. At this time, also add the appropriate amounts of inositol (usually 100 mg) and sucrose (usually 20–30 g/l).
4	Add the appropriate volume of the PGR stocks that you plan to use. (Details on how to make them are also found in Procedure 3.2.)
5	Adjust the pH using 0.1–1.0 N NaOH or HCl depending upon how high or low it is. For most media, the pH should be 5.4 to 5.8. Adjust the final volume (1 liter) with deionized, distilled water.
6	If you are going to use liquid medium, you can distribute the media into the tissue culture vessels you plan to use and autoclave. If you plan to use solid medium, weigh out and add the agar or other gelling agent to the liquid. If you are going to use tissue culture vessels (tubes or boxes), melt the agar by putting it on a heat plate while stirring, or use a microwave. You may have to experiment with settings because microwaves differ in output. Once the media is distributed into the tissue culture vessels, these are capped and placed in the autoclave for sterilization. Autoclave for 15 min at 121°C. If a large quantity is being autoclaved, longer times may be necessary. If you plan to distribute the medium into tissue culture vessels after autoclaving (such as into sterile, disposable petri dishes), cap the vessel with a cotton or foam plug and cover the plug with aluminum foil. When the medium is cool enough to handle (about 60°C), it can be moved to the sterile transfer hood, uncapped, and poured into dishes.

follow the same steps and substitute the macro- and micronutrient stocks that you have made for the desired medium. Omit the agar to produce a liquid medium for use in suspension culture. PGRs can be customized for the medium of your choice, whether it is intended for initiation of callus, shoots, roots, or some other purpose.

Adjusting the pH of the medium is an essential step. Plant cells in culture prefer a slightly acidic pH, generally between 5.3 and 5.8. When pH values are lower than 4.5 or higher than 7.0, growth and development *in vitro* is in general greatly inhibited. This is probably due to several factors, including PGRs, such as IAA and gibberellic acid, becoming less stable; phosphate and ion salts precipitating; vitamin B_1 and pantothenic acid becoming less stable; reduced uptake of ammonium ions; and changes in the consistency of the agar (the agar becomes liquified at a lower pH). Adjusting the pH is the last step before adding and dissolving the agar and then distributing it into the culture vessels and autoclaving. If the pH is not what it should be, it can be adjusted using KOH to raise pH or HCl to lower it (0.1–1.0 N). While NaOH can be used, it can lead to an undesirable increase in sodium ions. The pH of a culture medium generally drops by 0.3 to 0.5 units after it is autoclaved and then changes throughout the period of culture, due to both oxidation and to the differential uptake and secretion of substances by growing tissue.

MAKING STOCK SOLUTIONS OF THE MINERAL SALTS

Mineral salts can be prepared as stock solutions 10 to 1000 times (10× to 100×) the concentration specified in the medium. Mineral salts are often grouped into two stock solutions, one for macro-elements and one for microelements, but unless these are kept relatively diluted (10×), precipitation can occur. In order to produce more concentrated solutions, the preferred method is to group the compounds by the ions they contain, such as nitrate, sulfate, halide, phosphorous (P), boron (B), molybdenum (Mo), and iron, and make them up as 100× stocks. Table 3.3 lists the stock solutions for MS medium made at 100× final concentration, which means that 10 ml of each stock is used to make one liter of medium. Some of the stock solutions require extra steps to get the components into solution (for example, NaFeEDTA stock), or may require making a serial dilution to obtain the amount of a trace component for the stock (sulfate and halide stocks).

Sometimes the amount of a particular component needed for a tissue culture medium is extremely small so that it is difficult to weigh out the amount even for the 100× stock solution. Because such small quantities of a substance cannot be weighed accurately, a serial dilution technique is used. The following example illustrates how a serial dilution can be used to obtain the correct amount of a component (in this case $CuSO_4 \cdot 5H_2O$) of the medium for its appropriate stock solution.

The stock solution calls for 2.5 mg of $CuSO_4 \cdot 5H_2O$ as a part of the sulfate stock of MS medium. Make an initial stock solution by placing 25 mg of $CuSO_4 \cdot 5H_2O$ in 100 ml of deionized/distilled water. After mixing thoroughly, use 10 ml of this solution, which will contain the desired 2.5 mg, and place it into the sulfate stock. This procedure required only one serial dilution but any component can be subjected to one or more dilutions to obtain the desired amount.

Once a stock solution is made, it should be labeled as in the example below to avoid error and to prevent inadvertently keeping a stock solution too long.

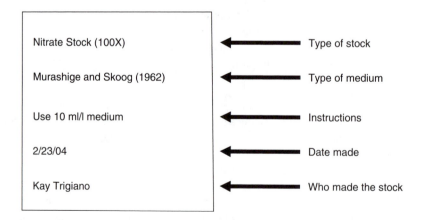

MAKING STOCK SOLUTIONS OF THE PGRS AND VITAMINS

Vitamins and PGR stock solutions can be made up in concentrations 100 to 1000× what is required in the medium. Determine the desired amount for 1 liter of medium, the volume of stock solution needed to deliver that dosage of vitamin or PGR, and the volume of stock you wish to make.

Procedure 3.2 gives examples of how to make vitamin and PGR stock solutions. Many of the PGRs require special handling to get them into solution. You will also find this information in Procedure 3.2.

STERILIZING EQUIPMENT AND MEDIA

Tissue culture media, in addition to providing an ideal medium for the growth of plant cells, are also an ideal substratum for the growth of bacteria and fungi. For this reason it is necessary to sterilize the media, culture vessels, tools, and instruments, and to surface disinfest the explants as well. The most commonly used means of sterilizing equipment and media is autoclaving at 121°C with a pressure of 15 psi for 15 min, or longer for large volumes. Glassware and instruments are usually wrapped in heavy-duty aluminum foil or put in autoclave bags. Media (even those that contain agar) are in a liquid form in the autoclave, requiring a slow exhaust cycle to prevent the media from boiling over when pressure is reduced. Larger volumes of media require longer autoclave times. Media should be sterilized in tissue culture vessels with some kind of closure, such as caps or plugs made of nonabsorbant cotton covered by aluminum foil. This way, they do not become contaminated when they are removed from the autoclave for cooling. Use of racks that tilt the tissue culture tubes during cooling can give a slanted surface to the agar medium. These can be purchased or made from scratch using a little ingenuity.

Some components of the medium may be heat labile or altered by the heat so that they become inactive. These are usually added to the medium after it has been autoclaved, but before the medium has solidified. It is filtered through a bacteria-proof membrane (0.22 µm) filter and added to the sterilized medium after it has cooled enough in order not to harm the heat-labile compound (less than 60°C), and then thoroughly mixed before distributing it into the culture vessels. A rule of thumb is to add the filtered material at a point when the culture flask is just cool enough to be handled without burning one's hands. Some presterilized filters are available; these fit on the end of a syringe for volumes ranging from 1–200 ml. More elaborate disposable assemblies are also available.

PREPARING THE STERILE TRANSFER HOOD

Explants are transferred to the sterile tissue culture medium in a laminar flow transfer hood. A transfer hood is equipped with positive-pressure ventilation and a bacteria-proof high-efficiency particulate air (HEPA) filter. The laminar flow hoods come in two basic types. Generally, air is forced into the cabinet through a dust filter and a HEPA filter, and then it is directed either downward (vertical flow unit) or outward (horizontal flow unit) over the working surface at a uniform rate. The constant flow of bacteria- and fungal-spore-free filtered air prevents nonfiltered air and particles from settling on the working area, which must be kept clean and disinfected. The simplest transfer cabinet is an enclosed plastic box or shield with a UV light and no airflow. A glove box can also be used, but both of these low-cost, low-technology options are not convenient for large numbers of transfers.

In the transfer hood, you should have ready some standard tools such as a scalpel (with a sharp blade), long-handled forceps, and sometimes a spatula. Occasionally, fine-pointed forceps, scissors, razor blades, or cork borers are needed for preparing explants.

Many people prefer doing sterile manipulations on the surface of a presterilized, disposable petri dish. Other alternatives that have worked well are stacks of standard laboratory-grade paper towels. These can be wrapped in aluminum foil and autoclaved. When the top sheet of the stack is used, it can be peeled off and discarded, leaving the clean one beneath it exposed to act as the next working surface. Another alternative is to use ceramic tiles. These can also be wrapped in aluminum foil and autoclaved, but tiles with very slick or very rough surfaces should be avoided.

Procedure 3.3
Making Stock Solutions for Vitamins and Plant Growth Regulators

Step	Instructions and Comments
1	To make a stock solution for nicotinic acid, look at how much is required for 1 liter of medium. In this example, we are going to assume we are making MS medium, so we will need 0.5 mg.
2	It would be convenient to be able to add the 0.5 mg of nicotinic acid by adding a volume of the nicotinic acid stock that corresponded to 1 ml. If 0.5 mg of nicotinic acid must be in 1 ml of the stock solution, then (multiplying by 100) use 50 mg for 100 ml of stock solution. One milliliter may be dispensed into 1.5 ml eppendorf tubes and may be frozen (–20°C).
3	Now prepare a PGR stock — IBA for a rooting medium. If 1.0 mg/l of IBA is require for rooting medium, then our solution must contain 1 mg in each milliliter of the IBA stock. We must weigh 100 mg of IBA for the 100 ml of IBA stock solution. IBA is not very soluble in water, so first dissolve it in a small amount (~1 ml) of a solvent such as 95% ethanol, 100% propanol, or 1 N KOH. Swirl it to dissolve, and then slowly add the remainder of the water to a final volume of 100 ml. Label the stock solution and add 1 ml for every liter of medium to be made. Store in a brown bottle in the refrigerator (4°C). Note: Many growth regulators need special handling to dissolve. Indoleacetic acid, indolebutyric acid, naphthaleneacetic acid, 2,4-D, benzyladenine, 2-iP, and zeatin can be dissolved in either 95% ethanol, 100% propanol, or 1 N KOH. Kinetin and ABA are best dissolved in 1 N KOH, but thidiazuron will not dissolve in either alcohol or base, so a small amount of dimethylsulphoxide must be used. When using only 1 ml of each stock, only very small amounts of the solvent are added to the medium; this will minimize toxic effects.
4	Now let us make a stock solution of IBA, but this time we want to have a 5 micromolar (μM) solution of IBA in the medium. Look up the molecular weight of IBA in Table 3.2 and find that it is 203.2. Using the same rationale in the two examples above, we wish to have 5 μM delivered to the medium by using 1 ml of IBA stock solution. So 100 ml of stock solution must have 500 μM or 0.5 millimoles (mM). If 1 mole (M) is 203.2 g, 1 mM is 203.2 mg, and 0.5 mM IBA × 203.2 mg/1 mM = 101.6 mg of IBA needed for 100 ml of IBA stock solution. To prepare the solution, weigh out the IBA, dissolve it in 1 N KOH as described above, label the stock solution, and store in the refrigerator (4°C) in a brown bottle. Add 1 ml of the IBA stock solution to deliver 5 μM for every liter of the medium.

In Procedure 3.4 is a suggested protocol to follow for preparing the sterile transfer hood. It also contains tips for keeping your work surface clean, eliminating contamination, and avoiding burns when flaming your instruments.

STORAGE OF CULTURE MEDIA

Once culture media have been made, distributed into tissue culture vessels, and sterilized by autoclaving, they can be stored for up to one month, provided they are kept sealed to prevent excessive evaporation of water from the medium. They should also be placed in a dark, cool place to minimize degradation of light labile components such as IAA. Storage at 4°C prolongs the time

Procedure 3.4 Getting Under the Hood	
Step	Instructions and Comments
1	Turn on the transfer hood so that positive air pressure is maintained. This ensures that all of the air passing over the work surface is sterile. You should feel air flowing against your face at the opening. Make sure there are no drafts such as open windows or air conditioning vents that may interfere with the air flow coming out of the hood.
2	Use a spray bottle filled with 70% (v/v) ethanol to spray down the interior of the hood. Do not spray the HEPA filter! This is more effective than absolute alcohol for sterilizing surfaces, perhaps because 70% (v/v) ethanol denatures DNA. You can also use a piece of cheesecloth saturated with 70% (v/v) ethanol to help distribute the ethanol more uniformly. Allow it to dry. To maintain the cleanliness of the interior, anything that is now placed inside the hood must be sprayed with 70% (v/v). This includes the alcohol lamp, the slide staining jar filled with 80 to 95% (v/v) ethanol for flaming the instruments, a rack for the tools, racks of the tissue culture vessels containing medium, and all of the presterilized wrapped bundles containing your working surface (tiles, paper towels, ashtrays, etc.) and your tools (forceps, spatula, scalpel, etc.).
3	When you are ready to begin, remove jewelry and wash your hands thoroughly with soap and water. Just before placing your hands in the hood, spray them down with 70% ethanol.
4	You may now open your work surface by peeling back the heavy duty aluminum foil, exposing the surface of the tile (or alternative surface). Never block the air flow across the surface coming from the filter unit. Also, do not pass your hands across the surface of the tile. Talking while you are in the unit also compromises sterility. If you must talk, turn your head to one side. If you have long hair, tie it up so that it does not dangle onto your work surface.
5	Keep any open sterile containers as far back in the hood as you can. When you open containers that have been sterilized, keep your fingers away from the opening. If you open a glass container or vessel, as general rule, pass the opening through the flame. This creates a warm updraft from the vessel, helping to prevent contamination from entering it.
6	Flaming instruments can be hazardous if you forget that ethanol is flammable. When you are flame sterilizing an instrument, such as forceps, dip the instrument into the jar containing the 80% ethanol. When you lift it out, keep the tip of the instrument at an angle downward (away from your fingers) so that any excess alcohol does not run onto your hand (Figure 3.1). Then pass the tip through the flame of the alcohol burner and hold the instrument parallel to the work surface. When the flame has consumed the alcohol, let the tool cool on the rack until you are ready to use it. Never, never place a hot tool back into a jar containing 80% ethanol. I know an experienced scientist who momentarily forgot this simple rule, and the resulting fire burned his hand and singed the hair off his forearm!

that media can be stored, but condensate may form in the container and encourage contamination. By making media 5 to 7 days in advance, you allow time to check for any unwanted microbial contamination before explants are transferred onto the medium. However, media for certain sensitive species or operations, or media that contain particularly unstable ingredients, must be used fresh and cannot be stored.

SURFACE DISINFESTING PLANT TISSUES

Just as the media, instruments, and tools must be sterilized, so must the plant tissue be disinfested before it is placed on the culture medium. Many different materials have been used to surface disinfest explants, but the most commonly used are 0.5–1% (v/v) sodium hypochlorite (commercial bleach contains 5% sodium hypochlorite), 70% alcohol, or 10% hydrogen peroxide. Other methods include using a 7% saturated solution of calcium hypochlorite, a 1% solution of bromine water, a 0.2% mercuric chloride solution, and a 1% silver nitrate solution. If these more rigorous techniques are used, precautions should be taken to minimize health and safety risks, especially with the solutions containing heavy metals.

The type of disinfestant used, the concentration, and the amount of exposure time (1–30 min) all vary depending upon the sensitivity of the tissue and how difficult it is to disinfest. Woody or field-grown plants are sometimes very difficult to disinfest and may benefit from being placed in a beaker with cheesecloth over the top and placed under running water overnight. In some cases, employing a two-step protocol (70% ethanol followed by bleach) or adding a wetting agent such as Tween 20 or detergent helps to increase the effectiveness. In any case, the final step before trimming the explant and placing it onto the sterile medium is to rinse it several times in sterile, distilled water to eliminate the residue of the disinfesting agent.

FINAL WORD

Tissue culture is much like good cooking. There are simple recipes and then there is "haute cuisine." Cooking is also a very rewarding activity, particularly when the end result is delicious. By following the procedures outlined in this chapter and in the succeeding chapters of the book, you should find that with care and attention to detail, you will be a "chef extraordinaire," and your tissue culture ventures will be successful!

LITERATURE CITED AND SUGGESTED READINGS

Ammirato, P.V. 1974. The effects of abscisic acid on the development of somatic embryos from cells of caraway (*Carum carvi* L.). *Bot. Gaz.* 135:328–337.

Coffin, R., C.D. Taper, and C. Chong. 1976. Sorbitol and sucrose as carbon source for callus culture of some species of the Rosaceae. *Can. J. Bot.* 54:547–551.

Epstein, E., K.-H. Chen, and J.D. Cohen. 1989. Identification of indole-3-butyric acid as an endogenous constituent of maize kernels and leaves. *Plant Growth Regul.* 8:215–223.

Gamborg, O.L., R.A. Miller, and K. Ojima. 1968. Nutrient requirements of suspension cultures of soybean root cells. *Exp. Cell Res.* 50:151–158.

Kochba, J. et al. 1978. Stimulation of embryogenesis in *Citrus ovular* callus by ABA, Ethephon, CCC and Alar and its supression by GA$_3$. *Ztg. Pflanzenphysiol.* 89:427–432.

Lloyd, G. and B. McCown. 1980. Commercially feasible micropropagation of mountain laurel, *Kalmia latifolia*, by use of shoot tip culture. *Int. Plant Prop. Soc. Proc.* 30:421–427.

Murashige, T. and F. Skoog. 1962. A revised medium for rapid growth and bio-assays with tobacco tissue cultures. *Physiol. Plant.* 15:473–497.

Nissen, S.J. and E.G. Sutter. 1990. Stability of IAA and IBA in nutrient medium to several tissue culture procedures. *HortScience* 25:800–802.

Nitsch, J. P. and C. Nitsch. 1969. Haploid plants from pollen grains. *Science* 163:85–87.

Oka, S. and K. Ohyama. 1982. Sugar utilization in mulberry (*Morus alba* L.) bud culture. In: *Plant Tissue Culture*. (Ed.) A. Fujiwara. Proc. 5th Int. Cong. Plant Tiss. Cell Cult., Jap. Assoc. Plant Tiss. Cult., Tokyo, Japan, 67–68.

Roberts, D.R., B.S. Flinn, D.T. Webb, F.B. Webster, and B.C. Sutton. 1990. Abscisic acid and indole-3-butyric acid regulation of maturation and accumulation of storage proteins in somatic embryos of interior spruce. *Physiol. Plant.* 78:355–360.

Schenk, R.V. and A.C. Hildebrandt. 1972. Medium and techniques for induction and growth of monocotyledonous plant cell cultures. *Can. J. Bot.* 50:199–204.

Singha, S. 1982. Influence of agar concentration on *in vitro* shoot proliferation of *Malus* sp. 'Almey' and *Pyrus communis* 'Seckel'. *J. Am. Soc. Hort. Sci.* 107:657–660.

Skoog, F. and C.O. Miller. 1957. Chemical regulation of growth and organ formation in plant tissues cultured *in vitro*. *Symp. Soc. Exp. Biol.* 11:118–131.

White, P. R. 1963. *The Cultivation of Plant and Animal Cells.* 2nd ed. Ronald Press, New York.

4 Histological Techniques

Robert N. Trigiano, Kathleen R. Malueg, Kimberly A. Pickens, Zong-Ming Cheng, and Effin T. Graham

CHAPTER 4 CONCEPTS

- Histological examination of plant tissues using paraffin or plastic media involves the following steps: fixing, dehydrating, infiltrating, embedding, sectioning, and staining of samples.

- Histological techniques can reveal structural details and developmental processes in plants.

- Histological techniques are useful for investigating and confirming various responses of plant tissues to *in vitro* manipulations.

INTRODUCTION

Plant histology can be defined as the study of the microscopic structures or characteristics of cells and their assembly and arrangement into tissues and organs. Several histological techniques are commonly used for examining plant tissues, each providing somewhat similar gross information, but differing in resolution of details and the media in which the samples are prepared. These include techniques for brightfield and fluorescent microscopy, in which specimens can be prepared for cutting into thick sections (15–40 micrometers [µm]) either without a stabilizing medium (fresh sections), in cryofluids (frozen sections), or embedded in paraffin-like materials or in various formulations of plastic. Other techniques that employ electron microscopy either do not require a specialized embedding medium for specimen preparation (scanning electron microscopy) or use samples embedded in plastics (transmission electron microscopy) that can be cut into ultrathin sections (65–100 nanometers [nm]). Explanation of the details involved in all of these techniques is beyond the scope of this chapter.

One might reasonably ask, why include a chapter on histology in a plant development and biotechnology book? In many situations, histological techniques provide essential information that may not be evident upon mere visual inspection. Much of our understanding of the *in vivo* and *in vitro* developmental processes that are presented throughout this book has resulted from detailed histological research. For example, somatic embryos (Chapter 14) can be produced on the surface of a leaf explant, but may be so morphologically aberrant that they are unrecognizable. By using histological techniques and scrutinizing anatomical features, the scientist can see more readily the characteristics of somatic embryos. Another example for using histological techniques is to investigate the origin of specific structures — i.e., adventitious shoots and roots (Chapter 11), embryos, and so on — that develop in culture. Histological development can be studied over time by periodically sampling tissues, or the result can be examined in the mature structure. One must always be mindful that growth of tissues is dynamic and is changing from moment to moment, whereas histological sections, for example, are static and fixed in time, and only present a very narrow glimpse of the developmental process. Nevertheless, the origin of tissues and organs may be convincingly inferred from serial observation of many samples and sections.

The field of plant histology and microtechnique is quite broad and cannot be conveyed in one simple chapter. Therefore, the intent of this chapter is to summarize briefly the essentials and to provide a primer of sorts for preparing specimens from plants and tissue cultures for histological examination using elementary paraffin and scanning electron microscopy techniques. We also provide a brief account of staining nuclei and chromosomes to complement material presented in Chapter 17, "Haploid Cultures." The techniques presented herein are considerably easier and avoid many of the toxic materials found in more traditional protocols. For a broader view of general histological methodologies, students should consult one or more of the following references: Berlyn and Miksche, 1976; Sass, 1968; Jensen, 1962; and Johansen, 1940.

EQUIPMENT AND SUPPLIES

The equipment and materials needed to complete a paraffin histological study are listed below:

- Indelible marker (Scientific Products [SP], Atlanta, GA), or #2 lead pencil
- 500 or 1000-ml plastic beaker
- Aluminum pan
- Rotary microtome with disposable blade holder (SP)
- Warming oven (58–60°C)
- Two slide warming trays (40 and 50°C)
- Water bath (50°C)
- Hot/stir plate
- 1% agar or agarose in a 150-ml beaker
- Disposable microtome blades (SP)
- Paraplast® embedding medium (SP)
- Disposable plastic base molds and embedding rings (SP)
- 5-ml microbeakers (SP)
- Alcohol lamp and metal spatula
- Screw-capped Coplin jars
- Histochoice® fixative (Amresco, Inc., Solon, OH) plus 20% ethanol
- Disposable snap cap plastic 10-ml specimen vials
- Wooden applicator sticks
- Clean glass slides and coverglasses
- Camel or sable hair artist's brush
- Test tube racks
- Eukitt or similar coverglass resin (EMS, Fort Washington, PA)
- Microclear® (Micron Environmental Industries, Fairfax, VA)
- 30%, 50%, 75%, 95% (all aqueous) and 100% isopropanol
- 0.2% aqueous solution of Alcian blue 8GX (Sigma) 200 mg/100 ml of water.
- Quick-mixed hematoxylin (Graham, 1991). Stock A: 750 ml distilled water, 210 ml propylene glycol, 20 ml glacial acetic acid, 17.6 g aluminum sulfate, and 0.2 g sodium iodate; Stock B: 100 ml propylene glycol and 10 g certified hematoxylin (Sigma). Stain is prepared by adding 2 ml of Stock B to 98 ml of Stock A.

GENERAL PROTOCOL FOR PARAFFIN STUDIES

The procedure described next will work well with most tissues, including those from stock plants as well as those cultured *in vitro*. Moreover, the procedure has several important advantages over more traditional paraffin protocols. First, it circumvents the use of toxic aldehydes (formaldehyde) and heavy metals (chromium and mercury) as fixatives. Second, it avoids the use of specimen/slide adhesive agents, such as Haupt's, that require the use of formaldehyde and may produce staining artefacts. Third, it does not use toxic substances, such as xylene, for deparaffinizing sections. Fourth, sections are stained directly through the paraffin, greatly reducing the number of operational steps in the process. The advantages listed above combine to make preparing tissue for histology and slides a safe and less intimidating process for both students and instructors alike. There are circumstances where more traditional fixation and staining protocols are warranted, however, and one is included in this chapter.

FIXATION

The first step in the protocol is to identify typical specimens (i.e., embryos on leaf sections), remove them from the culture dishes, and place them in a small petri dish containing water. Most explants are too large to be adequately fixed, and usually the subject (embryos) occupies a very small area. The specimen should then be carefully trimmed using a razor blade or a scalpel so that the subject and a small area of surrounding tissue remain. In each specimen vial, five or fewer specimens are placed in about 5 ml of Histochoice®, a nonaldehyde, nontoxic fixative amended with 20% ethanol. If possible, the open vials should be aspirated (–1 atmosphere [atm] or less) for about 30 min to remove air from the samples and promote infiltration of the fixative. For very small or very delicate specimens, we suggest that fixation without aspiration be tried first. The vials are closed and the tissue allowed to remain in the fixative at room temperature for a minimum of 24 h; samples may be stored at room temperature for weeks or months without degradation of structures. However, replace the original fluid with fresh fixative after 24 h. If long-term storage is needed, we suggest using screw-capped rather than snap-capped vials.

DEHYDRATION AND INFILTRATION

After fixation, the samples must be dehydrated and placed in a solvent compatible with paraffin. Traditionally, this has been accomplished with graded series of ethanol and transitioning to tertiary butyl alcohol. This process can be considerably shortened by using isopropanol to dehydrate the tissue and dissolve the paraffin. Histochoice® fixative can be removed from the vials with a long thin pipette and poured down the drain with plenty of water. Specimens are dehydrated using 30%, 50%, 75%, 95%, and three changes of 100% isopropanol — each for about 10–30 min. After the last change of pure isopropanol, fill the vials about one-quarter full with fresh 100% isopropanol, add a few pellets of Paraplast® embedding medium, loosely recap the vials, and incubate at 60°C. Periodically over the next several hours, swirl the contents and add a few more pellets of Paraplast® to each of the vials. At the end of the day, remove the caps from the vials to allow the isopropanol to evaporate. At this time, fill a 500 or 1000-ml plastic beaker with Paraplast® pellets, place several pellets into a number of base molds contained in an aluminum pan, and incubate in the 60°C oven. The next morning all of the isopropanol should be dissipated, the specimens completely infiltrated with Paraplast®, and the pellets in the beaker and base molds should be melted.

CASTING SPECIMENS INTO BLOCKS

Using a sharp razor blade, whittle a flattened paddle-shaped end on several wooden applicator sticks and store in a vial with the flattened end immersed in molten paraffin in the oven. Transfer a specimen into the molten paraffin contained in a base mold and position it so that the sectioning plane of the specimen is parallel to the bottom of the base mold. For example, if a transverse section of a stem is required, the stem should be placed perpendicular to the base mold; the long axis of the stem should be sticking straight up in the molten paraffin. If the stem is positioned flatly in the base mold, longitudinal sections will result. With more complex specimens, such as single somatic embryos or shoots growing from basal tissue, this orientation step becomes more critical.

It is important to visualize the exact desired section plane at this step. Once the sample is oriented correctly, quickly transfer the base mold to a cool surface — an inverted aluminum pan containing ice works well. The paraffin surrounding the bottom of the base mold will quickly become cloudy (congealed) and specimen will be immobilized. Rapidly attach an embedding ring to the base mold and fill the apparatus with molten paraffin from the plastic beaker. This is a very vulnerable point in the process. If the immobilizing paraffin is allowed to congeal too far, the added molten paraffin will not bond with it and the block will split apart. At this time, a label consisting of white paper and text identifying the specimen written in pencil (do not use ink) may be inserted partially into the molten paraffin. The casted blocks may now be placed in the refrigerator, or they may be floated in a large beaker containing ice water. The base mold may be removed after the block has hardened.

MICROTOMY AND MOUNTING SECTIONS ON SLIDES

A word of caution before beginning to section. The stainless steel knives and disposable razor blades that are used to cut sections are both extremely sharp and cut flesh very easily. Exercise care while working with the microtome. If the microtome is left unattended, place a note on the knife holder warning in bold print that a knife or blade has been installed. The rotary action arm should be kept locked at all times except when sectioning.

One of the advantages of using embedding molds is that the cutting face of the block is square. If the block were sectioned as is, the resulting "ribbon of very large sections" would be straight. However, seldom does the specimen occupy the entire block face. Therefore, it is often desirable to trim away some of the excess paraffin from around the sample. The horizontal edges of the block must be parallel with each other. To put it another way, the sides of the block must be of equal length. The block edges must also be kept parallel to the knife edge. If the ribbon of sections curves to the right or left, the side of the block opposite the direction of curvation is too long. Remove the block from the chuck and retrim so that the sides are parallel and of equal length. Do not trim the block while mounted in the chuck over the knife!

Mount a new disposable blade in the blade holder. Only a small portion of the blade should be clear of the carriage. Typically, new blades are coated with oil and may be cleaned with a small wad of tissue moistened with a little Microclear®. Carefully wipe the blade perpendicular to the blade surface with the tissue and allow to completely dry. Set the microtome to cut 10 to 12-µm sections. Now mount the specimen block in the chuck and orient it so the bottom of the cutting face is parallel to the edge of the blade. Position the specimen at the level of the knife and carefully and gently slide the knife holder toward the block until they are very close. With the utmost care, turn the wheel clockwise, causing the specimen to advance 10 µm per revolution. The first sections may not include the entire block face; continue advancing the block until a complete section is obtained. Lock the advancing wheel into position, carefully remove the partial sections, and clean the knife as before. After the Microclear® has evaporated, continue to operate the microtome and, after five or more sections are produced, place a camel hair brush underneath the ribbon for support while it is being held away from the knife holder. A 30 to 50-section ribbon can be produced and easily manipulated. Lay the ribbon out on a clean piece of dark paper and, using a razor blade, divide it into segments (sections) that do not exceed about 75% of the nonfrosted length of a glass microscope slide.

Modern Superfrost® slides afford the following well-established advantages:

1. They can be used directly from the manufacturer without cleaning.
2. The glass surface bonds with tissue sections without an adhesive coating.
3. The special labeling pen resists all common histological solvents (Graham and Trentham, 1998).

If these are unavailable in your laboratory, prewash glass microscope slides in a mild dish-washing detergent solution, rinse well with tap water, and rinse with distilled water. Allow the slides to dry in test tube racks with an aluminum foil base. Washed and dried slides can be stored indefinitely in slide boxes. Label the frosted portion of the slide with appropriate information using an indelible marker or lead pencil and cover the specimen surface of the slide with distilled water. Moisten the tip of a brush, touch it to the surface of the first section to the left, and lift the entire ribbon off of the paper. Now, touch the first section to the right onto the water and lay the rest of the ribbon down so that air is not trapped underneath the sections. Place the slide on the 50°C warming table to expand the sections. After about 10 min, set the slide on a sponge and slightly tilt it to drain off most of the water. Affix the sections to the glass by placing the slide on the 40°C warming table for 1–16 h, or overnight. The sections are now ready to be stained.

STAINING SECTIONS

The staining procedure shown in Procedure 4.1 is adapted from Graham and Joshi (1995). An alternative to this staining process is offered by Graham and Trentham (1998). This method (Procedure 4.2) is very efficient and provides a triple stain in a single solution. More traditional fixing and staining procedures are provided in Procedure 4.3. Do not become discouraged if your results are less than perfect the first time. Do not give up — try again! As with any other technique, the operator becomes more skilled after some experience.

IMMOBILIZATION OF SPECIMENS FOR PARAFFIN SECTIONING

This sample preparation technique involves immobilization and precise orientation of specimens between two layers or a "sandwich" of agar. It offers several advantages over more conventional methods of preparation, especially when working with small or delicate tissues. Small samples, e.g., embryos or callus, are often lost or damaged during the dehydration sequences when fluids are removed and replaced. Immobilizing the specimens in agar makes it nearly impossible to lose samples, and the agar minimizes damage from handling. However, the real advantage to the technique is that the sample can be oriented in the agar before fixation, eliminating the need to position it while casting the block.

This procedure is adapted primarily from Hock (1974). Prepare 1% water agar (1 g agar to 100 ml of water in a tall 150-ml beaker) on a hot/stir plate. After the agar is melted, store in a 50°C water bath. Dip a clean glass microscope slide into the molten agar, and then place it in a 100×15 mm petri dish containing moistened filter paper. The remainder of the immobilization procedure is diagrammatically represented in Figure 4.1. Select a specimen and place it on top of the cool and hardened agar. With a pasteur pipette, cover the sample with a small amount of molten agar. The position of the sample is now secure. Draw an arrow in pencil on a piece of white paper about the width of the sample. Using a stereomicroscope if necessary, and fine forceps, place the arrow directly opposite or behind the desired cutting face of the tissue and tack it into position using molten agar. Cover the entire preparation with a thin coating of molten agar. Once the agar has hardened, trim any excess agar away using a razor blade. The specimen can be removed from the slide by slipping the razor blade under the bottom of the agar sandwich and placing it in fixative. Follow the previously described protocols for fixation of tissue and infiltration with paraffin.

Instead of using an embedding mold, use a 5-ml plastic microbeaker to cast the specimen into a paraffin block. Draw a straight line on the bottom of the beaker with an indelible pen. Fill the container with molten paraffin and transfer the sample to the beaker, so that the sample is parallel but slightly behind the paper, and the arrow is positioned orthogonally (at a right angle) to the line on the bottom (Figure 4.2). Place the microbeaker on a cold surface and label with a paper slip as described above.

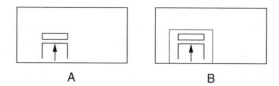

FIGURE 4.1 Immobilization. (A) The specimen (stem) is placed on an agar-coated slide, then covered with molten agar. A white slip of paper with an arrow drawn in pencil is also affixed with agar. (B) After the agar has hardened, the preparation may be trimmed and removed from the glass slide (see dotted lines).

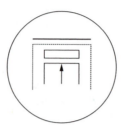

FIGURE 4.2 Top view of positioned specimen cast in a plastic microbeaker.

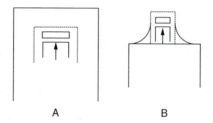

FIGURE 4.3 Mounting the dissected specimen (A) to a flat block cast using an embedding mold (B). Note the paraffin reinforcement along the sides of the specimen.

Procedure 4.1
Staining through Paraffin with Alcian Blue and Hematoxylin

Step	Instructions and Comments
1	Preheat the alcian blue solution in a closed Coplin jar and place the slides in the stain for 1 h at 50°C.
2	Gently rinse the slides in a stream of cold tap water and carefully blot dry the remaining water droplets.
3	Transfer the slides to Coplin jars containing hematoxylin stain for 15 min at room temperature. Rinse and dry as described above and place on the 40°C slide table for a minimum of 30 min.
4	The paraffin in the sections is removed by means of three 5-min soaks in Microclear® followed by three 5-min changes of 100% isopropanol.
5	After the slides have air-dried, pool some resin (Eukitt) in the center of the slide and gently lower a coverglass so that the resin spreads evenly over the surface without air bubbles. The resin may be cured overnight on a 40°C warming table.

<table>
<tr><td colspan="2" align="center">Procedure 4.2
Triple Stain Using Alcian Blue, Safranin O, and Bismarck Brown</td></tr>
<tr><td align="center">Step</td><td align="center">Instructions and Comments</td></tr>
<tr><td align="center">1</td><td>Prepare a 0.1 M acetic acid-sodium acetate buffer, pH 5.0, and three 1% stock solutions of alcian blue 8 GX, safranin O, and Bismarck Brown, using 0.1 g of dye powder dissolved in 100 ml of absolute ethanol.</td></tr>
<tr><td align="center">2</td><td>To 100 ml of acetate buffer, add 5 ml of alcian blue, 2 ml of safranin O, and 1 ml of Bismarck Brown stock solutions.</td></tr>
<tr><td align="center">3</td><td>Extract paraffin from sections by moving slides through four baths of Microclear®, then through five baths of 100% isopropanol, to remove the paraffin solvent. Finally, place the slides on a rack to air-dry.</td></tr>
<tr><td align="center">4</td><td>Immerse slides in the triple stain. Coloring of tissues will be evident within 15 min, but optimum staining differs widely among various specimens. Progress of staining is monitored easily by rinsing a slide briefly with distilled water, blotting it dry with soft tissue paper, and observing under a compound microscope. The slide may be returned to the stain solution until satisfactory staining is achieved. We suggest reexamining the slide at 15 min intervals.</td></tr>
<tr><td align="center">5</td><td>When staining is sufficient, dry slides for 1 h on a 40°C slide warmer and affix a coverslip as described in Procedure 4.1, Step 5. The results of the triple stain are the following: polysaccharide cells are blue; lignified walls, nuclei, and chloroplasts are red; and cuticle (if present) is brown or yellow-brown.</td></tr>
</table>

DISSECTION AND MOUNTING OF SPECIMENS

The block of paraffin in which the specimen was cast cannot be mounted on the microtome. Therefore, it is necessary to dissect the specimen and a small amount of the surrounding block and mount it on a block formed by an embedding mold. Usually there are some old blocks around from previous projects; newly cast blocks may also be used. Before the microbeaker is removed, take a sharp razor blade and cut through the beaker and score the paraffin in the block underlying the line on the microbeaker. This will ensure that the cutting face is identified even if the specimen and arrow are not clearly visible in the block. Dissect the specimen from the block. Create a flat field on the face of the embedding mold block. Melt a thin layer of the embedding block using the flat side of a spatula heated with an alcohol burner. Quickly attach the side opposite of the cutting face of the specimen block to the embedding mold block. Reinforce the attachment by melting small chips of paraffin with a hot spatula tip along all four sides of the specimen block (Figure 4.3). The block may be sectioned and slides prepared as previously described.

PREPARATION OF SPECIMENS FOR SCANNING ELECTRON MICROSCOPY (SEM)

The sections prepared for paraffin histology present a two-dimensional (flat) interior view of a small area of a much larger three-dimensional object. In contrast, SEM allows for three-dimensional topical or internal views of an entire specimen. Samples for SEM do not need to be embedded in paraffin, nor do they need to be stained. Moreover, the resolving power of a scanning electron microscope is much greater than that of a compound light microscope. However, in many ways, preparation of samples for SEM is technically easier than for paraffin histology. *Note*: methodologies for critical point drying and sputter coating are not presented here. These processes should be performed by a competent SEM technician or by a person familiar with these techniques.

	Procedure 4.3 **Traditional Fixing and Staining Schedule**
Step	Instructions and Comments
1	Fix tissue in 50% FAA (5 ml of 37% formaldehyde, 5 ml of glacial acetic acid, and 90 ml of 55% ethanol) for a minimum of 24 h. **Caution**: Do this operation under a fume hood, and avoid breathing vapors or making contact with the skin.
2	Decant the 50% FAA from the specimen jars. Begin dehydration and paraffin infiltration of the samples with 50% isopropanol as described under the dehydration and infiltration section. Cut sections and mount on clean glass slides as previously described.
3	The following is a description of a triple stain that works well for most plant tissues. Place slides with sections in carrier and immerse in the following solutions, contained in Coplin jars, for a minimum of 5 min each: 100% Microclear®; 100% Microclear®; 100% Microclear®; 50% Microclear® and 50% absolute ethanol; 100% ethanol; 95% ethanol (you may skip this step); 70% ethanol; and 50% ethanol.
4	Safranin O (4 g safranin O; 100 ml water; 100 ml 95% ethanol; 4 g sodium acetate) for 30 min to 2 days. It is very difficult to overstain with safranin. Wash slides with several changes of water, until water is no longer pink. Do not let running water directly contact slides. Immerse slides in crystal violet (1 g crystal violet with 100 ml of water) for 30–90 sec. It is very easy to overstain. Immediately wash in water, as with safranin, until water is no longer purple (water may remain slightly violet).
5	Dehydrate sections for 5 min each in the following solutions: 50% ethanol; 70% ethanol; 95% ethanol; and 100% ethanol. Place slides in fast green (1.2 g fast green; 80 ml methyl cellusolve; 80 ml methyl salicylate; 80 ml absolute ethanol) for 30–90 sec. **Caution**: It is easy to overstain. Immediately immerse slides in 100% ethanol for 10 sec to remove most of excess stain, and then transfer to a new Coplin jar containing 100% ethanol for 5 min.
6	Transfer slides through the following solutions for a minimum of 5 min each: 50% Microclear® and 50% absolute ethanol; 100% Microclear®; 100% Microclear®; and 100% Microclear®. Coverslips may now be added to the slides as described previously.

The following materials will be needed to prepare samples for SEM:

- 3% glutaraldehyde in a 0.05 M potassium phosphate buffer with a pH of 6.8; store in the refrigerator[*]
- 1% osmium tetroxide in the same buffer (optional); store in a shatter-proof bottle in the freezer[*]
- Aluminum mounting stubs[*]
- Double-sided sticky tape
- A 0.05 M potassium phosphate buffer with a pH of 6.8; prepare by dissolving 0.87 g K_2HPO_4 and 0.68 g KH_2PO_4 in 200 ml of water
- A graded series of ethanol or acetone (10%, 30%, 50%, 75%, 95%, and 100%)

---

[*] May be purchased from Electron Microscopy Sciences, Fort Washington, PA.

Note: glutaraldehyde and osmium tetroxide are extremely toxic and should only be handled when wearing gloves and in a certified fume hood. For additional information, consult material safety data sheets.

Follow the steps in Procedure 4.4 to prepare samples for SEM.

Procedure 4.4 **Preparation of Specimens for Scanning Electron Microscopy (SEM)**	
Step	Instructions and Comments
1	Fix and aspirate small pieces of tissue in cold (4°C) buffered solution of glutaraldehyde for at least 24 h.
2	Dispose of the fixing solution in a waste bottle and rinse specimens at least three times with cold (4°C) phosphate buffer.
3	With great care and working in a fume hood, dispense enough of the cold, buffered osmium tetroxide (osmium) solution to cover the samples and recap the vials. After 2 h, dispose of the osmium in a properly labeled hazardous waste bottle, and rinse the tissues with three changes of cold buffer. In many instances, the tissue will be blackened after postfixing with osmium. This is normal.
4	Dehydrate the tissue for 30 min in each member of the graded ethanol or acetone series except 100%. The final steps of dehydration are three changes of room temperature absolute ethanol or acetone over 30 min. The samples may be stored in absolute ethanol for a few days until they can be critically point-dried by a SEM technician. *Note:* We suggest that, if the specimens cannot be dried within a few days, they should be stored in 75% ethanol or acetone.
5	Have a SEM technician critically point-dry the specimens. Place a piece of clear double-sided tape onto an aluminum stub and mount the dried sample on the tape. The samples are ready to be gold–palladium coated and viewed with the aid of a technician.

CHROMOSOME COUNTING

Determination of the chromosome number of a plant species may be important for research in genetics, evolution, and taxonomy. The chromosome number is generally conserved within a species, and taxonomical assumptions can be made according to this number. Chromosome counting is also a necessary means for determining the ploidy level from plants derived from anther and ovule cultures or chromosomal doubling by mitotic inhibitors (see Chapter 17).

The equipment and materials needed for study are listed below.

- Onion bulb and moist sand in a large plastic beaker
- Acetocarmine stain (45 ml glacial acetic acid, 55 ml of distilled water, and 0.5 g carmine)*
- Fixative (75 ml of absolute ethanol and 25 ml of glacial acetic acid) and 70% ethanol
- Microscope slides and coverslip
- Compound microscope equipped with oil immersion lens
- Alcohol lamp, blotting paper, and pencil with rubber eraser

---

* Combine in beaker covered with aluminum foil and gently boil solution for 5 min under a fume hood. Filter solution through filter paper (this may take several days) and store in refrigerator. The stain should be dark red.

Follow the protocol listed in Procedure 4.5 to stain chromosomes of onion.

Procedure 4.5 Staining Chromosomes in Onion Root Tip Squashes	
Step	Instructions and Comments
1	Place an onion bulb with flat sides on moist sand contained in a large beaker. Bulb cultures may be incubated at room temperature on the lab bench.
2	New roots should emerge from the base of the bulb after about 10–14 days (depending on temperature). No later than mid-morning (10 AM), excise about 1–2 cm root tips, remove sand with water and store in fixative for 2 h. Roots may be stained immediately in acetocarmine or stored in 70% ethanol until needed.
3	Trim roots to about 0.5 cm and place in acetocarmine stain for 20–120 min. Transfer the tissue to a glass slide along with acetocarmine stain. Gently pass the slide through the flame from an alcohol lamp. **Caution**: Do not allow the stain to boil away; add more to prevent drying. Cover the preparation with a coverslip. Place several layers of tissue paper or blotting paper over the specimen and squash the tissue flat by pressing evenly and firmly with the eraser end of a pencil. Try not to move the coverslip.
4	Examine the preparation with a high dry or an oil immersion lens for even spreads of chromosomes and mitotic figures. The preparation may be made permanent by sealing the edges of the coverslip with nail polish.

LITERATURE CITED

Berlyn, G.P. and J.P. Miksche. 1976. *Botanical Microtechnique and Cytochemistry.* Iowa State University Press, Ames, IA.

Graham, E.T. 1991. A quick-mixed aluminum hematoxylin stain. *Biotech. Histochem.* 66:279–281.

Graham, E.T. and P.A. Joshi. 1995. Novel fixation of plant tissue, staining through paraffin with alcian blue and hematoxylin, and improved slide assembly. *Biotech. Histochem.* 70:263–266.

Graham, E.T. and W.R. Trentham. 1998. Staining paraffin extracted, alcohol rinsed and air dried plant tissue with an aqueous mixture of three dyes. *Biotech. Histochem.* 73:178–185.

Hock, H.S. 1974. Preparation of fungal hyphae grown on agar coated microscope slides for electron microscopy. *Stain Technol.* 49:318–320.

Jensen, W.A. 1962. *Botanical Histochemistry.* W. H. Freeman, San Francisco.

Johansen, D.A. 1940. *Plant Microtechnique.* McGraw-Hill, New York.

Sass, J.E. 1968. *Botanical Microtechnique.* 3rd edition. Iowa State University Press, Ames, IA.

5 Photographic Methods for Plant Cell and Tissue Culture*

Dennis J. Gray

CHAPTER 5 CONCEPTS

- Plant cell and tissue culture is based on visual observation, making photographic documentation necessary.

- Digital imaging technology is rapidly replacing film.

- Proper equipment is essential in obtaining acceptable photographic documentation.

- Details of composition often determine the difference between acceptable and unacceptable photography.

INTRODUCTION

Plant cell and tissue culture is based on direct visual observation, unlike other disciplines, such as molecular genetics, virology, and physiology, where the results from experiments are often abstract. With *in vitro* culture, the confirmation that the experimental response has occurred is based primarily on visual proof. For example, the scientist determines whether or not cultured tissue has undergone embryogenesis or organogenesis by looking for the presence of somatic embryos or organs. Thus, the fundamental documentation of plant cell and tissue culture research is often a photograph.

As an essential component of cell and tissue culture research, accurate photography is needed in order to satisfy the doctrine of scientific research, which demands that experimental methods and results be documented and repeatable so that they can be confirmed and built upon by others. Unfortunately, many research reports are plagued by poor or inadequate photographic technique. Due to the pivotal nature of photography in plant cell and tissue culture research, poor photographic documentation severely detracts from or, in many cases, jeopardizes the credibility of a research report. This is unfortunate because high-quality photography is not difficult, given proper equipment and supplies.

In-depth instruction in photographic techniques is beyond the scope of this book and will not be attempted. Instead, we will assume that a general knowledge of photography and microscopy exists or is readily attainable. In this chapter, the use of equipment, supplies, and techniques required for basic photographic documentation will be described. The field is changing rapidly due to the availability of high-resolution digital cameras and accessories. Therefore, requirements for digital photography are described, along with those for film. Although advanced techniques that utilize photography, such as confocal microscopy, are becoming more common, they remain highly specialized and will not be included in this chapter. Similarly, scanning and transmission electron photomicroscopy routinely are used in plant cell and tissue culture research, but are generally taught as separate courses and will not be discussed here.

---

* Florida Agricultural Journal Series No. R-10164

DIGITAL VS. FILM

Until very recently, the camera of choice for most biological photography was one that utilized 35-mm film. Although available for many years, digital cameras remained prohibitively expensive or of unacceptably low resolution. Advances in digital imaging technology coupled with mass production of components have resulted in affordable cameras with useful resolution and light capturing abilities. The extreme usefulness of such cameras has convinced many to abandon film altogether because tedious development is eliminated and all steps of photography from taking the picture to producing a high-quality print can now be accomplished on the desktop computer.

The resolution of affordable digital cameras, as measured by the number of pixels available to capture an image, has now exceeded 3 megapixels, a threshold level for "publishable" photographs. Currently, cameras are available that exceed 6 megapixels, with image enhancement software increasing the virtual resolution to over 12 megapixels. While the resolution of a 3-megapixel image does not approach that of film, the apparent differences are often undetectable unless images are significantly enlarged.

Along with advancements in digital camera technology, the development (and drop in price) of compact memory devices for storing large image files, and the availability of fast computers to process images and high-resolution inkjet printers to produce prints, have made digital photography a viable option for biologists.

For macrophotography, almost any of the high-resolution "consumer" cameras are sufficient. However, for microphotography, where the camera is attached to a microscope, a digital camera back (where the lens can be detached) is essential. These so called "prosumer" (i.e., profes-sional/consumer) cameras tend to cost more, but are still affordable. In choosing a digital camera back, one should ensure that the lens mount can be attached to the microscope. Another detail to consider is that the camera should have as large an image sensor as possible. Currently, the largest sensors in prosumer cameras are still smaller than those in 35-mm cameras; this limits the image size projected by the lens system (also typically optimized for 35 mm) onto the sensor. In addition, among sensors of a given resolution, the largest will have the greatest light sensitivity, making low-light and fluorescence photomicroscopy more convenient. This is because the indi-vidual pixels of larger sensors are larger and permit more light to strike them when recording an image.

Another convenience of digital photography is the ability to utilize software to modify and process the image. There is an abundance of software available, often bundled with the camera. The most commonly used software is Adobe Photoshop™, but there are many others, too numerous to mention. Similarly, many printers are available, and these have been increasingly optimized for photographs. Typically, an inkjet type printer offers the highest quality at the lowest cost. Although printing a full-page color photograph with high-resolution color can be costly on such a printer, due to price of paper and ink, it is still less expensive than high-quality photo enlargements, especially when the time it takes to make enlargements is considered.

TYPES OF PHOTOGRAPHY USED IN PLANT CELL AND TISSUE CULTURE RESEARCH

Several different photographic techniques are required in order to document various aspects of plant cell and tissue culture. Because objects of interest vary greatly in size and type (from a cell to a culture vessel), but often are very small, equipment for both macro- and microphotography is required.

FIGURE 5.1 Typical setup of specimen in petri dish on stereomicroscope. Note fiber optic illuminator bundles positioned on each side of the specimen.

MACROPHOTOGRAPHY

When cultures contained in whole petri dishes or groups of petri dishes must be photographed, a simple digital or 35-mm single-lens reflex (SLR) camera equipped with a close-up lens and mounted on a copy stand is preferable. In this instance the photographic field might range from 2 cm (a large callus culture) to 10 cm (a petri dish) to 40 cm (a group of petri dishes).

MICROPHOTOGRAPHY USING STEREOMICROSCOPES

For characterization of callus cultures, small shoots, nonzygotic embryos, and other objects that range in size from 0.5 mm to 1.5 cm, stereomicrography is most appropriate. A stereomicroscope or "dissecting scope" allows the surface morphology of a specimen to be examined. Because specimens often possess much surface relief, the ability to photograph a wide depth of field is essential in stereomicrography. Depth of field can be adjusted on those stereomicroscopes that incorporate an iris diaphragm between the objective lens and the eyepieces. Depth of field increases as the iris diaphragm is closed (see the example below). Stereomicroscopes often use a separate light source. The ideal type of light source for stereomicrography consists of a fiber optic illuminator with a bifurcated fiber bundle. This arrangement allows for great flexibility in lighting, allowing an optimum view of the specimen to be obtained (Figure 5.2).

MICROPHOTOGRAPHY USING COMPOUND MICROSCOPES

A compound microscope is used when smaller objects such as early-stage embryos, entire cells, or the internal structure of cells and tissues must be examined. A compound microscope can resolve objects from the 10 μm to 1.5-mm size range, depending on optical quality. A compound microscope differs from a stereomicroscope in that it utilizes light that is transmitted through the specimen. Therefore, the specimen must be thin and lucent enough to allow for light to pass. Typically, histological processing of tissue is employed to produce lucent sections (see Chapter 4), or cells and tissues that can be pressed thin enough for light to pass through are utilized. Compound microscopes are configured as either "upright," in which the specimen is below the objective lens, or "inverted," in which the specimen is placed above the objective lens. Upright microscopes are the most common, typically have the highest resolving power, and are used to view specimens mounted on microscope slides with cover slips. Inverted microscopes are used to examine cells and tissues in liquid, such as suspension cultures or protoplasts, since they can view specimens through the bottom of the culture vessel. Inverted microscopes are indispensable in protoplast research, where living cultures must be monitored *in situ*.

FOCUSING AND CROPPING THE IMAGE

Methods of adjusting the camera to obtain sharply focused photographs differ from camera to camera and between macrophotography and the various types of microphotography. Stereomicrography is generally considered the method in which it is most difficult to focus accurately.

Focusing the subject on a copy stand is usually not difficult, and amounts to following the directions for a given digital or 35-mm SLR camera. One note of caution, however, is to make sure that the camera angle is the same as that of the copy stand bed in order to maintain a consistent focal plane across the field of view. This parameter is frequently out of alignment, but can be measured and adjusted by using a small bubble level (obtained from a hardware store) to compare and adjust the angle of the camera and that of the copy stand bed.

Focusing the subject through microscopes is more difficult, since most photomicrography systems must be adjusted to synchronize the optical focal plane (i.e., the image that the eye sees) with the film focal plane. Because everyone's eyes focus differently, users must learn to adjust the camera to their own preference. Focal systems vary between photomicroscopes (individual owner's manuals should be consulted), but often consist of a photoreticle (an inscribed pattern of black lines visible through one eyepiece) that must be brought into sharp focus along with the specimen by adjusting both the microscope focus and the focusing ring on the individual eyepiece. Once this is accomplished, the optical and film focal planes are synchronized so that focusing the specimen through the eyepiece results in the camera also being in focus. It is recommended that those who normally wear corrective lenses use them during photomicrography in order to obtain a consistent sharp focus.

Generally, the specimen should be magnified to fill most of the viewing area. A common mistake is to produce a photograph of a large empty field with only a small specimen in the center. This is a waste of film or digital memory, and it does little to illustrate the specimen.

COMPOSITION

While proper equipment and materials are essential in photography, the composition of the subject in the viewing area often makes the difference between a poor photograph and an excellent photograph. As mentioned above, the specimen should be carefully cropped so as to fill most of the viewing area. Also, the orientation and lighting of the specimen is critical. The following example illustrates the importance of depth of field adjustment and specimen orientation in obtaining the best possible micrograph with a stereomicroscope.

Figure 5.1 shows the typical setup of a stereomicroscope and fiber optic illuminator with a bifurcated fiber bundle. A petri dish, containing cotyledon bases of *Cucumis melo* L. on agar, which were cultured to produce somatic embryos, is on the microscope stage. The petri dish lid is removed briefly during each exposure. The microscope stage has a black surface, which generally provides better contrast. The fiber bundles are positioned to provide optimum lighting of the specimen, as determined by viewing the specimen through the microscope while maneuvering the bundles. The illuminator is turned to the maximum power setting if color film is to be used. The specimen within the dish is adjusted on the agar so as to show the desired view (in this case, the embryos) and the magnification is adjusted to fill the viewing area to the greatest extent possible.

DEPTH OF FIELD

With stereomicrography, depth of field, which can be adjusted on those microscopes that possess an iris diaphragm between the objective and eyepiece lenses, is of critical importance. Figure 5.2(A) shows a typical photomicrograph obtained with the iris diaphragm set to full open. The focus was set on the apical dome and the foreground of the cotyledon (large arrow). Note that the background area of the cotyledon (smaller arrow) is out of focus, as is the globular stage embryo in the foreground (smaller arrow). This demonstrates the limited focal plane obtained under these conditions, due to reduced depth of field. An improved view of the specimen is shown in Figure 5.2(B), which was obtained by adjusting the iris diaphragm on the stereomicroscope so that it was nearly closed, thus increasing the depth of field. No other adjustments were made. Note now that both the foreground and background are in focus, resulting in a superior micrograph.

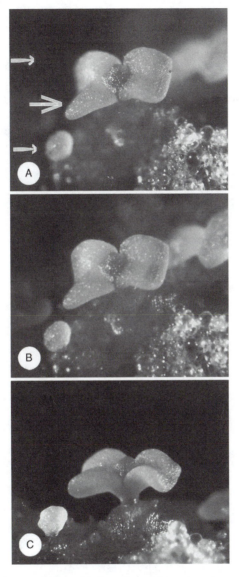

FIGURE 5.2 Optimization of depth of field and orientation of *Cucumis melo* somatic embryos for photography with a stereomicroscope. (A) Specimen photographed with poor depth of field. Note area that is in focus (large arrow) vs. out of focus areas in the foreground and background (small arrows). (B) Specimen photographed with increased depth of field, obtained by closing the stereomicroscope iris diaphragm. Note that both foreground and background are now in focus. (C) Same embryos, as in A and B, demonstrating improved photograph obtained by careful orientation of the specimen.

ORIENTATION

Specimen orientation, particularly in the case of stereomicroscopy, makes a dramatic difference in the clarity of documentation. Figure 5.2(C) shows the same embryos as in Figures 5.2(A) and 5.2(B), but here the specimen was positioned to show the embryos from the side. In this orientation, much more of the important morphological attributes of the larger embryo can be seen, including the attachment of the embryo to cotyledon tissue. Note also that the morphology of the subtending globular-stage embryo can be better determined.

CONCLUSION

Photography is an important tool in tissue and cell culture research. Due to the ranges in size and detail of tissue and cell culture specimens, several different types of photography are usually needed to provide adequate documentation. While a photograph is relatively simple to obtain, proper photographic and microscopic equipment and supplies are required to obtain high-quality photographs. With equipment in place, photographic technique becomes of paramount importance. Details of composition often determine the difference between acceptable and unacceptable photography.

6 Elements of *In Vitro* Research

Michael E. Compton

CHAPTER 6 CONCEPTS

- The most commonly used form of hypothesis for *in vitro* research is the null hypothesis. This popular hypothesis states that all treatments will elicit a similar response on all plant tissues.

- An experimental unit (EU) is the smallest unit that receives a single treatment — the culture vessel, in most tissue culture experiments. The observational unit (OU) is the object being observed or measured. However, an explant may be considered as both the EU and OU in cases where one explant is cultured per vessel.

- The completely randomized, randomized complete block, incomplete block, and split plot designs are most commonly used for *in vitro* research. These designs differ in their randomization schemes and experimental requirements.

- Many researchers find it beneficial to culture multiple explants in a culture vessel. This conserves resources and time, but leads to a situation in which multiple measurements are made for each replicate (e.g., culture vessel). Statistically this is referred to as subsampling or repeated measures.

- The nature of the data influences the type of statistical procedure that should be used. Most observations made in plant tissue culture experiments result in data that can be classified as continuous measurements (shoot length, root length, or embryo size); as counts (the number of organs per explant or callus [shoots, embryos, or protocorm-like bodies]); as binomials (response or no response, expressed as percentages); or as multinomials (shoot, root, callus, or no response). The statistical methods used to evaluate the treatment effects vary for each type of data observation.

INTRODUCTION

Research projects begin with an idea aimed at solving a specific problem. From that first idea blossom a set of planned experiments, wherein treatments are devised and tested in the controlled conditions under which the problem typically exists. Data are gathered from observations made during experimentation and evaluated, using statistical analysis, to ascertain the effectiveness of each treatment in correcting the problem.

This chapter is an introduction to the basic elements of plant tissue culture research. More specifically, it demonstrates the experimental designs and statistical methods that are used to analyze and interpret data as well as to present results in tables and graphs.

0-8493-1614-6/05/$0.00+$1.50

METHODOLOGIES ASSOCIATED WITH SCIENTIFIC RESEARCH

ESTABLISHING THE HYPOTHESIS AND THE EXPERIMENTAL OBJECTIVES

In science, a logical and step-by-step approach is best for problem solving. We examine what is already known by reading literature and combining that information with knowledge gained from prior experiences to formulate a hypothesis and a set of experimental objectives that can be tested objectively through experimentation. Writing these basic elements of research down as part of the planning process is important. If the hypothesis and objectives cannot be outlined, it is unlikely that they can be tested experimentally. The most commonly used form of hypothesis for *in vitro* research is the "null hypothesis." This popular hypothesis states that all treatments will elicit a similar response on all plant tissues. Application of the null hypothesis can be illustrated by using a situation in which different cytokinins (zeatin, zeatin riboside, 2-iP, BAP, kinetin, and TDZ) were tested to evaluate their ability to stimulate adventitious shoot organogenesis in petunia (*Petunia* × *hybrida*) leaf explants. In this case, the null hypothesis would read: Adventitious shoot regeneration in petunia is similar among leaf explants incubated in media containing equal concentrations of zeatin, zeatin riboside, 2-iP, BAP, kinetin, and TDZ. Rejection of the null hypothesis through statistical analysis implies that the experimental tissues reacted differently to the imposed treatments. Statistical procedures used to test research hypotheses will be discussed later in this chapter.

SELECTING THE MAIN COMPONENTS AND THE EXPERIMENTAL DESIGN

Once the hypothesis and the experimental objectives have been outlined, the main components of the experiment can be selected. This includes selection of the treatments and supportive materials required for the experiment. Treatments are conditions that characterize the population of interest and whose effect on the population can be measured and tested experimentally (Lentner and Bishop, 1986). Treatments may include a myriad of conditions, including culture vessels and their contents (medium salt formulations, type or concentration of plant growth regulators [PGRs], vented versus unvented vessel enclosures, and so on); environmental conditions outside of the culture vessel (temperature, light source, quality, intensity, or photoperiod); or factors inherent to the explant, such as genotype or stock plant physiology (explant size or age, stock plant age or maturation stage, preconditioning, and so on). Researchers may choose to examine one treatment factor at a time or test several factors simultaneously (e.g., stock plant genotype and PGR concentration). The latter is often appropriate for *in vitro* studies because the way that the factors interact with each other can be examined, and time and materials can be conserved by conducting one experiment rather than two (Compton, 1994). Regardless of the number of treatments selected, it is important that they are chosen according to the experimental objectives and that they evoke a wide range of responses, so that an optimum level can be identified.

Supportive materials are items present under the conditions in which the problem is observed, but are not treatments. In plant tissue culture studies, these materials may be culture vessels, medium components, growth regulator types and concentrations, states of medium solidification, growth environments (light photoperiod and intensity), or any other items important for satisfactory explant growth. Supportive materials should not interfere with the treatment effects on plant tissues. Knowledge of previous experiments, whether successful or not, provides beneficial information when selecting treatments and supportive experimental materials.

The experimental units, observational units, and experimental design are chosen once the treatments and supportive materials are selected. By definition, an experimental unit (EU) is the smallest unit that receives a single treatment. This is the culture vessel, in most tissue culture experiments. The observational unit (OU) is the object being observed or measured. Explants are the most common OU in tissue culture studies, because most studies are devoted to discovering treatments that optimize explant development and performance. However, an explant may be considered both the EU and the OU, in cases where one explant per vessel is cultured (Compton and Mize, 1999). The number of EUs required per treatment is determined once the EUs and OUs are selected. Replication occurs when more than one EU is evaluated for each treatment, and is necessary to estimate treatment effects and experimental error. The number of replicates needed per treatment is determined by the amount of variability inherent in the data and the degree to which treatment effects must be separated (Zar, 1984). At least three replicates per treatment are required to calculate means and measure error. Most researchers rely on prior experience, review of literature, and speculation to determine the number of replicates to be used. Insufficient replication can reduce the ability of statistical procedures to detect treatment differences (Lentner and Bishop, 1986). Kempthorne (1973) outlines how information from previous studies can be used to estimate the number of replicates required for an experiment.

The completely randomized (CR), randomized complete block (RCB), incomplete block (IB), and split plot (SP) designs are most commonly used for *in vitro* research. These designs differ in their randomization schemes and experimental requirements. Randomization allows researchers to assign treatments fairly and to account for variation associated with experimental treatments and supportive materials, as well as factors outside of the experiment. When selecting an experimental design, it is important to examine the supporting materials and the population of interest for possible variation and to choose a design that is simple to employ, efficient in regard to measuring treatment effects and residuals, and in accord with the experimental objectives (Compton and Mize, 1999).

The CR design is the one that is most commonly used for *in vitro* research because it is easy to use and has a randomization scheme with no set pattern. This allows researchers to maximize the replicate number and to utilize either equal or unequal treatment replication (Little and Hills, 1978). This is important because unequal replicate numbers often occur in plant tissue culture studies due to contamination or death of explants. The CR design can be illustrated using an example in which the effectiveness of six types of cytokinins on adventitious shoot organogenesis from *Petunia* × *hybrida* leaf explants was examined. Culture vessels were test tubes containing 20 ml of Murashige and Skoog (MS) medium, supplemented with 4 ml of either zeatin, zeatin riboside, 2-iP, BA, kinetin, or TDZ. One explant was cultured in each vessel, and ten vessels were established for each treatment. The randomization scheme for this experiment, with vessels randomized in a CR design, is shown in Figure 6.1(A).

Several techniques can be used to assign explants to treatments. Many researchers simply place culture vessels containing experimental media in the laminar flow hood and randomly select a vessel while preparing explants. However, it is best to establish a specific randomization scheme beforehand by drawing numbers representing treatment replicates from a hat and diagraming the results on paper. Culture vessels may be placed in the hood and used as diagramed in Figure 6.1(A). A more efficient method is to use a statistical analysis software package to establish the randomization scheme.

Statistically, the CR design is most efficient because it allows researchers to maximize degrees of freedom (DF) for error, which is important for detecting treatment differences during statistical analysis (Little and Hills, 1978). However, this design has some problems. One pitfall of the CR design is that the EUs must be homogeneous for data analysis to be effective. Situations in which EUs or OUs are highly heterogeneous cause increased variability in the experiment and decrease the likelihood that treatment differences will be detected during statistical analysis (Little and Hills, 1978).

Designs that employ blocking are used when a high degree of heterogeneity exists among the EUs. Unevenness or inconsistency among environmental factors in the growth room, differences among stock plants or leaves from which explants are obtained, variability among technicians, and differences among batches of *Agrobacterium* or bombardments with a particle gun are a few factors that introduce heterogeneity. When using blocking, only factors that are measurable can be blocked.

The RCB design assembles replicates of each treatment into blocks. This creates a high level of uniformity among treatments within a block, reducing nontreatment variation (Little and Hills, 1978). It is important that all explants in a block are as uniform as possible for the design to be

A.

Completely Randomized

1	5	4	5	3	6	4	5	3	4
5	3	2	1	5	1	2	6	1	2
2	6	4	3	4	5	3	4	3	2
6	4	1	1	2	6	4	2	1	1
2	3	5	6	6	2	1	5	5	3
3	1	6	2	4	5	6	3	4	6

B.

Randomized Complete Block

1	2	3	4	5	6	7	8	9	10
1	2	4	3	5	6	1	4	3	5
2	4	6	1	6	2	4	3	6	2
3	1	3	6	3	4	5	1	5	3
4	5	2	5	1	3	6	2	1	4
5	6	5	4	2	1	2	5	4	6
6	3	1	2	4	5	3	6	2	1

C.

Incomplete Block

1	2	3	4	5	6	7	8	9	10	11	12	13	14	15
1	2	2	3	4	5	1	6	2	6	4	5	3	4	1
2	3	4	6	6	4	3	5	1	3	6	2	5	3	4
3	4	1	5	2	1	6	3	5	4	2	1	4	5	6
4	5	5	2	1	6	2	1	3	2	5	6	6	1	3

FIGURE 6.1 Randomization schemes for completely randomized (A), randomized complete block (B), and incomplete block (C) designs, each with six treatments. Each treatment was replicated ten times in all designs. *Note*: The number of blocks and the number of times that any two treatments appear in the same block for the incomplete block design were determined using the formula $kb = tr$ and $\lambda(t - 1) = r(k - 1)$, from Lentner and Bishop (1986).

effective. Explants from the same leaf or stock plant often have similar regeneration competence, and each leaf can constitute a block. The petunia study could be used to illustrate the RCB design, with explants from one leaf establishing replicates of each treatment in a block. In this case, ten leaves were needed to produce enough blocks to meet the requirement of ten replicates per treatment (Figure 6.1B). Using the RCB in this situation would maximize uniformity within blocks by utilizing the homogeneity present in each leaf.

Situations often exist in which there is too little source material to produce the number of explants required to establish complete blocks, each with one replicate of each treatment. In these cases the IB design may be the best choice (Compton and Mize, 1999). Like the RCB design, this design uses blocking to maintain homogeneity, but differs in that not all treatments are present in each block. The petunia example could be modified to illustrate use of the IB design by limiting the number of explants obtained from each leaf to four. In this case, there would not be enough leaf material to establish complete blocks with all six treatments represented. Therefore, 15 incomplete blocks would be required to replicate each treatment 10 times (Figure 6.1C).

The number of blocks, the number of replicates, and the number of times that any two treatments appear together in a block are determined using a special formula: $kb = tr$ and $\lambda(t-a) = r(k-1)$ (Lentner and Bishop, 1986). Information regarding the use of IB designs in plant tissue culture was published by Kuklin et al. (1993). Although the IB is an excellent design for *in vitro* studies, an RCB design is often preferred when there is adequate explant material available to form complete blocks (Compton and Mize, 1999).

Another design that utilizes blocking is the SP design, which is best used when two treatment factors with different degrees of variation are examined simultaneously (Little and Hills, 1978). The SP design uses two separate levels of randomization for each factor. The treatment factor with the greatest degree of variation is assigned to large plots called main plots, and the treatment factor with less variation is assigned to subplots. Main plots can be completely randomized or arranged in blocks.

It is common that the youngest fully expanded leaves are most competent for adventitious shoot organogenesis. Given this circumstance, researchers must often use leaves from several different stock plants to obtain enough explant material for experiments. This often induces additional variability into the experiment. In the petunia example, the researcher believes that stock plants from which leaves are harvested for explantation, as well as cytokinin type, influence adventitious shoot regeneration. Based on this hypothesis, an experiment was designed in which the influences of stock plant and cytokinin type on shoot regeneration were examined simultaneously. Because greater variation was expected to be introduced into the experiment from the five stock plants than from cytokinin type, the former was assigned to main plots. Stock plants were replicated three times, with main plots completely randomized and subplot treatments randomized within each main plot, resulting in 15 replicates of each cytokinin type (Figure 6.2). Although SP is the most efficient experimental design to use when examining two treatment factors with different degrees of influence on explant response, the setup, data recording, and data analysis and interpretation are more complicated than with the other designs.

It is impossible to be absolutely sure if blocking will be required when designing experiments. The need for blocking can be determined by conducting a small preliminary experiment and testing for blocking when conducting the statistical analysis. Kempthorne (1973) outlines a method for determining if blocking is necessary.

Subsampling

Many researchers find it beneficial to culture multiple explants in a culture vessel. This conserves resources and time, but leads to a situation in which multiple measurements are made for each replicate (e.g., culture vessel). Statistically this is referred to as subsampling or repeated measures (Compton and Mize, 1999). Subsampling is not an experimental design, but a modification that can be used with any of the designs discussed above. Subsampling can be illustrated by modifying the petunia example so that 2 explants are cultured in petri dishes containing 25 ml of MS medium, supplemented with 4 µ*M* of zeatin, zeatin riboside, 2-iP, BA, kinetin, or TDZ. Petri dishes containing explants were randomized using a CR design. Subsampling may be used with blocking designs when warranted. Culturing multiple explants in one vessel not only saves resources and space, but maximizes the number of OUs and minimizes variation associated with explants by

Plant Genotype	1_1	3_1	4_1	2_1	5_1	....	4_3	3_3	2_3	5_3	1_3
	①	④	②	③	⑤		④	①	⑥	②	③
	③	⑥	⑤	⑥	①		③	④	①	⑥	⑤
Cytokinin Type	⑤	②	③	①	④		⑤	②	③	④	⑥
	②	⑤	⑥	④	②		①	③	④	①	①
	⑥	③	④	⑤	⑥		⑥	⑤	②	③	④
	④	①	①	②	③		②	⑥	⑤	⑤	②

FIGURE 6.2 Randomization scheme for a split plot design in which main plots are completely randomized and subplots are randomized within each main plant. Stock plants were assigned to main plots. Each stock plant (1–5) was replicated three times. Cytokinin types were assigned to subplots: zeatin (①); zeatin riboside (②); 2-Isopentenyl adenine (2-iP) (③); BA (④); kinetin (⑤); and thidiazuron (TDZ) (⑥).

making multiple measurements per vessel (Mize and Winistorfer, 1982). One disadvantage of subsampling is that the statistical analysis is slightly more complicated than when subsampling is not used (Compton and Mize, 1999).

Determining How Many Times an Experiment Should Be Replicated

Researchers often wonder how many times an experiment should be repeated. Most statisticians and researchers believe that an experiment must be conducted at least twice to validate the results (Lentner and Bishop, 1986). The number of times that the experiment is to be conducted should be decided during planning. Repeating experiments does not guarantee a similar outcome, even if the same stock plants are used. Repeating experiments may introduce time as a nontreatment factor, if changes in stock plant physiology have occurred with increased stock plant age or time in the growing season. This can lead to differences in experimental results. Researchers generally have two options when differences between runs are observed. The entire experiment can be repeated, which can be costly, or each run can be examined as a block or main plot. The best option would be first to look at the data from the two runs and determine if similar trends exist. If the trends are similar, the experiment should not be repeated, but the circumstances that caused the differences should be examined and explained. If the results obtained in the two runs contrast, the experiment should be repeated. If differences among runs persist, the possibility of seasonal effects on the results should be examined experimentally. A decision should be made during the planning phase about what will be done if differences between runs of the experiment are observed.

Choosing the Best Data and Planning the Statistical Analysis

The data recorded should help evaluate treatment effects according to the experimental objectives. If inappropriate data are recorded, it is unlikely that the researcher will be able to assess explant response and address the experimental objectives. Therefore, it is important to discuss which observations should be recorded with colleagues and statisticians during planning.

Most researchers peruse relevant literature and combine the information gathered with previous experience to determine which observations to measure in their experiments. In the petunia illustrations used throughout this chapter, the number and percentage of explants producing shoots, the number of shoots per explant (or all explants in a petri dish), and shoot length are observations that most researchers would record to evaluate explant competence. However, to get the maximum information from the project, the researcher may wish to make additional observations, such as the number and percentage of explants that produced callus, the amount of callus produced per explant (by weight or volume), and the ploidy level of regenerants, as well as histological observations of explants, periodically during the experiment to ascertain the mode of shoot regeneration (direct or indirect; see chapters on histology [Chapter 4] and shoot regeneration [Chapters 11 and 12]).

Observations made during the experiment are recorded on data sheets to facilitate data processing. This activity can lead to mistakes if the sheets are not well organized. Designing data sheets the way that observations will be entered into the computer helps to prevent mistakes when transcribing data. Many statistical software packages will organize and print data sheets for you. Data sheets should be designed during planning to ensure that the researcher is considering observations that measure treatment effects.

The nature of the data influences the type of statistical procedure that should be used. Most observations made in plant tissue culture experiments result in data that can be classified as continuous measurements (shoot length, root length, or embryo size); as counts (the number of organs per explant or callus [shoots, embryos, or protocorm-like bodies]); as binomials (response or no response, expressed as percentages); or as multinomials (shoot, root, callus, or no response) (Mize et al., 1999). The statistical methods used to evaluate the treatment effects vary for each type of data observation. Outlining sources of variation and DFs during planning can help identify data analysis procedures that are most appropriate for the type of data recorded and the experimental objectives. Most researchers use analysis of variance (ANOVA) to evaluate data. While ANOVA is well suited to continuous data, it is not suitable for analyzing count, binomial, or multinomial data without prior manipulation.

CONDUCTING THE EXPERIMENT

Experimentation may begin once planning is completed and thoroughly reviewed. Personal bias must be avoided during all phases of the experiment. When setting up an experiment, bias can be avoided by reliance on the structure of the experimental design (Mize et al., 1999). Do not prepare all explants for one treatment at the same time. Instead, arrange culture vessels according to the randomization scheme selected, and place explants in vessels according to the design.

It is also important to avoid any circumstances that introduce error into the experiment. Many mistakes occur when researchers become fatigued. The time required for each phase of the experiment should be estimated during planning. Tasks that are taxing should be staggered to avoid worker strain. Mistakes may also occur when explants are transferred to fresh media. Carefully arrange culture vessels (both new and currently used) in the laminar hood in a manner that reduces the opportunity for mistakes. Be careful to label each culture vessel with the correct information. If using codes, be sure to record keys to your codes in your laboratory notebook. There are many avenues for introducing operator error into your experiment. Take the precautions necessary to encourage safe and accurate experimentation.

RECORDING DATA

As when starting an experiment, you should rely on the structure of the experiment when recording observations. For example, record observations of replicates in one block before moving to the next block. This helps to avoid bias. Be meticulous and take your time when recording data. A sloppy technique introduces error and inconsistency. Take breaks to avoid fatigue when recording data. Many observations are stressful on the eyes, making it important to rest periodically.

Data should be checked for errors after they are recorded. Erroneous data entries lead to incorrect treatment evaluation and interpretation (Mize et al., 1999). Make sure the data of each treatment fall within the expected boundaries. Data outside these boundaries are called outliers and may result from errors in reading, recording, or transcribing data, or may be values obtained from plant tissues with inherent problems or genetic mutations. Outliers are often discarded. However, researchers should exercise caution when deleting data, as personal bias may be unintentionally introduced and may affect the outcome of the experiment. Several statistical procedures can be used to detect outliers; these are helpful for reducing bias when editing (Barnett and Lewis, 1984). Treatment codes should be examined while checking data. Do not assume that codes were entered correctly. Errors associated with entering codes can be examined by calculating the mean for the coded variables. In the petunia example, obtaining a mean of 3.5 for the cytokinin treatment (coded 1 through 6) indicates that the code was entered correctly for each observation. Editing data can be facilitated by printing data sheets and thoroughly reviewing them. This can be a time-consuming task but it is worth the effort.

DATA ANALYSIS AND INTERPRETATION

Data must be statistically analyzed and interpreted as planned. As discussed earlier, the nature of the data influences the results and the interpretation of the statistical analysis. Researchers should choose an analysis procedure that is best suited for the recorded observations and experimental objectives. When analyzing data, a general analysis is conducted first to determine if there are any differences between the treatment levels. Further analysis of treatment means is conducted if differences are detected by the general analysis (Compton, 1994; Mize et al., 1999).

Analysis of Continuous Data

Continuous data are normally distributed with treatments having similar variances (Mize et al., 1999). Because of this characteristic, ANOVA is well suited for analyzing this type of data. During ANOVA, the value of each observation is subtracted from the overall mean. Differences between these values are considered random error (residual) and used in the analysis (Mize et al., 1999). A model statement is created that identifies the dependent variables (observations recorded) and independent variables (treatments, treatment interactions) based on the experimental design that has been used. The influence of independent variables on the dependent variables is tested according to the model statement. A summary table is generated during ANOVA that provides the results of the model tested (Lentner and Bishop, 1986). Information in the summary table includes the DFs, the sources of variation (SS), the mean square error (MSE), the F-statistic (F), and estimates of the probability (P) that a similar result would be obtained the next time the experiment is conducted under similar conditions (Lentner and Bishop, 1986).

In the petunia regeneration example, the shoot length variable is considered continuous. Because a CR design was used in this study, cytokinin type and experimental error were identified as sources of variation (Table 6.1A). According to the results, the length of regenerated shoots was influenced by the type of cytokinin used. This is determined by looking at the F and P values. In ANOVA, F (12.46) is obtained by dividing the MS for cytokinin (360.67) by the MSE (28.9537). The MS value reflects the degree of variation associated with the source. Therefore, the cytokinin MS reflects the amount of variation caused by cytokinin treatments. The MSE measures variation due to culture vessels and any other nontreatment sources. The P value (the level of significance of the F value) is obtained by comparing the calculated F value against standardized values for ANOVA. P values equal to or less than 0.05 are considered significant.

TABLE 6.1
Analysis of Variance Table Comparing Petunia Shoot Organogenesis Data (Shoot Length) Using the Following Designs: Completely Randomized (CR); Randomized Complete Block (RCB); Incomplete Block (IB); and Completely Randomized with Subsampling (CR-SS)

Source	DF	SS	MS	F	P
A. Completely Randomized (CR) Design without Subsampling					
Cytokinin	5	1803.35	360.67	12.46	<0.000
Experimental error	54	1563.50	28.9537		
Total	59	3366.85			
B. Randomized Complete Block (RCB) Design					
Cytokinin	5	1803.35	360.67	12.69	<0.000
Block	9	284.35	31.5944		
Experimental error	45	1279.15	28.4256		
Total	59	3366.85			
C. Incomplete Block (IB) Design					
Cytokinin	5	1803.35	360.67	67.20	<0.000
Incomplete blocks	14	490.10	63.928		
Experimental error	40	1073.40	26.835		
Total	59	3366.85			
D. Completely Randomized with Subsampling (CR-SS) Design					
Cytokinin	5	1803.35	360.67	13.93	<0.000
Subsampling error	24	787.0	32.7917		
Experimental error	30	776.50	25.8833		
Total	59	3366.85			

Note: Explants were prepared from leaves of petunia plants grown in the greenhouse. Six cytokinin types were evaluated (zeatin, zeatin riboside, 2-iP, BA, kinetin, and TDZ). Explants were either cultured in 25 mm × 150 mm test tubes (1 per tube; CR, RCB, and IB) or 100 mm × 15 mm petri dishes (2 per dish; CR-SS) containing 25 ml of watermelon shoot regeneration medium. There were ten replicate test tubes (CR, RCB and IB designs) or five petri dishes per treatment (CR-SS). Data are hypothetical.

This level is chosen by most researchers because 0.05 indicates that a similar result would be obtained 95% of the time that the experiment is repeated, signifying a high level of confidence for the outcome. Another reason for selecting 0.05 is that most journals require researchers to test at this level of significance.

ANOVA for experiments in which blocking is used must be conducted differently than when a CR design is used. When an RCB design is used, the block must be specified in the model statement (Compton, 1994). Comparing ANOVA summary tables for the RCB, IB, and CR designs illustrates the origin of DF and SS for blocking (Table 6.1A–C). The DF and SS calculated for cytokinin are identical for all experimental designs. However, DF and SS values for experimental error differ for the three randomization schemes. Adding the DF for block to the DF for experimental error in the RCB design results in the same value calculated for error DF in the CR design. (Compare experimental error rows in Tables 6.1A and 6.1B.) The same is true for the SS values. Specifying blocks in the model statement instructs ANOVA to separate variation associated with blocks from the experimental error, resulting in a lower MSE and increased F for the treatment variable. This increases the sensitivity of ANOVA, increasing likelihood that treatment differences will be detected.

Conducting ANOVA is slightly more complicated when an IB design is used. This is because two model statements must be generated. One statement is written that is similar to a CR design, while a second is written to generate a value for using incomplete blocks. Values calculated for treatment and incomplete blocks are used to construct an ANOVA table for calculating the F and associated P value for the treatment. In the petunia example, use of incomplete blocks removed variation from the experimental error more effectively than did the RCB design (compare Tables 6.1A and 6.1C), possibly because the use of smaller blocks (four treatments per block) improved homogeneity within individual blocks.

ANOVA is preferred by most researchers because it is easy to perform, can be used to evaluate data obtained from virtually all experimental designs, and generates an MSE value that is considered to be the best estimate of experimental error (Mize et al., 1999). ANOVA provides a good overall analysis because it accurately measures how treatments relate to each other when used in the proper circumstances.

Analysis of Count Data

Count data are not normally distributed because treatment variances are equal to the average response of the treatments (Zar, 1984). Because of their distribution, count data should be analyzed using Poisson regression (Mize et al., 1999). This statistical method is best suited for analyzing count data because it uses a logarithmic value of the mean counts, which normalizes the data during analysis. Poisson regression calculates a coefficient that is divided by the standard error to determine if the model is significant. In the petunia example, observations on the number of shoots per explant are considered count data. Poisson regression indicates that cytokinin type influenced the number of shoots per explant (p = 0.005; see Table 6.2).

ANOVA is often used by researchers to analyze count data. Unfortunately, ANOVA is unreliable when count values less than 10 are observed (Mize et al., 1999). Because the number of organs regenerated by explants in adventitious organogenesis and embryogenesis studies are typically low (less than 10 regenerants), use of the Poisson regression is usually warranted. In experiments where more than 10 organs are regenerated per explant (e.g., micropropagation studies using shoot tips), ANOVA and Poisson regression yield similar results (Mize et al., 1999). Zar (1984) believes that ANOVA should not be used to analyze count data regardless of how large values are. Therefore, it may be in the best interest of the researcher to transform count data in all situations.

Analysis of Binomial Data

Response data — for example, the ability of explants to respond to a treatment — are generated to evaluate the influence of treatments on regeneration competence. These data have a binomial distribution, causing treatment variances to be dependent on explant success (Mize et al., 1999). Percentages calculated by these observations are often considered important because researchers are interested in identifying treatments that optimize explant response. The ability to optimize explant response can translate directly to increased profits for plant tissue culture businesses

---

TABLE 6.2
Poisson Regression Summary Table Petunia Shoot Organogenesis Data (Percentage of Explants with Shoots)

Predictor Variables	Poisson Regression Coefficient (A)	Standard Error (B)	Poisson Regression Statistic (A/B)	P Value
Constant	1.120	0.150	7.45	<0.0000
Cytokinin	0.101	0.036	2.79	0.0052

TABLE 6.3
Logistic Regression Summary Table Petunia Shoot Organogenesis Data (Number of Shoots per Explant)

Predictor Variables	Logistic Regression Coefficient (A)	Standard Error (B)	Logistic Regression Statistic (A/B)	P Value
Constant	1.1296	0.7435	1.74	0.0813
Cytokinin	0.0573	1.9580	0.29	0.7969

micropropagating house plants or biotechnology researchers interested in using recombinant DNA technology to insert transgenes into plant tissues.

Most researchers use ANOVA to analyze response data. This is unfortunate because, as stated earlier, ANOVA assumes that data are normally distributed and that treatment variances are equal. Because treatment variances of response data are dependent on the success of responses, ANOVA results are unreliable. Logistic regression is the statistical procedure most suited for analyzing response data because the procedure does not produce separate estimates of experimental error, but calculates a special coefficient and standard error value to derive a test statistic and thus to determine statistical significance (Mize et al., 1999).

In the petunia example, the percent regeneration observations were binomially distributed. Analyzing these data using logistic regression revealed that the percentage of explants that produced shoots was influenced by the type of cytokinin used (Table 6.3). Again, a mean comparison test must be used to determine specific differences among the cytokinin treatments.

Before analyzing response data, it is important to determine if there are treatments in which the response did not vary. This usually occurs when all or none of the explants responded. These values should either be changed or deleted before analysis (Mize et al., 1999). If a decision is made to change the values, zeros should be changed to a slightly higher value (0.000001) and 100% should be reduced slightly (0.999999). It is important to decide during planning if values are going to be altered or deleted. You should also indicate if treatments were dropped or values altered when writing reports and manuscripts.

ANALYZING EXPERIMENTS USING SUBSAMPLING

Most researchers culture several explants in a culture vessel to save time and resources. This is called subsampling. But researchers often go on to analyze the data as if subsampling was not used. This action is incorrect because subsampling reduces the variation introduced into the experiment by culturing multiple explants per vessel (Compton and Mize, 1999). Two methods can be used to analyze observations wherein subsampling is used. Observations from each explant in a dish may be combined, resulting in one value per vessel. In this situation the model statement used in the general analysis (ANOVA) would be written as if subsampling was not used (Lentner and Bishop, 1986). An alternative method would be to record observations from each explant in a culture vessel and enter them individually. However, the model statement should be written so that the DF and SS associated with subsampling are separated from the experimental error (Compton, 1994). This method is preferred by researchers wishing to document the variation associated with explants. If the subsampling error is not separated in this situation, the error DF, SS, and MS will be inflated, resulting in a smaller F value for the treatments of interests, which reduces the likelihood that treatment differences will be detected. In the petunia regeneration example, 24 DF and 787.0 SS were removed from the experimental error. This reduced the MSE from 28.9537 to 25.8833, resulting in a greater treatment F value (Table 6.1D). Subsampling can be used with single factor and factorial experiments and with any of the experimental designs discussed previously.

Methods for Comparing Treatment Means

Once a significant value is obtained in the preliminary statistical test (ANOVA, Poisson regression, or logistic regression), the researcher must elucidate specific differences among treatments. This is not a problem when there are only two treatments because the general test alone will be sufficient for determining statistical differences. However, most researchers test more than one treatment, making further analysis necessary. The easiest way to compare treatment means is to rank them in ascending or descending order and then pick the best treatment. The problem with this method is that variation within the treatments is not considered. It is important to consider treatment variation when comparing means because the means alone may not accurately represent how the OUs respond to the treatments. There are many mean separation procedures that account for within treatment variation, including the standard error of the mean, multiple comparison and multiple range tests, orthogonal contrasts, and trend analysis.

Standard Error of the Mean

The standard error (SE) of the mean is obtained by dividing the sample standard deviation by the square root of the number observations for that treatment (Zar, 1984). When most researchers use SE values for mean separation purposes, they use treatment means, calculating their respective SE values and the difference between paired values. Treatments are declared different if the collective values for the paired treatments do not overlap. Many researchers use SE values to compare treatment means. Problems occur because most researchers use SE values to compare the means of treatments that are ranked far apart. This use of SE overestimates treatment differences and violates the assumption of ANOVA that treatment variances are equal (Mize et al., 1999). This does not produce useful results because SE values increase with the numerical value of the data and do not accurately reflect population variance (i.e., treatments with large values have a larger SE than treatments with small values). When using SE values to compare treatment means, one SE value should be calculated from the MSE, from ANOVA or SE values obtained from Poisson or logistic regression (Mize et al., 1999). This use of SE is more likely to yield realistic results. However, mean comparison procedures that use the sample variance often provide more useful information.

Multiple Comparison and Multiple Range Tests

Multiple comparison and multiple range tests are statistical procedures that use the population variance in a formula to calculate a numerical value and to compare treatment means. Means are ranked as when SEs are used, and the difference between compared means is calculated. The calculated value is compared with the critical value, computed by the mean comparison statistic. If the difference between the compared means exceeds the computed statistical value, the treatments are considered statistically different. However, if the difference between treatment means is equal to or is less than the computed statistical value, the treatments are considered similar (Mize et al., 1999). When presenting means in a table or graph, means designated as different are assigned different letters, whereas treatments declared to be similar are assigned the same letter.

Most researchers use the terms "multiple comparison test" and "multiple range test" interchangeably. However, the two differ in their method of making comparisons. Multiple comparison tests use the same critical value to compare adjacent and nonadjacent means, whereas multiple range tests employ different critical values to compare adjacent and nonadjacent means (Compton, 1994). Multiple range tests are considered to be more accurate than multiple comparison tests because different values are used to compare means that are ranked far apart, making it less likely that errors will be made when comparing distantly ranked means. Examples of multiple comparison tests are Bonferoni, Fisher's Least Significant Difference (LSD), Scheffe's, Tukey's Honestly Significant Difference Test (Tukey's HSD), and the Waller-Duncan K-ratio T test (Waller-Duncan).

TABLE 6.4
Comparison of Results Obtained from Analyzing Treatment Means Using Least Significant Difference (LSD) and Tukey's Honestly Significant Difference (Tukey's HSD) Mean Comparison Tests

Cytokinin	Shoot Length (mm)	Results of Mean Comparison Tests	
		LSD[a]	Tukey's HSD[b]
Zeatin	17.6	a	a
2-iP	14.3	ab	ab
BA	11	bc	abc
Zeatin riboside	8.5	cd	bcd
kinetin	4.1	de	cd
TDZ	1.8	e	d

Note: Data were analyzed following ANOVA for a single factor completely randomized design.

[a]Means with the same letter are not significantly different according to LSD at 0.05.
[b]Means with the same letter are not significantly different according to Tukey's HSD at 0.05.

Examples of multiple range tests include Duncan's New Multiple Range Test (DNMRT), the Ryan-Einot-Gabriel-Welsh Multiple F-Test (REGWF), the Ryan-Einot-Gabriel-Welsh Multiple Range Test (REGWQ), and the Student-Newman-Kuels Test (SNK). The REGWQ is considered one of the best mean comparison tests to use. However, this procedure is not available in all computer software packages.

In the petunia example, Tukey's HSD and LSD were used to compare the effects of different cytokinins on elongation of regenerated shoots (Table 6.4). Normally only one comparison test is used to compare treatment means. However, two are used here to demonstrate differences due to procedures. Means were ranked from highest to lowest and means of all treatments were compared. Different results were obtained from the two tests. LSD sorted the means into five groups, while Tukey's HSD divided the means into four groups. Which procedure is correct? LSD is considered by many statisticians to be the most liberal mean comparison procedure, i.e., the one that is most likely to declare means different. This often leads researchers to declare means different that are similar. This is often true with means that are ranked far apart (e.g., highest versus lowest means). For this reason, LSD should not be used unless treatments are considered different in a general test (ANOVA, etc.). Tukey's HSD is considered more conservative than LSD, and produces more accurate results.

Results of mean comparison tests should be presented in tables or in bar graphs, since treatments are unrelated. If using bar graphs, letters assigned to specific treatments should be positioned above each bar (Figure 6.3).

Multiple comparison and multiple range tests should only be used when treatments are unrelated (Lentner and Bishop, 1986). In plant tissue culture studies, these are treatments in which different PGRs, genotypes, culture vessels, medium solidification agents, and so on, are tested. These procedures should not be used to compare means of treatments that are related, e.g., different concentrations of the same PGR, antibiotic, medium solidification agent, or activated charcoal (Compton, 1994; Mize et al., 1999).

Multiple comparison and multiple range tests can be used for count and binomial data. However, the data must be normalized first. Count data can be normalized using square root transformation (Compton, 1994). The arc sin transformation is typically used for binomial data. Once results of transformed data are calculated, the data are converted back to the original scale for presentation.

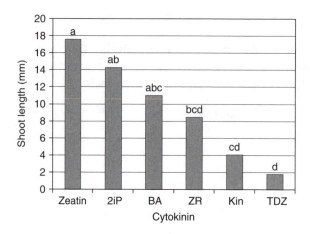

FIGURE 6.3 Effect of cytokinin type on elongation of adventitious shoots regenerated from leaf explants of *Petunia* × *hybrida* incubated in MS medium containing 4 μ*M* of zeatin, 2-Isopentenyl adenine (2-iP), benzyladenine (BA), zeatin riboside (ZR), kinetin (Kin), or thidiazuron (TDZ).

Orthogonal Contrasts

Multiple comparison procedures are most commonly used by researchers to detect differences between treatment means. However, there are mean comparison procedures that give more meaningful results. "Orthogonal contrasts" is a statistical procedure in which comparisons are made between treatments with similar characteristics (Compton, 1994; Mize et al., 1999). In plant tissue culture studies, these may be PGRs with similar activity, natural versus synthetic PGRs, or DNA constructs with the same promoter. Orthogonal contrasts differ from mean comparison tests in that more than two treatments can be compared in one test. However, the number of comparisons that can be made is restricted to the number of DFs for the treatment variable. Contrast statements are ususally performed as part of the ANOVA procedure (Compton, 1994). A contrast statement is written that specifies the treatments or group of treatments to be compared, and ANOVA calculates the DF, SS, and MS for the comparison. An F statistic is calculated by dividing the MS for the contrasts by the MSE and the level of significance (P value) is determined. One DF is used for each comparison.

In the petunia example, five contrasts of interest were planned before the experiment:

1. Natural cytokinins: zeatin, zeatin riboside, and 2-iP versus BA, kinetin, and TDZ
2. Cytokinins containing zeatin (zeatin and zeatin riboside) versus cytokinins not containing zeatin
3. BA (as a control) versus all other cytokinins
4. BA versus TDZ
5. TDZ versus all other cytokinins

Contrasts were conducted for all dependent variables as part of ANOVA. Results of orthogonal contrasts are usually presented in a table containing ANOVA computations (Table 6.5). Means of each contrast may be placed in a table or presented in the text, with the means and P values given for each contrast. Examining results for shoot length shows that shoot elongation was promoted when PGRs containing zeatin (Z and ZR) were used (P = 0.039). Likewise, natural cytokinins (Z, ZR, and 2-iP) promoted shoot length more than synthetic formulations of BA, kinetin, and TDZ (P = 0.0001). Shoot elongation was inhibited (P = 0.0008) when TDZ (1.8 mm) was used, as compared with any of the other cytokinin formulations (11.1 mm).

TABLE 6.5
Analysis of Variance Summary Table with Orthogonal Contrasts for Shoot Length from the Petunia Shoot Organogenesis Study

Source	DF	Mean Square	F Value	P Value
Cytokinin	5	1803.35	12.46	<0.00001
Experimental error	54	1563.50		
Corrected total	59			
Orthogonal contrasts				
Zeatin and ZR versus 2-iP, BA, kinetin, and TDZ	1	367.5	2.54	0.039
Zeatin, ZR, and 2-iP versus BA, kinetin, and TDZ	1	920.42	6.36	0.0001
BA versus zeatin, ZR, 2-iP, kinetin, and TDZ	1	25.230	0.17	0.9711
BA versus TDZ	1	423.20	2.92	0.0209
TDZ versus zeatin, ZR, 2-iP, BA, and kinetin	1	720.75	4.98	0.0008

Note: Data were analyzed using ANOVA for a single factor completely randomized design.

Orthogonal contrasts can be used to analyze binomial or count data given that the data are transformed before analysis. Information gained from orthogonal contrasts is often much more useful than outcomes of multiple comparison or multiple range tests. Orthogonal contrasts are underutilized in plant tissue culture studies, possibly because researchers have not been exposed to the concept or because its usefulness is not well understood. See Lentner and Bishop (1986), Little and Hills (1978), or Zar (1984) for more information regarding the use of orthogonal contrasts.

Trend Analysis

Trend analysis is the most effective method for analyzing data of treatments consisting of various levels of a single factor (different concentrations of the same growth regulator) or increments in time (Kleinbaum et al., 1988). With these treatments, the primary objective is to identify a dose or time period that simulates optimal explant response. Models identifying specific trends (linear, quadratic, cubic) are tested in a step-by-step fashion, from simplest (linear) to most complex (cubic in most cases). The procedure is stopped when a nonsignificant trend value is identified, and the last significant trend is then considered the best fit for the data. Trend analysis uses SS, T, and R^2 values to indicate significant trends. Trends may be tested through regression analysis or through polynomial contrast statements in ANOVA (Compton, 1994).

Trend analysis can be illustrated by changing the cytokinin treatments evaluated in the petunia example from six cytokinin types to six concentrations of BA (0 [control], 2, 4, 6, 8, and 10 μM). Data recorded included the percentage of explants that produced shoots, the number of shoots per explant, and the length of regenerated shoots.

In the petunia example, trend analysis was used to examine the effect of cytokinin concentration on shoot length. Because no shoots were produced by explants cultured in media without BA, observations recorded for 0 BA were deleted before analysis. In addition, zero values from unresponsive explants were deleted from the remaining treatments because only shoot length values from responding explants were of interest. Based on information from responding explants, shoot length was dependent on BA concentration (Table 6.5) with the linear model ($\hat{Y}$ = 22.954 − 2.1014*BA) giving the best fit (P < 0.00001; R^2 = 0.86). Means of dosage treatments should be presented in graphs displaying the regression equation and the fitted line, R^2, as well as individual data values and confidence intervals. As with mean comparison tests, confidence intervals of treatments that overlap are considered similar. In the petunia example, confidence intervals for 2 and 4 μM overlap (Figure 6.4). However, confidence intervals for 6, 8, and 10 μM do not overlap

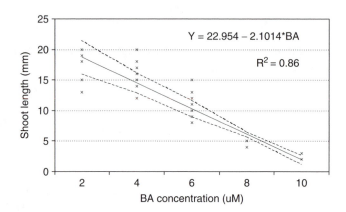

FIGURE 6.4 Influence of benzyladenine (BA) concentration on elongation of adventitious shoots regenerated from leaf explants of *Petunia* × *hybrida* incubated in MS medium containing 0, 2, 4, 6, 8, or 10 μ*M* BA.

with 2 and 4 μ*M*, and are considered significantly different from the former treatments. The conclusion would be that shoot elongation was inhibited with increasing concentration of BA above 4 μ*M*.

Trend analysis and polynomial contrasts may be used for evaluating optimum treatment levels, even if those levels were not directly tested but lie within treatment boundaries (Compton, 1994). Predicted values can be obtained using information generated from the analyses, allowing researchers to estimate the effects of treatments that were not evaluated.

Trend analysis can be used to evaluate binomial and count data, provided that the data are transformed prior to analysis. As with mean comparison procedures, data are converted back to the original scale for presentation.

CONCLUSIONS

This chapter has outlined the steps of experimentation and many of the procedures used in the analysis of plant cell culture data. It is important to remember that any well-conceived research project should be planned carefully on paper first. This helps to ensure that experiments will answer the desired questions. Be sure to minimize bias, fatigue, and errors during the course of experiments. These factors lead to untrue results that mislead researchers. Use the experimental design while conducting the experiment and recording and analyzing data. This helps to provide an objective, unbiased means of accurately evaluating treatment differences. Used properly, statistical analysis is a valuable tool to the scientific researcher, one that allows investigators to declare experimental results with confidence.

LITERATURE CITED AND SUGGESTED READINGS

Barnett, V. and T. Lewis. 1984. *Outliers in Statistical Data*. John Wiley & Sons, Chichester, UK.

Compton, M.E. 1994. Statistical methods suitable for the analysis of plant tissue culture data. *Plant Cell Tissue Organ Cult.* 37:217–242.

Compton, M.E. and C.W. Mize. 1999. Statistical considerations for *in vitro* research: I - Birth of an idea to collecting data. In Vitro *Cell. Dev. Biol.-Plant.* 35:115–121.

Kempthorne, O. 1973. *The Design and Analysis of Experiments*. Robert E. Krieger Publishing Co., Malabar, FL.

Kleinbaum, D.G., L.L. Kupper, and K.E. Muller. 1988. *Applied Regression Analysis and Other Multivariable Methods*. 2nd edition. PWS-Kent Publishing Co., Boston, MA.

Kuklin, A.I., R.N. Trigiano, and B.V. Conger. 1993. Incomplete block design in plant tissue culture research. *J. Plant Tissue Cult. Methods* 15:204–209.

Lentner, M. and T. Bishop. 1986. *Experimental Design and Analysis*. Valley Book Company, Blacksburg, VA.

Little, T.M. and F.J. Hills. 1978. *Agricultural Experimentation: Design and Analysis*. John Wiley & Sons, New York.

Mize, C.W. and Y.W. Chun. 1988. Analyzing treatment means in plant tissue culture research. *Plant Cell Tissue Organ Cult.* 13:201–217.

Mize, C.W., K.J. Koehler, and M.E. Compton. 1999. Statistical considerations for *in vitro* research: II - Data to presentation. In Vitro *Cell. Dev. Biol.-Plant.* 35:122–126.

Mize, C.W., and P.M. Winistorfer. 1982. Application of subsampling to improve precision. *Wood Sci.* 15:14–18.

Zar, J.H. 1984. *Biostatistical Analysis*. 2nd edition. Prentice Hall, Englewood Cliffs, NJ.

7 A Brief Introduction to Plant Anatomy

Robert N. Trigiano and Dennis J. Gray

CHAPTER 7 CONCEPTS

- Plant cells can be classified into the following basic types: meristematic and their immediate derivatives, parenchyma, collenchyma, and sclerenchyma.

- Simple tissues consist of one cell type, whereas complex tissues contain more than one of the basic cell types.

- The four plant organs are roots, stems, leaves, and flowers.

- Generally, the organization of monocots and dicots is similar, but they each have distinctive arrangements of cells and tissues in their organs.

- The four parts of flowers are sepals, petals, anthers, and pistils.

- The study of plant anatomy is useful for understanding plant development, determining the origin of cells and tissues *in situ* and *in vitro*, and identifying the mode of regenerated plants in culture.

INTRODUCTION

This chapter explores some of the anatomy, or internal organization (cells, tissues, and organs), of angiosperms. For simplicity, we first look at cell types, and then compare and contrast the anatomy of monocotyledonous (monocot) and dicotyledonous (dicot) tissues and organs. For the purposes of this book, we will consider angiosperms to have the following four organs: roots, stems, leaves, and flowers—fruit morphology and anatomy will not be discussed in this chapter. It is impossible to discuss adequately all the details of anatomy and development of these organs in this short chapter. Therefore, most treatments of cell types, tissues, and organs are described in broad terms, and students are cautioned that many exceptions to our generalizations can be found. Readers with greater interest in a more thorough exploration of anatomy, of the development of angiosperms, and of the comparison of angiosperms to other divisions of plants are directed to the botany and anatomy textbooks cited at the end of this chapter. Most of the material in this chapter is derived from Esau (1960) and Fahn (1990). Readers should also note that while plant anatomy and morphology are typically studied through the use of static materials, such as histological sections (Chapter 4), it is important to keep in mind that they represent growing, changing three-dimensional organisms. In this way, an understanding of plant development can be better achieved.

Plants, like other complex organisms, are constructed of cells, the basic unit of life. However, unlike animal cells, most plant cells are surrounded by a more or less rigid wall, composed of a variety of structural polymers, which may include pectin, cellulose, lignin, and hemicellulose. The wall may be relatively thin, as in many parenchyma cells, or rather thick, as in collenchyma and sclerenchyma cell types (see below). In fact, the structure of the plant cell wall imparts to some degree the function of the cell. Cells may have only a primary cell wall, which can be more or less accurately defined as the wall material deposited while the cell is increasing in size and having cellulose microfibrils that are laid down randomly or in more or less parallel orientations (Esau, 1960). The primary wall usually contains cellulose, hemicellulose, and pectic compounds, referred to as the middle lamella. The middle lamella acts as "cement" between adjacent cells. In contrast, secondary walls found in some cells are deposited over or integrated into the primary wall and middle lamella after the primary wall has been completed. The new cellulose microfibrils have a definite parallel orientation. Secondary walls containing cellulose and hemicellulose and lignin may or may not be present (Esau, 1960).

CELL TYPES

Let us consider the basic cell types of plants before examining the internal arrangement of cells into tissues and, in turn, tissues into organs. For the purposes of this chapter, plant cells will be classified into the following types and their variations: (1) meristematic, (2) parenchyma, (3) collenchyma, and (4) sclerenchyma. Note that most references consider the meristematic cells to belong to the parenchyma class.

MERISTEMATIC CELLS

Meristematic cells are very thin-walled cells that undergo mitosis to increase the length (apical meristem) or thickness (lateral meristem) of the organ. The meristematic initials (stem cells; see Chapter 13) reproduce themselves as well as form new cells, termed derivatives, that increase the body of the plant. These derivative cells usually continue to divide several times before any significant differentiation into other cell types occurs. The initials and their derivative cells constitute the apical meristem (Esau, 1960), which can be found at shoot (Figure 7.4A) and root (Figure 7.1A) tips. The stem apical meristem may be divided further into the tunica and corpus. The tunica (coat) is one to several cell layers thick and divides only by anticlinal divisions to increase the surface area of the tip and to surround the corpus (body), which consists of a number of cells that divide in different planes to increase the volume of the meristem. This two-part arrangement of the meristem is absent in roots. The shoot apical meristem can be divided into a peripheral zone, which gives rise to leaves, buds, and flowers (lateral organs), and the rib zone, which produces the stem tissues (see Chapter 13). Several cell layers distal to the apical meristem and in the rib zone, the procambium (vascular), ground meristem (cortex), and protoderm (epidermis) of the stem, are differentiated and give rise to the primary tissues of the plant body.

Primary growth of the plant is brought about from the activities of apical meristems and subsequent divisions and differentiation of the derivative cells into the tissues and organs of the plant. All tissues originating from a primary meristem are termed primary tissues; e.g., primary xylem and epidermis. Most monocots and some dicots complete their growth and development via primary growth only.

Secondary growth exhibited by many dicots is achieved through specialized lateral meristems. The vascular cambium (Figure 7.5A), located between the primary phloem and the primary xylem, produces cells that differentiate into additional vascular tissue—i.e., secondary phloem and xylem—which increases the girth of stems and roots. Another lateral meristem, the phellogen or cork cambium, is found near the exterior of stems and roots and arises in the primary cortex. It produces phellem and phelloderm cells that replace the epidermis and cortex, respectively, which are lost or crushed due to the expanding diameter of the root or stem. Collectively, this new tissue is called the periderm.

Parenchyma Cells

Parenchyma cells generally have the following characteristics:

They are typically nearly isodiametic (about as long as they are wide); however, cells may vary in shape, some being elongated or even lobed.

The primary cell wall of this cell type is relatively thin, and is composed mainly of cellulose and hemicellulose with a layer of pectic substances, the middle lamella, on the exterior of the primary wall. Note that some parenchyma cells, especially in vascular tissue, may develop a secondary wall or become sclerified with lignin (see sclerenchyma).

Parenchyma cells always have nuclei and functioning protoplasts (cytoplasm).

These cells are generally considered to be relatively undifferentiated (compared to sclerenchyma) and capable of resuming meristematic activities by dedifferentiation. Indeed, this cell type and tissue is involved in the development of adventitious roots and shoots, wound healing, and other activities. Note, though, that some parenchyma cells can be very differentiated and specialized in their function.

This cell type can be found through the body of the plant in primary and secondary tissues.

Collenchyma Cells

Collenchyma cells have the following characteristics:

They are typically more elongated than parenchyma cells, and are specialized to function as mechanical support for the plant. Collenchmya cells have soft, pliable, unevenly thickened primary walls composed mostly of cellulose with some pectin, but never lignin.

Collenchyma cells are similar to parenchyma cells in having nuclei and living protoplasm, and are capable of dedifferentiation and meristematic activity.

This type of cell is generally found in young stems (Figure 7.5D) and leaf petioles. It functions as flexible supporting tissue.

Sclerenchyma Cells

Sclerenchyma cells have the following characteristics:

Sclerenchyma cells can be long and thin (fibers; Figure 7.5A) or isodiametic to elongated (sclerids). They are involved in mechanical support and water conduction. This cell type is widely distributed in the four major plant organs. Specific cell types include fibers, vessel elements, trachieds, and sclerids (taking various forms, including astrosclerids [star-shaped], stone cells, and so on).

The hallmark of sclerenchyma cells is the deposition of a secondary wall on the interior of the primary wall. Proceeding from the outside of the cell inward, the wall layers encountered in these cells are the middle lamella, the primary wall, and lastly, the secondary wall. By definition, the secondary wall is laid down after the growth of the cell ceases. The wall is composed primarily of cellulose arranged in parallel fibers and usually lacks pectic components. If a cell becomes lignified, deposition of lignin starts at the middle lamella and proceeds inward.

Although many sclerenchyma cells are dead (i.e., they lack a protoplasm) at functional maturity, some types of cells may retain living protoplasm. However, the protoplasm of these cells appears to be physiologically nonfunctional or inactive (Esau, 1960).

Sclerenchyma cells are highly differentiated and are usually considered incapable of dedifferentiating and resuming meristematic activity.

THE INTERNAL ARRANGEMENT OF CELLS

Cells are organized into simple tissues (one cell type) and complex tissues (more than one cell type). For example, young ground and pith tissue found in stems is a simple tissue made up of parenchyma cells (Esau, 1960). Another example would be the collenchyma tissue found in the four corners of a mint stem (Figure 7.5D) or the "strings" in a celery stalk (petiole). An excellent example of a complex tissue is the secondary vascular tissues found in many dicots. This tissue contains representatives of both parenchyma and sclerenchyma cell types. Tissues, in turn, are organized into the four primary organs: roots, stems, leaves, and flowers. The remainder of this chapter will be devoted to illustrating the general arrangement of cells and tissues within these organs and giving a brief account of seed anatomy. Anatomical studies of tissue culture regeneration of organs are included where applicable.

ROOTS

Roots serve as the primary water- and mineral-absorbing organ of plants. They also act to anchor the plant in the soil, and may also function as storage organs and in vegetative (asexual) reproduction. Dicots typically have a persistent taproot, whereas with many monocots, the taproot is ephemeral and is replaced with a fibrous root system consisting of many adventitious roots.

   The anatomy of roots is extremely variable, but general models may be developed for both monocots and dicots. The apical meristem of most roots (Figure 7.1A) appears less conspicuous than shoot meristems (Figure 7.5B), which are arranged as a tunica and corpus. In many plants the root apical meristem gives rise to the cells of both the root cap and the primary meristems; however in some grasses, a group of cells, the calyptrogen, produces the cells of the root cap (Figure 7.1A). In many instances, a quiescent zone is located in the apical meristem. This region exhibits low mitotic activity, but cell division typically resumes distal to this zone. The most noticeable differentiation of cells is in the vascular tissue — primary phloem differentiates first, followed by primary xylem. Cells continue to mature elongation and enlargement. Root hairs, extensions of epidermal cells that increase the surface area to absorb water and minerals, are typically first seen behind the zone of mitotic activity and mark the maturation of the first xylem elements.

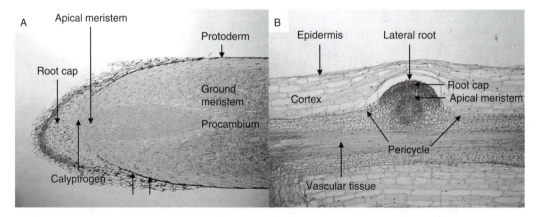

FIGURE 7.1 Root apical meristem and lateral root origin. (A) Near median longitudinal section through a young corn (*Zea mays*) root. The calyptrogen, in this case, gives rise to the cells of the root cap. In other roots, including those of many dicots, the root cap is derived from the apical meristem. The three meristematic areas are the protoderm (which gives rise to the epidermis), the ground meristem (cortex), and the procambium (vascular tissue). Arrows indicate muscilaginous wall substance. (Slide courtesy of Carolina Biological Supply Co., Burlington, NC.) (B) A young lateral (secondary) bean (*Phaseolus vulgaris*) root that originated from the pericycle. Note that as the lateral root develops, it pushes through and crushes the cortical and epidermal tissues of the primary root. The architecture of lateral roots is similar to that of primary roots.

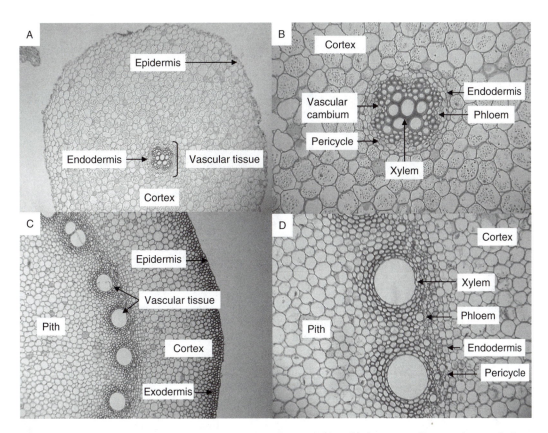

FIGURE 7.2 Dicot and monocot root structure. (A) Cross-section of a buttercup (*Ranunculus* species), a dicot root. Note the arrangement of tissues and the lack of pith in the center. (B) Higher magnification of the central area shown in Figure 7.1A. The stele includes all vascular tissue (xylem and phloem), the vascular cambium (discernible only by relative position in the young root), and the pericycle. It does not include the endodermis. Note the large xylem vessels in the center and the thickened walls (Casparian strips) of the endodermal cells. (C) Cross-section of a large corn (*Zea mays*), a monocot root. Note the central core of pith and the very large xylem vessels. (D) Enlargement of the vascular area of the cross-section shown in Figure 7.1C. (All slides provided courtesy of Carolina Biological Supply Co., Burlington, NC.)

Lateral or secondary roots greatly increase the absorptive area of the root system and usually are initiated from the pericycle, a layer or layers of cells in the vascular tissue or stele, at some distance behind the apical meristem (Figure 7.1B). Several adjacent pericycle cells divide to form a root primordium, and continued divisions force the developing root through and crush the endodermis, cortex, and epidermis of the primary root. Vascular tissue within the lateral root is connected to similar elements in the primary or parent root by differentiation of pericycle cells (Esau, 1960).

Most of the same cells and tissues are present in monocots and dicots; however, the arrangement of these tissues is somewhat different. A summary of the primary differences between the two divisions of angiosperms is presented in Table 7.1. The most conspicuous difference between dicot and monocot root anatomy is the arrangement of the primary vascular tissue. Most dicot roots lack central pith tissue; instead, the core of the root is occupied by large metaxylem vessels (Figures 7.2A and B). Additionally, the primary xylem is arranged in arches or arms, with the primary phloem located between the arms. A vascular cambium is located between the primary xylem and phloem in dicot roots. In contrast, many primary monocot roots have pith in the center of the primary root surrounded by many arms of vascular tissue (Figures 7.2C and D). Dicot roots generally exhibit secondary growth via vascular cambia located between the primary xylem and phloem, and

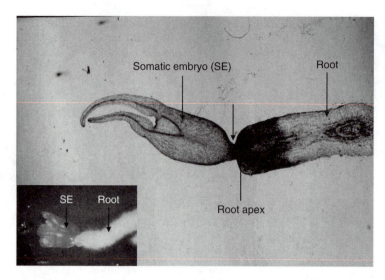

FIGURE 7.3 Somatic embryos of redbud (*Cercis canadensis*) that originated from root tissue. Near median longitudinal section through the somatic embryo and oblique section through the root, which developed from a callus culture (see inset). Close examination of serial sections revealed five distinct but fused somatic embryos from a common suspensor (arrow), which originated from the apical area of the root.

TABLE 7.1
A Summary Guide to Morphological and Anatomical Traits of Monocots and Dicots

Organ	Monocot	Dicot
Root	Usually fibrous	Usually a taproot
	Pith present	Pith lacking
	Lacking vascular and cork cambia or secondary growth	Vascular and cork cambia and secondary growth present
		Primary xylem typically arranged in "arches"
Stem	Pith lacking, but vascular bundles embedded in pith-like fundamental tissue	Pith present
	Vascular and cork cambia typically lacking; no secondary growth, although may have primary thickening meristems	Vascular and cork cambia typically present; secondary growth evident, manifested as rings of vascular tissue
Leaf	Typically blade-like with parallel venation	Variously shaped with net venation
	Leaf mesophyll generally undifferentiated into distinct layers	Mesophyll may be differentiated into spongy and palisade parenchyma layers
Flower and Seed	Flower parts typically in threes or in multiples of three	Flower parts typically in fours or fives or in multiples of four or five
	Embryo has one cotyledon	Embryo has two cotyledons

Note: Although this table provides very broad characterizations and contrasts of monocots and dicots, the reader is cautioned that there are many exceptions to these generalizations.

by cork cambia, which forms in the cortex (see discussion of stems below). Monocots lack lateral cambia or meristems.

Growth of plant tissues and organs in culture often present anomalies, which are difficult to understand. One such instance was seen in a culture of redbud (*Cercis canadensis*) cotyledons on which adventitious roots had formed. At the tip of one such root, a green, multilobed tissue had

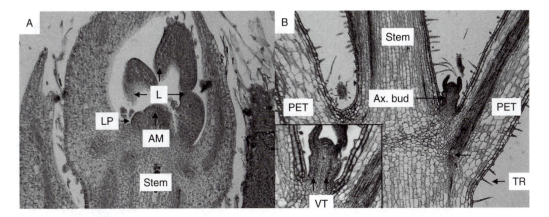

FIGURE 7.4 Apical and lateral (axillary) shoot meristems. (A) Longitudinal section through the stem tip of bean (*Phaeseolus vulgaris*). The apical meristem (AM) is surrounded by a number of very young leaf primordia (LP) and more developed leaves (L). (B) Longitudinal section through a node of catnip (*Nepeta cataria*). The axillary bud (Ax. bud) is located in the axil formed by the leaf petiole (PET) and the stem. Note the vascular tissue (VT) extending from the petiole and connecting to the vascular tissue of the stem (arrow). The AM is dome-shaped and the VT has differentiated (see inset). TR = trichome.

grown (Figure 7.3, inset). Upon histological sectioning, the unrecognizable tissue was determined to be five fused somatic embryos (see Chapter 14) that had been initiated from the apical meristem of the adventitious root (Figure 7.3).

STEMS

As noted before, the shoot meristem (Figure 7.4A) is organized into two parts, the tunica and the corpus. Just as in the root apical meristem, the primary meristems, the ground, the procambium, and the protoderm are also found in the shoot tip. However, the shoot tip is more complex than the root apex, as it differentiates leaves, axillary buds (Figure 7.4B), and flowers from the peripheral zone (see Chapter 13). The stem of the plant is divided into nodes (with leaves present) and internodes (with leaves absent). Leaves are defined by the presence of an axillary bud (Figure 7.4B and insert) lying between the main stem and the leaf petiole. The anatomy of axillary buds is equivalent to the original shoot tip. The bud has the same structure as that of the apical meristem, and serves to initiate branch or lateral growth and/or flowers under the proper environmental conditions.

The arrangement of tissues in stems is variable and very different than the arrangement found in roots. Just as in root anatomy, dicot and monocot stems are generally different from one another. These differences are summarized in Table 7.1. Dicots typically have primary vascular bundles (fasciculars) arranged in a ring around the central pith (Figures 7.5A and B). The primary xylem is located toward the pith, while the primary phloem is toward and contiguous with the cortex. Note that in some instances the primary phloem may lie on either side of the primary xylem. The fascicular may also have a cap of very conspicuous phloem fibers (Figure 7.5A). A portion of the vascular cambium (intrafascicular cambium) is located between the primary xylem and phloem, with another portion between the vascular bundles (interfascicular cambium). The activities of the vascular cambium produce secondary vascular tissues, which in due course surround the pith with a continuous ring of vascular tissue. The secondary vascular tissues eventually crush and obliterate the primary tissues, especially in perennials and woody plants. It is this continuous ring of vascular tissue that easily differentiates pith from cortex in the dicots.

Dicots also have another lateral or secondary meristem — phellogen or cork cambium. This cambium or meristem is formed in the cortex and is responsible for forming additional cortical cells (phelloderm) and replacement for the epidermis (periderm).

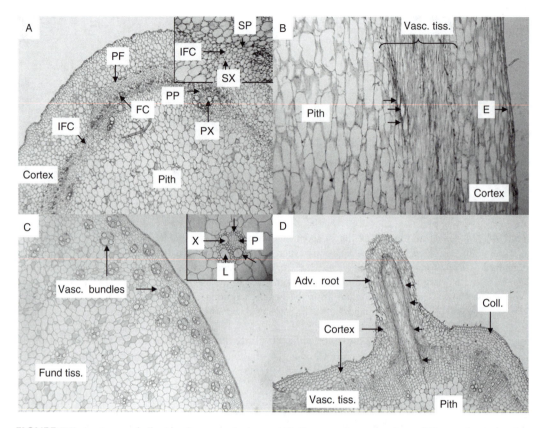

FIGURE 7.5 Anatomy of dicot and monocot stems. (A) Cross-section of a bean (*Phaeseolus vulgaris*) stem. Note the prominent pith and the cap of phloem fibers (PF). Secondary xylem (SX) and secondary phloem (SP) tissues, which originated from the interfascicular vascular cambium (IFC) and fascicular cambium (FC), are also common in dicot stems (inset). PP = primary phloem; PX = primary xylem. (B) Longitudinal section of a bean illustrates the arrangement of tissues in the stem. Arrows indicate xylem vessels with helical secondary wall patterns. E = epidermis. (C) Cross-section of an older corn (*Zea mays*) stem. Notice that vascular bundles are scattered throughout the stem and embedded in fundamental tissue (Fund. tiss.). The vascular bundles contain only primary xylem (X) and phloem (P) tissues (vascular cambium absent) with a prominent lacuna (L) or air space (inset). Arrows indicate sclerenchyma cells of the sheath. (Slide courtesy of Carolina Biological Supply Co., Burlington, NC.) (D) Cross-section through the stem of catnip (*Nepeta cataria*) and a longitudinal section through an adventitious root. The root, which originated near the primary phloem and vascular tissue (arrows), has differentiated and connected to the vascular tissue (Vasc. tiss.) in the stem. Coll = collenchyma tissue.

In monocot stems, the vascular bundles are scattered throughout the ground tissue or fundamental tissue, and as a result a pith and cortex are not discernable (Figure 7.5C). The vascular bundles typically are surrounded by a sheath of sclerenchyma cells, which helps support the stem (Figure 7.5C, inset). Additionally, many monocot stems, such as that of corn, have an abundance of small vascular bundles located just under the epidermis, imparting stiffness to the stem and helping the plant to withstand environmental stresses. Monocots lack vascular cambia and therefore do not have secondary growth.

An important feature of stems, especially in the production of shoots in tissue culture or cuttings, is the ability to form adventitious roots (Figure 7.5D). These roots have their origin in parenchyma cells lying near the vascular (phloem) tissue. They grow in a manner similar to that of lateral roots.

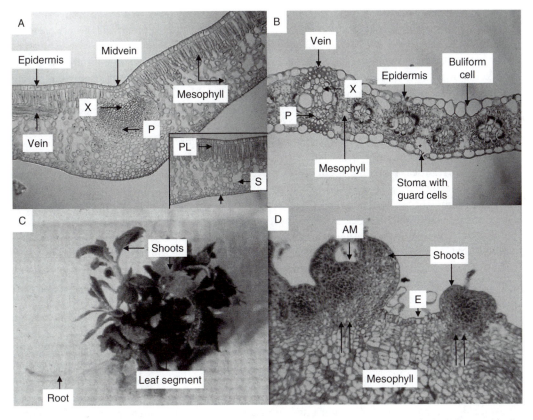

FIGURE 7.6 Leaf anatomy. (A) Cross-section of a dicot leaf (privet: *Ligustrum* species). A large midvein, with xylem (X) toward the top (adaxial) surface and phloem (P) oriented toward the abaxial (bottom) surface, is shown. The mesophyll is differentiated into palisade (PL) and spongy (S) layers, with stomata and guard cells (arrow) on the bottom surface (see inset). (B) Cross-section through a corn (*Zea mays*) leaf, a typical monocot. Notice that the epidermis contains specialized buliform cells and that the mesophyll tissue is not differentiated into layers. Stomata and guard cells are associated with large substomatal cavities (*). (Figures 7.6A and B are courtesy of Carolina Biological Supply Co., Burlington, NC.) (C) Adventitious shoots and roots forming directly from a mum leaf segment cultured on medium containing cytokinin. (D) Cross-section through the leaf segment and longitudinal sections through several adventitious shoots shown in Figure 7.6C. Note the lack of an intermediate callus and the vascular connections (arrow pairs) between the adventitious shoots and the leaf mesophyll. E = epidermis; AM = apical meristem.

LEAVES

Leaves are the primary photosynthesizing organs of angiosperms. Most leaves are relatively thin and flat (with a large surface area compared to volume) and are adapted for capturing light and facilitating gas and water exchange with the atmosphere. There are, of course, many exceptions to the above statement. Leaves can exhibit extensive, often unusual, modifications depending on the environment, coupled with genetic inputs and evolution. Leaves are found at nodes of the stem and are variously arranged (phyllotaxy): opposite (two leaves at the node), alternate (one leaf at the node), and so on. Some plants have leaves that are very difficult to recognize as leaves, whereas others are very obvious. Generally, leaves have three parts—blades or lamina, petioles, and stipules—although it is not unusual for many leaves to lack the petiole (sessile leaf) and/or the stipules. The blade may be simple and variously shaped, or compound, with the leaf divided into leaflets.

All leaves have an axillary bud associated with them. The bud is located in the axis formed by the petiole and stem. Compound leaves do not have buds at the base of the branch points or where leaflets join the common axis. Many dicot leaves exhibit net venation patterns with major and minor veins (Figure 7.6A), whereas monocots typically have a parallel venation arrangement with most veins of more or less equal size (Figure 7.6B). Note that there are many exceptions to this general rule. Anatomically, the vascular tissue in the node will exhibit a leaf gap, or parenchyma tissue in the vascular cylinder of the stem where the leaf traces (vascular tissue connecting the leaf to the stem) are bent toward the leaf petiole (Esau, 1960).

Leaves typically do not exhibit secondary growth and therefore have only primary tissues. Leaf tissue can contain all three of the basic cell types discussed above. The epidermis of leaves has many special features, including a cuticle to prevent wall loss, various trichomes (hairs), guard cells and accessory cells to regulate water and gas exchange, and other cells with specialized functions (Figures 7.6A and B). The leaf mesophyll of many dicots is differentiated into one or more palisade layers toward the adaxial (top) surface of the leaf, and the spongy mesophyll toward the abaxial (bottom) surface. The palisade cells are elongated (shoebox-shaped) and arranged "tightly" and orthogonally to the leaf surface. The cells of the spongy mesophyll tissue are isodiametric to slightly elongated and are arranged "loosely" with large intercellular spaces (Figure 7.6A). The mesophyll of monocot leaves is typically not differentiated into layers and has only spongy parenchyma (Figure 7.6B). Vascular tissue exists as bundles running through the mesophyll of the leaf. Xylem tissue is usually toward the top of the leaf and the phloem toward the bottom of the leaf. In many plants, especially the grasses and some other monocots, the vascular tissue may be surrounded by a sheath of sclerenchyma cells (Figure 7.6B).

Young, partially expanded leaves are a commonly explanted tissue for plant regeneration studies (Figure 7.6C). Shoots, roots, and somatic embryos arise from the explant, but it is sometimes difficult to visually distinguish one structure from another and to determine from which tissue each regenerated structure originated. A quick and simple anatomical study can help resolve these issues (Figure 7.6D). In the case of regeneration from a chrysanthemum leaf segment, the regenerated tissues were shoots originating from mesophyllic tissue. Not all cases are as clear-cut as this example, and a more extensive examination of serial sections of the tissue must be made in order to determine both the origin and identity of the regenerated tissue (somatic embryo versus shoot; see Chapters 12 and 14).

FLOWERS

Flowers are organs unique to the angiosperms. They are incredibly diverse in morphology and range from very showy and conspicuous to very bland and almost imperceptible. They may occur singly, or they may be arranged in different types of inflorescences (multiple flowers on a common axis). Flowers contain the sexual reproductive apparatus of plants. Both female and male sexes may be contained in the same flower (monoecious, "one house") or in two separate flowers or plants (dioecious, "two houses") for each of the sexes (pistillate and staminate flowers). Meiosis and alternation of generations occur in the ovule (female) and the anther (male). The male and female gametophytes are greatly reduced, consisting only of a few cells.

For the sake of simplicity, we will consider only perfect flowers, wherein all flower parts, including both male and female structures, are present. The four basic units in a flower are sepals, petals, stamens, and pistils (Figures 7.7A, B, and C). The floral parts are arranged as whorls on the receptacle (Figure 7.7D). Working from the outside inward at the base of the receptacle, the sepals — usually green, leaf-like structures — make up the first whorl; these are collectively termed the calyx. The next layer is composed of the petals, which can be green, but are generally white or colored. The petals are usually larger than the sepals and are collectively termed the corolla. The petals and sepals taken together are also called the perianth. In some cases, when the sepals and petals are similar or almost indistinguishable, they are called tepals (Figure 7.7C).

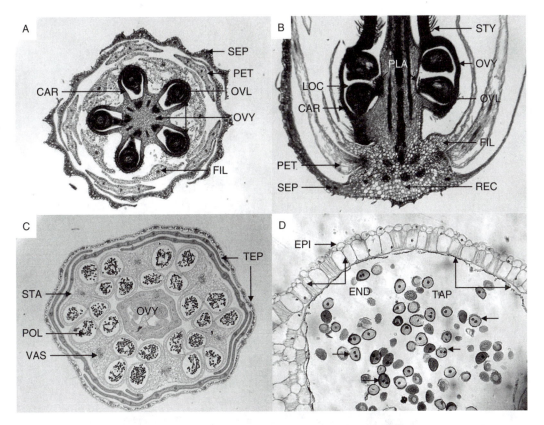

FIGURE 7.7 Flower anatomy. (A) Cross-section through a *Geranium* species (Stork's-bill) flower. Progressing from the outside inward, sepals (SEP), petals (PET), and filaments (FIL; the stalks of the anther) and ovary (ovy) are borne. Occupying the center of the flower is the ovary (OVY), which is composed of five carpels (CAR) in which two ovules (OVL) are borne. *Note*: only a single ovule can be seen in each locule (space) with this section. (B) Longitudinal section through the Stork's-bill flower seen in Figure 7.7A. In this view, two carpels (CAR) and locules (LOC) of the ovary (OVY) are visible, but each contains only two ovules (OVL), of which only one will mature. Note the trichomes (hairs) on the style (STY) portion of the ovary. The sepals (SEP), petals (PET), and filaments (FIL) are inserted below the ovary on the receptacle (REC). (C) Cross-section of a Lily (*Lilium* species) flower. In this view, six stamens (STA) containing pollen (POL) are shown. The vascular tissue (VAS) in the filament and four lobes of each anther are evident. Sepals and petals are collectively termed tepals (TEP) for lily flowers. OVY = ovary. (D) Higher magnification of an anther of a lily. The anther wall is composed of the epidermis (EPI), the endothecium (END), and the tapetum (TAP), which in this case has degenerated and is only represented by the remnants of the cells. Many pollen grains are present — some in the binucleate stage (arrows). (Figures 7.7C and 7.7D are courtesy of Carolina Biological Supply Co., Burlington, NC.)

The stamens, comprising the male portion of the flower, make up the next set of structures found in the flower. A stamen consists of the filament or stalk (Figures 7.7A and B), which terminates in the anther, and which contains pollen (Figures 7.7C and D). The pistil, or female structure, which is more or less flask-shaped, occupies the center of the flower and consists of a swollen base, the ovary, and a smaller diameter stalk (style) that ends in a somewhat swollen tissue, the stigma (Figure 7.7B — stigma not shown). Dicots typically have sepals, petals, and stamens in whorls of four or five or in multiples of four or five (Figure 7.7A), whereas monocots generally have these structures in groups of three or multiples of three (Figure 7.7C). Pistils may be composed of a single carpel (a structure analogous to a leaf rolled along the long axis and bearing ovules on the inner surface) or of many carpels (Figure 7.7A). One or more ovules may be attached to a common

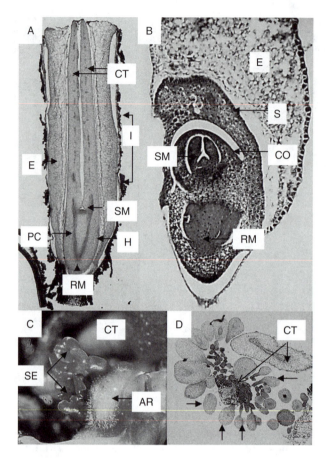

FIGURE 7.8 Zygotic and somatic embryos. (A) Near median longitudinal section through an immature seed of the dicot, *Cercis canadensis* (redbud). Note the well-developed pair of cotyledons (CT), shoot meristem (SM), and hypocotyl (H); the root meristem (RM) is inconspicuous. E = endosperm, I = integument, and PC = provascular tissue. (B) Near median longitudinal section through a seed of the monocot, *Dactylis glomerata* (orchardgrass). The shoot (SM) and root (RM) meristems, the single cotyledon (S = scutellum) and the first leaf (CO = coleoptile) are well developed. E = endosperm. (Photo courtesy of Springer-Verlag, *Protoplasma* 110:121–128.) (C) Somatic embryos (SE) of redbud (*Cercis canadensis*) that formed on a cotyledon (CT) cultured on medium containing 2,4-D. AR = adventitious root. (D) Section through the cotyledon (CT) and somatic embryos (arrows) shown in Figure 7.8C. Note the well-developed suspensors on some of the somatic embryos.

surface or placenta in each carpel (Figure 7.7B). Microspore mother cells (2N) in the anther and one megaspore mother cell (2N) in each ovule undergo meiosis (a reduction division) followed by mitosis to form the male and female gametophyte, respectively. When mature, the male gametophyte consists of three cells or nuclei, whereas the female gametophyte typically has eight cells or nuclei. The gametophytes are often placed in tissue culture in hopes of obtaining haploid plants. One must be cautious because many diploid plants may arise from the tapetum, the inner layer of cells lining the anther (Figure 7.7D) or the nucellus, which is a maternal cell layer delimiting the female gametophyte.

Pollination occurs when pollen (containing the male gametophyte) is transported via wind, insect, or another vector to the stigma. If the pollen is compatible with the female tissue and the stigma is receptive, the pollen grain will germinate by a germ tube and grow downward through the style toward the female gametophyte in the ovule. The pollen tube contains two sperm nuclei (each N), of which one fuses with the polar nuclei to form the primary endosperm nucleus (3N or another polyploid number), while the other unites with the egg (N) to form the zygote (2N). Thus, the sporophytic phase (2N) is restored. The primary endosperm nucleus divides and produces endosperm, and the zygote divides, producing the suspensor and embryo. The embryo develops into a bipolar structure, possessing both shoot and root meristems and exhibiting bilateral symmetry. The dicot zygotic embryo has two cotyledons (Figure 7.8A) and the monocot embryo has one cotyledon (scutellum; Figure 7.8B). The integuments (seed coat) covering the ovule harden, and the ovule is now a seed (Figure 7.8A). Endosperm may be completely absorbed by the embryo, or may remain outside the embryo.

Immature zygotic embryos or embryo parts (hypocotyls and cotyledons) are often used in experiments in which somatic embryogenesis is the goal. These embryos are excellent sources of relatively undifferentiated, totipotent cells that are capable of producing somatic embryos and regenerating entire plants. In many cases, embryo tissues are exposed to high concentrations of synthetic auxins, and somatic embryos are produced (Figure 7.8C). Histological sections of the materials provide evidence of somatic embryo formation (bipolar structures without vascular connections to the explant) and the origin of the somatic embryos (Figure 7.8D).

CONCLUSION

We have presented a very brief account of the anatomy of angiosperms, and we hope that we have whetted your appetite for further study of this area of plant science. The study of plant anatomy allows us to understand more fully the process of plant development from the union of egg and sperm, to the formation of the zygotic embryo, to the development of the various tissues in the embryo and the mature plant — all fascinating processes. The study of anatomy also allows us to understand, to some degree, the origin and development of regenerated tissue from *in vitro* culture. Aspects of the regeneration of plants *in vitro* are of keen interest to researchers conducting studies of genetic transformation (on the use of ballistics versus *Agrobacterium*-mediated methods, see Chapter 19); evaluating the quality of regenerated plants; and determining the mode of regeneration (e.g., somatic embryogenesis versus shoot formation).

LITERATURE CITED AND SUGGESTED READING

Dickison, W.C. 2000. *Integrative Plant Anatomy*. Academic Press, San Diego, CA.

Esau, K. 1960. *Anatomy of Seed Plants*. John Wiley & Sons, New York.

Fahn, A. 1990. *Plant Anatomy*. 4th edition. Pergamon Press, New York.

McDaniel, J.K., B.V. Conger, and E.T. Graham. 1982. A histological study of tissue proliferation, embryogenesis, and organogenesis from tissue cultures of *Dactylis glomerata*. L. *Protoplasma* 110:121–128.

Raven, P.H., R.F. Evert, and S.E. Eichhorn. 1999. *Biology of Plants*. 6th edition. W.H. Freeman and Co., New York.

Wilson, C.L., W. E. Loomis, and T. A. Steeves. 1971. *Botany*. 5th edition. Holt, Rinehart & Winston, New York.

8 Plant Growth Regulators in Plant Tissue Culture and Development

Victor P. Gaba

CHAPTER 8 CONCEPTS

- Regulation of developmental processes in plant tissue culture generally requires the addition of plant growth regulators (PGRs) to the medium.

- Endogenous PGRs are those produced by the plant, and exogenous PGRs are those applied to the plant, often through incorporation in culture media.

- Explant response to PGRs depends on the genotype (species and/or cultivar).

- Explant response to PGRs depends on the state of the plant tissue, the organ explanted, growth conditions, and the nutrition of the source plant.

- Explant response to PGR depends on class (auxin, cytokinin, gibberellin, abscisic acid, or ethylene), particular chemical structure and concentration.

- Major PGRs, such as the cytokinins and auxins, are especially important in the control of development in plant tissue culture.

- Factors other than PGRs also control *in vitro* growth and development.

INTRODUCTION

Plant growth and developmental processes, such as germination, stem elongation, leaf growth and development, flowering, fruit set, and fruit growth and ripening, are controlled by plant growth regulators (PGRs), commonly but incorrectly called "plant hormones." For general review of these processes, see recent plant biology textbooks (Mauseth, 1991; Raven et al., 1992; Salisbury and Ross, 1992; Taiz and Zeiger, 2002); more focused reviews of PGRs are contained in Davies (1995) and Arteca (1996). The interesting stories of PGR discovery are covered in these textbooks and here in Chapter 2.

Study of PGR function can be complex because several PGRs typically work in concert with each other and their concentrations in the plant change with time, season, and developmental stage. Some PGRs are synthesized locally by cells for their own consumption, whereas others are synthesized in one organ and transported to other parts of the plant for a specific action, in which case their activity might be thought of as being similar to a plant "hormone." Those produced by the plant are referred to as endogenous PGRs. However, as described in this chapter, natural and synthetic versions of many PGRs have been identified and commonly are supplied to intact plants and tissue cultures; these are termed exogenous PGRs.

The concentration of endogenous PGRs in tissue is determined by the relative rates of synthesis (or inflow) and the rate of degradation (consumption) or conjugation. Exogenous PGRs are applied to plants in the field to obtain important agricultural responses, such as the defoliation of cotton or the growth or ripening of fruit such as grapes and peaches. In plant tissue culture, small explants are transferred onto culture media. Often such explants are too small (or not able) to make the PGRs needed for a developmental or growth response. However, by supplementing the medium with exogenous PGRs it is possible to stimulate the responses that are desired from plant tissues, including overwhelming the effect of endogenous PGRs present in the explant. Different PGR analogues have different effects, depending on chemical structure, plant tissue, and genotype. Sometimes it is necessary to mix different PGRs of the same or different classes to obtain the results desired, or to give sequential treatments with different PGRs.

Although the role of PGRs in plant tissue culture is critical, PGRs interact within important biological determinants. The most important factor in plant tissue culture is genotype; in other words, whether the plant material has the genetic ability to produce the response desired. For example, some species or cultivars regenerate shoots, or produce somatic embryos, with relatively little exogenous stimulation, whereas others do not. Additionally, the physiological state of the plant material will also determine the biological response. Therefore certain plant parts (explants) will be more responsive than others at a given time. The cotyledons (seed leaves) of some species regenerate shoots well, whereas the true leaves or stems do not. In some cases, the physiological state of explants can be altered by pretreatments with another PGR.

A great deal is known about these factors (genotype and explant/state), but very little is understood of how they control response. The concentration of a PGR is often very important, in that too little will not produce the desired response and too much will often produce an unwanted result. In either case, a less than optimum response occurs. PGR concentration response curves are usually hyperbolic. The concentration of exogenous PGRs is controlled by the rate of uptake into the tissue from the medium, the rate of transport through the tissue, and metabolism (degradation or conjugation with a monosaccharide) in the tissue.

Growth media are optimized for the period of treatment. This period starts with a high concentration of PGRs, which gradually declines with time due to degradation (caused by lability at growth temperatures, light, and so on) and uptake. The initial high concentration of PGRs stimulates certain activities, such as bud initiation, which are then permitted to elongate by reduced concentration. The optimum concentration to use is therefore the best-integrated concentration over the period of growth. The addition of activated charcoal to culture media will also alter the effective concentration of PGRs, due to adsorption. Optimization of the response will take all these factors into account. Note also, that while our discussion here is about morphogenetic development, PGRs also influence other *in vitro* processes such as secondary-product metabolism (see Chapter 24).

THE FIVE MAJOR GROUPS OF PGRS

There are five major groups of endogenous PGRs: auxins, cytokinins, gibberellins, abscisic acid, and ethylene. Most of these groups have multiple roles in plant growth and development, and are employed to manipulate plants in tissue culture.

AUXINS

The first isolated PGR was endogenous (natural) auxin, indolyl-3-acetic acid (commonly called IAA) (Figure 8.1A). IAA is rapidly degraded both in the medium and in the plant; therefore, more stable chemical analogues of IAA are often substituted. These are generically called auxins, due to their similar biological activity to IAA. Synthetic auxins such as 2,4-dichlorophenoxyacetic acid (2,4-D) (Figure 8.1B), 3-indolyl butyric acid (IBA) (Figure 8.1C), and 1-naphthaleneacetic acid (NAA) (Figure 8.1D) are frequently used. Note the similarity of the chemical structures of these auxins (Figure 8.1).

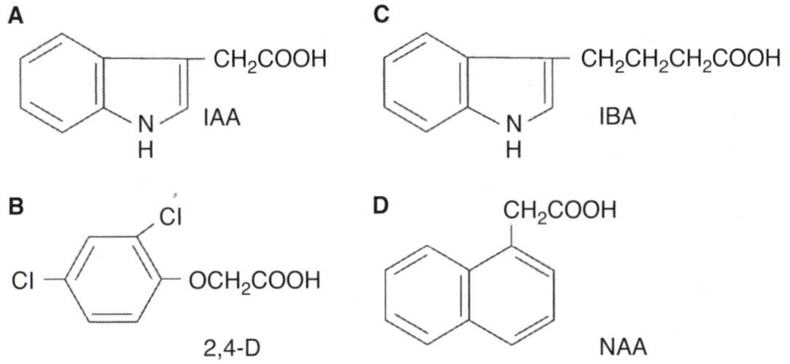

FIGURE 8.1 Chemical structures of some auxin-type PGRs. (A) Indolyl-3-acetic acid (IAA); (B) 2,4-dichlorophenoxyacetic acid (2,4-D); (C) indolyl-3-butyric acid (IBA); (D) 1-naphthaleneacetic acid (NAA).

Auxins have multiple roles in tissue culture, according to their chemical structure, their concentration, and the plant tissue being affected. Auxins cause the production of callus and roots as well as the extension growth of stems. Auxins generally stimulate cell elongation, cell division in cambium tissue, and, together with cytokinins (the next PGR to be discussed), stimulate differentiation of phloem and xylem. Additionally, a high concentration of an exogenous auxin can induce somatic embryogenesis. The essential function of auxins and cytokinins is to reprogram somatic cells that had previously been in a determined state of differentiation. Reprogramming causes dedifferentiation and then redifferentiation into a new developmental pathway. Thus, a cell that had been destined to develop into part of a leaf, for example, might become embryogenic, producing somatic embryos. The mechanism by which auxin causes dedifferentiation is not understood. High concentrations of exogenous auxin can be toxic, in large part because they stimulate production of concentrations of ethylene, which can cause growth inhibition (see below).

Example: Rooting of *In Vitro* Shoots, the Requirement for Auxin

Often it is necessary to root *in vitro*-derived shoots in order to convert the shoot to a functional plant that can be transferred to a greenhouse (see Chapters 11 and 12). Rooted shoots generally grow faster in culture. In many plant species (e.g., *Nicotiana* species, cucumber, squash, *Acacia* species, and some cultivars of rose), shoots generated *in vitro* root easily after transfer from the regeneration medium, which usually contains a high level of cytokinin, to a medium without PGRs. Cytokinins inhibit rooting and this treatment effectively lowers their level. The wounding treatment involved in preparation of shoots for rooting can stimulate endogenous auxin synthesis. However, many economically important species lack the ability to root easily. The reluctance to root could be due to a high level of cytokinin persisting in stem tissues from previous treatments. In such a case, rooting can be induced by transfer of the shoots once or twice to medium without PGRs to allow cytokinin levels to drop. Alternatively, shoots may be reprogrammed by treatment with auxin to produce roots if they have too much endogenous cytokinin. In such instances, exogenous auxin is required to overcome the cytokinin effect and induce rooting. Auxin treatment *in vitro* is analogous to the treatment with auxin rooting compound practiced by gardeners. Auxin appears to induce root initiation within a few hours of application. Most commonly IAA (0.6–60 μM), IBA (2.5–15 μM) and/or NAA (0.25–6 μM) are used to promote rooting *in vitro*. Not all of the above auxins have the same efficiency in rooting, the effectiveness varying with plant species, although often the type of auxin used for rooting is not important. It is important to note that the concentration of auxin that gives the greatest number of roots may not be the concentration that produces the greatest survival rate on subsequent hardening, as some roots induced directly by exogenous auxin are of low quality, as shown in rooting studies of *Acer* shoots using IBA or *Eugenia* shoots with NAA. A greater than optimum concentration of auxin often causes callus production and a reduction in root growth and quality. For instance, 60% of *Bougainvillea* shoots rooted after growth on media containing NAA followed by 2,4,5-trichlorophenoxyacetic acid (2,4,5-T, an analogue of 2,4-D), but with problematic callus production. However, a mixture of IBA and 2,3,5-T enabled more efficient rooting without callus production. Occasionally mixtures of auxins, usually at lower concentrations than those used singly, can promote root initiation whereas individual auxins have no activity, as shown with olive, *Eucalyptus*, *Hevea* and *Vitis*. High concentrations of auxin can be required to induce rooting, which may cause undesirable side effects such as growth inhibition of induced roots and disturbing subsequent plant growth. In such cases, it is sometimes possible to administer the elevated auxin treatment as a "pulse," such that the shoot is incubated with auxin for several days prior to transfer to a medium without PGRs, to permit root growth and development. This approach has been demonstrated for olive, apple, and some cultivars of coffee. In certain cases (e.g., some apple cultivars) auxin has been applied by placing the base of the shoot in a high concentration of auxin for an hour or two prior to rooting on a medium without PGRs.

Example: Auxins in Somatic Embryogenesis

Somatic or nonzygotic embryogenesis (see Chapter 14) is a process whereby plant somatic cells can be induced to form somatic embryos, very similar to sexually formed embryos in seeds, by the application of PGRs. Embryogenesis offers an alternative regeneration pathway to that of direct organogenesis from somatic tissues. Somatic embryogenesis is typically induced by auxins, sometimes in combination with cytokinins. Embryogenesis is good example of the ability of auxins to reprogram the development of somatic cells. Because somatic embryogenesis is discussed in detail in this book (Chapters 14 and 15), discussion here will be limited to the role of PGRs in somatic embryogenesis.

Induction of embryogenesis generally requires a high concentration of an artificial chlorinated auxin. The auxin 2,4-D is very commonly used at this stage. Alternatively, other artificial auxins such as dicamba or IBA might be used. More rarely IAA or NAA have been used to induce embryogenesis. Chlorinated auxins are difficult for the plant to metabolize and therefore may maintain the exogenous auxin at a high concentration. The function of auxins is to induce cells to become embryogenic and subsequently to promote repetitive cell division of embryogenic cell populations, while the high concentration of auxin prevents cell differentiation and embryo growth (Gray, 1996). Diverse auxins at different concentrations may produce variation, resulting in the production of abnormal embryos. The issue of abnormal embryo development is critical in production of a new embryogenic method and varies in severity from species to species. Consequently, when possible, embryo development occurring after transfer to a PGR-free medium often produces somatic embryos of more normal morphology. Abscisic acid (ABA) is often used to promote embryo maturation of somatic embryos. The important role of auxins is emphasized in a study of embryogenesis in fir (*Abies normanniana*), where the transition from embryo proliferation to maturation is problematic (Find et al., 2002). The transition can be promoted by the use of an antiauxin (a compound that blocks auxin action); this reduces proliferation and promotes maturation of many high-quality embryos, but only in the presence of ABA. An auxin transport inhibitor reduces proliferation of embryos without improving maturation, and an auxin synergist (which enhances auxin action) increases embryo proliferation, again without inducing maturation (Find et al., 2002).

CYTOKININS

Cytokinins cause cell division, as the name suggests. Such cell division can lead to shoot regeneration *in vitro*, by stimulating the formation of shoot apical meristems and, subsequently, shoot buds. The term "cytokinin" is used for compounds with similar biological activity. Additionally, the cell division caused by cytokinins can produce undifferentiated callus. Generally, a high concentration of cytokinin will block root development. Cytokinins can cause release of shoot

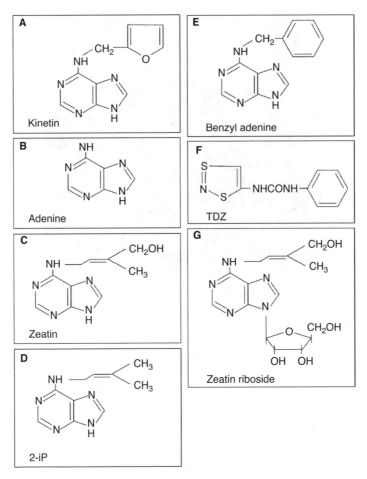

FIGURE 8.2 Chemical structures of some cytokinin-type PGRs. (A) Kinetin; (B) adenine; (C) zeatin; (D) N⁶-(2-isopentyl) adenine (2-iP); (E) 6-benzyladenine (BA); (F) thidiazuron (TDZ); (G) zeatin riboside.

apical dominance, thereby stimulating growth of lateral buds and resulting in multiple shoot formation. The first cytokinin isolated was kinetin (Figure 8.2A); shortly after its discovery, adenine (Figure 8.2C) was identified as possessing considerable cytokinin activity, as were many adenine derivatives (Figure 8.2). Zeatin (Figure 8.2C) and N⁶-(2-isopentyl) adenine (2-iP; Figure 8.2D), two other naturally occurring cytokinins, are often used in tissue culture. Additionally, the synthetic cytokinin PGRs 6-benzyladenine (BA) (also known as 6-benzyl amino purine [BAP]; Figure 8.2E) and thidiazuron (TDZ; Figure 8.2F) are commonly used in plant tissue culture. Zeatin riboside is an example of a PGR naturally complexed with a sugar molecule, altering the biological properties of the PGR (Figure 8.2G).

Example: The Use of Cytokinins in Micropropagation

Micropropagation is the mass vegetative production of plants *in vitro* for the purpose of commercial plant production. Micropropagation is performed by shoot multiplication followed by rooting (generally also *in vitro*), or more rarely by somatic embryogenesis. We will discuss the process using shoot multiplication, a process used industrially to produce large numbers of pathogen-free plants for horticulture, where cost of the final product is important (see Chapters 27 and 28). In designing a micropropagation process, it is best to avoid unstable forms of multiplication such as

FIGURE 8.3 Gibberellin and cytokinin effects on growth of *Nicotiana tabacum* (tobacco) shoots. Tobacco shoots grow and root easily on Murashige and Skoog (1962) basal medium (left). Cytokinin (5 μ*M*) enhances branching and multiple shoot growth and callus development at the stem base, and blocks rooting (center). Gibberellin (3 μ*M*) added to the medium induces stem extension and thin long leaves and reduces root formation (right). Bar = 10 mm.

through callus, due to the possibility of somaclonal variation. Therefore, the major regeneration pathway utilized is axillary shoot multiplication, noted for its genetic stability (Chapter 11). Shoot multiplication is induced by the application of exogenous cytokinins in the growth medium. The applied cytokinin breaks shoot apical dominance (in dicotyledonous plants), stimulating the growth of axillary buds into shoots (Figure 8.3), and often induces the growth of additional subsidiary buds adjacent to the new axillary shoots. Too high a concentration of the applied cytokinin can cause many small shoots to grow, which fail to develop. There is also a notable specificity of action in the type of cytokinin compound used; e.g., kinetin, BA, and zeatin were not effective in the instigation of shoot cultures of *Browallia viscosa*, whereas 2-iP was successful. Better *Gerbera* shoot quality was obtained with kinetin, although more rapid propagation was achieved with BA. TDZ can also be used for micropropagation. Notably, cytokinin treatments such as utilized in micropropagation inhibit rooting, and therefore a necessary part of such a project is rooting of the shoots produced in an efficient manner.

GIBBERELLINS

Gibberellins are another class of PGRs with nearly 100 different variants identified from various sources, although all are based on the same gibbane structure. The activity of exogenously applied gibberellin varies with the gibberellin type and the plant species treated. The gibberellin commonly applied in plant tissue culture is GA_3, also known as gibberellic acid (Figure 8.4A). Although gibberellins have important roles in the life cycle of plants (controlling stem length, flowering, and fruit set), in tissue culture GA_3 has generally been used to stimulate either shoot elongation or the conversion of buds into shoots. GA_3 interferes with bud initiation at very early stages of meristem formation, and thereby may reduce shoot production *in vitro* if given to plant tissue cultures at the shoot bud initiation stage. Similarly, GA_3 generally reduces root formation and embryogenesis *in vitro*. Therefore, for practical usage, it is necessary to optimize the stage-specific effects of GA_3.

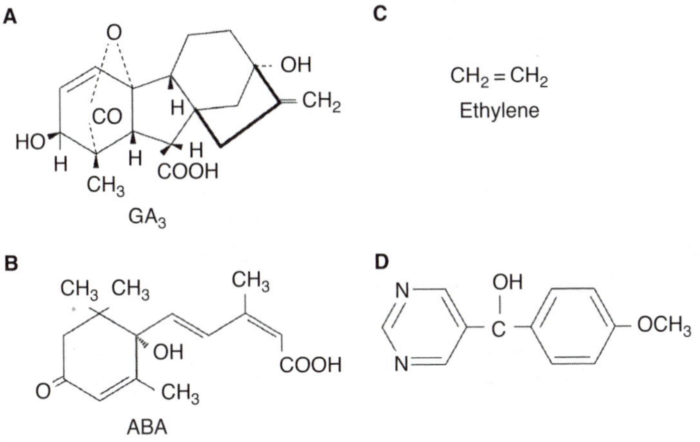

FIGURE 8.4 Chemical structures of PGRs of varied actions. (A) Gibberellic acid (GA_3); (B) abscisic acid (ABA); (C) ethylene; (D) ancymidol.

Example: Effect of Gibberellins on Stem Extension

Gibberellins were first discovered because of their dramatic stimulating effect on plant elongation growth. Gibberellins are primarily used in plant tissue culture to stimulate cell elongation, thereby producing elongated shoots. Stimulation of shoot elongation by gibberellin can be observed easily with tobacco shoots (Figure 8.3). The use of gibberellins in elongation medium is a valuable method to convert buds to functioning shoots. For instance, in some varieties of melon and squash, following induction of shoots on medium with BA (5 μM), shoot elongation is best stimulated by a medium with a reduced cytokinin concentration (0.5 μM) and the addition of GA_3 (1.5–3 μM). The reduction of the cytokinin concentration permits buds to elongate, aided by the cell elongating effect of the GA_3. However, the use of gibberellin at the earlier stage of bud induction often reduces the number of buds produced (Gaba et al., 1996). Other unwanted side effects of the use of gibberellins can include the elongation of leaf structures such as petioles and lamina, the excessive elongation of shoots, and the reduction of root production (Figure 8.3).

Example: The Use of Antigibberellins

Antigibberellins are artificial chemicals that block the activity of enzymes in the gibberellin synthesis pathway. As gibberellins are important for elongation growth, antigibberellin compounds are known as growth retardants. Antigibberellins, such as ancymidol (Figure 8.4D) are used in agriculture as dwarfing compounds that prevent the lodging of small grain crops, for the maintenance of golf greens with short green turf and for dwarfing of ornamental potted plants. Additionally, growth retardants are used to promote fruit ripening. In plant tissue culture, antigibberellins stimulate embryogenesis in *Citrus* and bud formation in melon (Gaba et al., 1996), both processes being inhibited by GA_3. Antigibberellins also stimulate *in vitro* formation of *Gladiolus* cormels, shoot regeneration in *Albizzia*, bud multiplication in *Philodendron*, and potato microtuber induction.

Abscisic Acid

Abscisic acid (ABA; Figure 8.4B) has been mainly used in plant tissue culture to facilitate somatic embryo maturation. In certain species, embryos may not mature adequately for lack of exogenous ABA. Endogenous ABA functions similarly during seed development in monocotyledonous and dicotyledonous plants and conifers. ABA induces the formation of the essential LEA proteins found at late stages of embryogenesis in both somatic and sexual embryos. Occasionally, ABA might stimulate some regeneration processes and, rarely, may also reduce the production of somatic embryos. ABA is the only natural member of its class.

ETHYLENE

Ethylene is the only gaseous natural PGR (Figure 8.4C). Although ethylene is primarily known in plant biology for its effects on fruit ripening, it is naturally produced by all higher plant organs in a controlled fashion. In plant tissue culture, endogenously produced ethylene can accumulate in a closed vessel to levels that are deleterious to plant growth and development. The result in ethylene-sensitive species is a typical syndrome of reduced stem elongation, restricted leaf growth, premature leaf senescence and, sometimes, increased growth of axillary buds. Enhanced ethylene concentrations can stimulate callus formation, while reducing bud and shoot regeneration. Somatic embryogenesis can be affected by ethylene concentration — low concentrations stimulate, while high concentrations inhibit. Rooting and bulb formation may also depend on an adequate concentration of ethylene in the culture vessel. Ethylene is the only member of its class.

Example: Ethylene Toxicity in Plant Tissue Culture

Plant tissue culture is generally performed in enclosed vessels, and the concentration of ethylene in the vessel increases from the gas produced by plant activity. Ethylene concentration increases with time in culture (Zobayed et al., 2001). The biological effect depends on how closed (or how "leaky") the culture vessel is, and on the sensitivity of the plant material. Closing the vessel tightly can be problematic, as the escape of the ethylene is reduced and the concentration of the ethylene increases. Again the response to the PGR ethylene is biphasic: explants in culture need a certain low level to function (depending on species, genotype, and activity), but too high an ethylene concentration is deleterious. Endogenous ethylene has an important role in shoot and root growth and differentiation, and adventitious root induction in plant tissue culture. The effects of ethylene can be observed by overaerating tissue culture vessels (flushing too frequently with sterile air), which reduces growth and development. There is a strong effect of ethylene on melon stem growth and rooting (Figure 8.5). Endogenous ethylene causes leaf crinkling and folding and reduces tendril growth (Figure 8.5B) after a month in culture. After explants have been maintained for two growth cycles (two months) in tubes that are sealed tightly, thus preventing endogenous ethylene escape, a more extreme response is observed (Figure 8.5C). In this case, the new internodes do not elongate, the original stem is swollen, apical dominance is broken, leaves become greatly reduced in size, multiple leaves are produced, and the explant does not root (Figure 8.5C). This syndrome is similar to the classic "triple response" of etiolated pea seedlings, whereby ethylene inhibits stem elongation and stimulates lateral internode expansion. A diagnostic of ethylene responses is that ethylene can be countered by the application of silver ions, which specifically block the ethylene receptor molecule. Silver thiosulfate incorporated in the medium counters the effect of ethylene, restoring leaf color and enlargement and root development, and partially restoring internode elongation (Figure 8.5D). A rare genotype-dependent toxicity to silver reduces melon internode elongation compared to the control (Figure 8.5D).

There are other treatments for excess ethylene in culture vessels. Vessels that are not tightly closed will permit the gas to escape, or an ethylene-absorbent material (such as potassium permanganate) can be placed inside. On the other hand, the use of larger vessels or fewer explants per vessel can reduce problems with ethylene. Sometimes plant material can be crowded more closely in vessels, resulting in more explants per vessel and thereby increasing the ethylene concentration to obtain the desired biological response. Alternatively, ethylene-releasing materials can be added

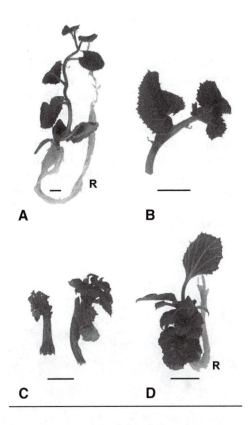

FIGURE 8.5 Ethylene effects *in vitro* growth of melon shoots. (A) Control shoots of melon. (B) Leaf crinkling and folding after a month in culture. (C) Response of melon shoots to endogenous ethylene for two months. New internodes do not expand; the original stem is swollen, especially at the base; leaves are greatly reduced in size; and roots do not develop. (D) Silver thiosulfate blocks ethylene effects, but reduces melon internode elongation. R = roots. Bars = 10 mm.

to the culture vessel, or a precursor of ethylene biosynthesis or air containing a defined concentration of ethylene can be introduced to the culture vessels. Potato is particularly sensitive to ethylene in tissue culture. Production of small virus-free potato tubers (microtubers) on *in vitro* grown potato plants is an important industry. Excess ethylene in the culture vessels reduces stem and leaf growth, and limits the production of microtubers (Perl et al., 1988; Zobayed et al., 2001).

PGR COMBINATIONS

The general responses of combinations of cytokinin and auxin are as follows. A high ratio of auxin to cytokinin induces root formation in shoots, callus initiation in monocotyledonous plants, and the initiation of somatic embryogenesis (with TDZ or NAA rather than IAA). Intermediate ratios of auxin to cytokinin induce adventitious root formation from callus and callus initiation from dicotyledonous plants. Low ratios of auxin to cytokinin induce adventitious shoot formation and axillary shoot production in shoot cultures. Therefore, cytokinin and auxin interactions are crucial in the control of many growth and developmental processes *in vitro*.

EXAMPLE: THE RATIO OF EXOGENOUS CYTOKININ TO AUXIN DETERMINES THE
PATTERN OF REGENERATION

In a classic paper, Skoog and Miller (1957) demonstrated how cytokinin and auxin interact to produce a variety of different morphogenetic responses, depending on the concentration of the PGRs and the ratio of the concentrations of exogenous auxin to cytokinin. Skoog and Miller exploited the ease with which tobacco tissues can be manipulated in culture. In general, they showed that there is cytokinin and auxin activity at low concentrations, and that when PGRs are combined at low concentrations, the activity is enhanced. Auxins and cytokinins can also have separate (opposite) actions at intermediate concentrations. However, combinations of auxin and cytokinin at high concentrations reduce the specific effects of either PGR, causing callus formation.

We have imitated the experiment of Skoog and Miller (with some alterations), in a factorial experiment of regeneration from tobacco leaf explants on three concentrations (0, 0.5, and 5 μM) each of cytokinin (BA) and auxin (NAA) (Figure 8.6). Treatment without exogenous PGR (cytokinin or auxin), the control, causes little enlargement of the explant or regenerative development (Figure 8.6).

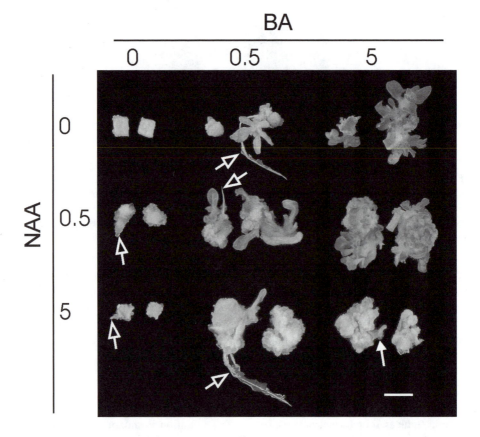

FIGURE 8.6 Regeneration from *Nicotiana tabacum* (tobacco) leaf explants on combinations of benzyladenine (BA) and 1-naphthaleneacetic acid (NAA) in Murashige and Skoog (1962) medium, after 31 d in culture. Factorial arrangement of concentrations of BA and NAA of 0, 0.5, and 5 μM was applied. For each treatment two explants are shown, representing the extremes of development achieved. After 31 d in culture, the PGR concentration in the treatment with 5 μM each of BA and NAA declined sufficiently to permit the regeneration of a few small shoots, marked here with solid arrows. Roots are marked with open arrows. Bar = 10 mm.

Low concentrations of either BA or NAA (0.5 μ*M*) incite limited enlargement of the explant and regeneration of many shoots on BA. The low concentration of NAA caused some regeneration of roots as well as consistent callus formation. Root regeneration on 0.5 μ*M* BA occurs occasionally because the tissue has a high concentration of endogenous auxin. Shoot regeneration occurs on 0.5 and 5 μ*M* BA at high efficiency. A high concentration of cytokinin (5 μ*M* BA) causes extensive shoot and leaf development from greatly enlarged tobacco leaf explants without root production. A high concentration of NAA (5 μ*M*) stimulates moderate explant enlargement and less root and callus formation than with the low NAA concentration. The mixture of 0.5 μ*M* each of BA and NAA disproportionately promotes explant enlargement and shoot development, although not to the extent of 5 μ*M* BA. The high concentration of cytokinin (5 μ*M* BA) with a low concentration of NAA (0.5 μ*M*) consistently provides the greatest level of explant enlargement and shoot regeneration, with no root production. The high concentration of auxin (5 μ*M* NAA) with a low concentration of BA (0.5 μ*M*) stimulates extensive explant enlargement, with much callus production, with much reduced shoot and root production compared to 0.5 μ*M* each of BA and NAA. High concentrations of both BA and NAA (5 μ*M* each) induce callus formation and explant enlargement, but less than with 5 μ*M* NAA and 0.5 μ*M* BA. The high combined PGR concentrations (5 μ*M* each) block shoot regeneration until the final days in culture, when the PGR concentration has declined with time in light permitting the development of a few small buds (Figure 8.6).

EXAMPLE: THE RELATIONSHIP BETWEEN SOMATIC EMBRYOGENESIS AND SHOOT ORGANOGENESIS

As we have discussed above, the effects of PGRs depend on the applied concentration. Here we will see the concentration-dependent effect of auxin in two separate morphogenetic processes: shoot organogenesis (the formation of shoot buds and shoots) and somatic embryogenesis. In an interesting experiment, Tabei et al. (1991) examined the response to a wide range of auxin concentrations on morphogenesis of cotyledons from mature melon (*Cucumis melo* L.) seeds. Fixed levels of cytokinin (0.44 μ*M* BA) and sucrose (3%) in the medium were used. Initial experiments tested 3 auxins (2,4-D, NAA, and IAA) for the ability to cause embryogenesis in melon and, surprisingly, IAA was found to be the most effective.

The use of IAA also gave rise to the fewest abnormal embryos. The response to a wide concentration range of IAA was tested (Figure 8.7). Adventitious shoots were formed at a low (control) level without embryogenesis in the absence of IAA; shoot formation did not increase with

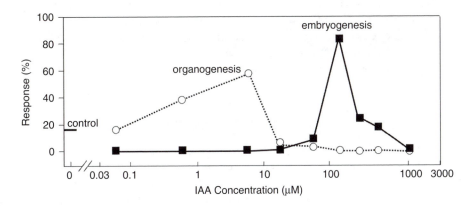

FIGURE 8.7 The effect of auxin on the induction of organogenesis and embryogenesis in melon cotyledons after Tabai et al. (1991). Organogenesis and embryogenesis are stimulated by different concentrations of IAA in the presence of 0.44 μ*M* BA. The control level of regeneration without IAA is marked.

addition of 0.057 μM IAA. However, an increase in organogenesis occurred with the addition of 0.57 μM IAA. Maximum organogenesis was observed at 5.7 μM IAA. Increasing the IAA concentration threefold to 17 μM reduced the organogenic regeneration to a level below that observed in the control without IAA; embryogenesis now occured in 1% of explants. A further threefold increase of IAA concentration (to 57 μM) produced a higher embryogenic response, permitting little organogenesis; therefore, the shoot-inducing effect of the BA was blocked almost entirely. However, at 57 μM IAA both modes of regeneration occurred simultaneously, albeit at low levels. Optimum embryogenic response was observed at 143 μM IAA. The embryogenic response was maintained, although reduced, at high IAA concentrations (285 and 570 μM), and declined to zero at 1710 μM (Figure 8.7).

This elegant experiment demonstrates several general points. First, there is usually a response curve for each PGR activity that rises with increasing concentration and then declines. This is evident for both adventitious shoot regeneration and somatic embryogenesis (Figure 8.7). Second, both adventitious shoot regeneration and embryogenesis are apparently controlled by the ratio between the cytokinin and the auxin, as discussed earlier. Third, the concentration ranges for adventitious shoot regeneration and somatic embryogenesis do not overlap, strongly suggesting that these morphogenic processes are controlled by different physiological pathways. Fourth, the concentration range of IAA over which embryogenesis occurs is narrower than the concentration range permitting organogenesis. Similar responses to auxin concentration have also been reported for adventitious shoot regeneration and somatic embryogenesis in eggplant, cucumber, and soybean.

OTHER GROWTH REGULATORS

Other small molecules are known to be important regulators of plant development. The responses to these compounds have not been investigated as thoroughly as the major PGRs and their use in tissue culture is relatively untested. However, these will be increasingly important in the future. Brassinosteroids are a group of chemicals of the steroid class (currently there are about 60), which generally promote stem elongation (e.g., shoot regeneration from cauliflower hypocotyls segments), inhibit root growth, and stimulate ethylene synthesis. Salicylic acid and its derivatives regulate some of the defense responses of plants, including the inhibition of ethylene synthesis. Jasmonic acid and its methyl ester inhibit many physiological processes of growth and germination; stimulate fruit ripening, senescence, tuberization, and abscission; and have functions in plant defenses. Jasmonic acid inhibits *in vitro* shoot organogenesis from *Pinus radiata* cotyledons. Polyamines are found in all cells and tissues and affect growth and development at low concentrations. Polyamines are important in cell division and growth and have many effects on plants.

It is important to recall that there are other factors that regulate plant growth and development *in vitro*, such as sucrose concentration, type of carbohydrate source, light (wavelength, intensity, photoperiod), darkness, and temperature. These factors interact in different ways with applications of the major PGRs.

ACKNOWLEDGMENTS

Contribution from the Agricultural Research Organization, the Volcani Center, Bet Dagan, Israel, No. 503/03. The author thanks Ms. Sima Singer for her skilled technical assistance. This work was supported by research grants to the author from the Office of the Chief Scientist of the Ministry of Agriculture (Israel), and from the United States–Israel Binational Agricultural Research and Development Fund (BARD). The author thanks Dr. B. Steinitz and Dr. K. Kathiravan for critical comments on the manuscript.

LITERATURE CITED

Arteca, R. 1996. *Plant Growth Substances: Principles and Applications*. Chapman & Hall, New York.

Davies, P. J. 1995. *Plant Hormones: Physiology, Biochemistry and Molecular Biology*. Kluwer Academic Publishers, Dordrecht, the Netherlands.

Find, J., L. Grace, and P. Krogstrup. 2002. Effect of anti-auxins on maturation of embryogenic tissue cultures of Nordmanns fir (*Abies normanniana*). *Physiol. Plant.* 116: 231–237.

Gaba, V. et al. 1996. Ancymidol hastens *in vitro* bud development in melon. *Hort. Sci.* 31:1223–1224.

Gray, D.J. 1996. Nonzygotic embryogenesis. In: *Plant Tissue Culture Concepts and Laboratory Exercises*. (Ed.) R.N. Trigiano and D.J. Gray. CRC Press, Boca Raton, FL.

Mauseth, J. D. 1991. *Botany: An Introduction to Plant Biology*. W.B. Saunders, Philadelphia, PA.

Murashige, T. and F. Skoog. 1962. A revised medium for rapid growth and bioassays with tobacco tissue cultures. *Physiol. Plant.* 15: 473–498.

Perl, A., D. Aviv, and E. Galun. 1988. Ethylene and *in vitro* culture of potato: suppression of ethylene generation vastly improve protoplast yield, plating efficiency and transient expression of an alien gene. *Plant Cell Rep.* 7:403–406.

Raven, P.H., R.F. Evert, and S.E. Eichhorn. 1992. *Biology of Plants*. Worth, New York.

Salisbury, F.B. and C.W. Ross. 1992. *Plant Physiology*. Wadsworth, Belmont, CA.

Skoog, F. and C.O. Miller. 1957. Chemical regulation of growth and organ formation in plant tissues cultured *in vitro. Symp. Soc. Exper. Biol.* 11, 118–131.

Tabei, Y., T. Kanno, and T. Nishio. 1991. Regulation of organogenesis and somatic embryogenesis by auxin in melon, *Cucumis melo* L. *Plant Cell Rep.* 10: 225–229.

Taiz, L. and E. Zeiger. 2002. *Plant Physiology*. 3rd edition. Sinauer Associates, Inc., Sunderland, MA.

Zobayed, S.M.A., J. Armstrong, and W. Armstrong. 2001. Micropropagation of potato: evaluation of closed, diffusive and forced ventilation on growth and tuberization. *Ann. Bot.* 87: 53–59.

9 Software and Databases as Tools for Analyzing Nucleic Acid and Protein Sequences*

Zhijian T. Li

CHAPTER 9 CONCEPTS

- Advances in molecular genetics have resulted in the rapid accumulation of vast amounts of plant genomic and EST sequence data and related functional information. Numerous databases designed to accommodate and categorize specific sequence data have now become accessible to plant scientists for their exploration.

- Desktop computer software packages provide highly integrated and multifunctional tools for individual researchers to investigate and manage sequences of genes, genetic elements, or proteins of interest and to gain access to a variety of databases.

- Desktop computer software packages also play an indispensable role in DNA manipulation and construction of plasmid vectors for use in molecular study of genes and genetic elements, using transgenic plants and genetic engineering of plants for crop improvement.

INTRODUCTION

Computer software designed for analysis of nucleic acid and protein sequence data is an indispensable tool in molecular biology and genetic engineering research. Concerted efforts in genome sequencing and gene discovery have resulted in large amounts of DNA sequence data and the development of databases containing enormous numbers of records. However, the majority of this sequence data functionally belong to so-called hypothetical genes or gene fragments that code for proteins of unknown functions, and DNA sequences and motifs with unknown regulatory capabilities. Hence, scientists are faced with the immense task of analyzing these sequence data in order to identify useful genes and genetic elements.

In addition, techniques of DNA manipulation and gene transfer are being utilized to introduce foreign genes and promoters into plants in order to study their functions, interactions, and modes of action. Foreign genes are also routinely introduced into crop plants to achieve improved agronomic performance. These plant transformation and genetic engineering operations necessitate the use of computer software programs to assist the precise manipulation of nucleic acid sequences and the construction of plasmid vectors.

Desktop computer programs for nucleic acid and protein analysis have been developed to facilitate the accomplishment of these challenging tasks. Using these programs, researchers can access data from publicly available databases to investigate genes, genetic elements, or proteins of

---

* Florida Agricultural Journal Series No. R-10165

interest. Such computer programs allow access to these databases from virtually any laboratory in the world. Computer programs also can be used to analyze nucleic acid sequence identity, predict functionality, and guide sequence manipulations.

A wide range of analytical functions are provided in commercially available DNA sequence analysis software packages for desktop computers. These packages typically support programs for:

- Editing of sequence data for sequence assembly
- Restriction of enzyme and proteolytic analysis
- Gene and protein identification
- Sequence alignment
- Promoter motif discovery
- Database or network searching for similarity analysis
- Oligonucleotide primer design
- Plasmid map construction for gene cloning

These software packages are user-friendly, highly versatile in functionality, and readily operable on different computer platforms. Many packages are updated frequently with enhancements. In this chapter, computer software packages that support DNA sequence editing, analysis, manipulations, and plasmid vector construction for genetic engineering projects are introduced. Examples of analyzing nucleic acid and protein sequences using the Vector NTI Advance suite of software by InforMax Inc. (Bethesda, MD) are demonstrated. However, programs specifically designed for contig assembly and related sequence analysis in DNA sequencing projects, such as the Sequencher program from Gene Codes Co. (Ann Arbor, MI), and network-based software programs such as the GCG Wisconsin Package running on the UNIX server operating system from Accelrys Inc. (San Diego, CA), will not be covered.

GENOMIC AND PROTEOMIC DATABASE RESOURCES

Currently, three major international nucleic acid databases are organized by the International Nucleotide Sequence Database Collaboration (http://www.ncbi.nih.gov/projects/collab). These include the following:

- GenBank of the National Center for Biotechnology Information (NCBI), Bethesda, MD
- The European Molecular Biology Laboratory (EMBL/EBI) Nucleotide Sequence Database, Hinxton, UK
- The DNA Data Bank of Japan (DDBJ), Mishima, Japan

These databases can be searched for sequences, which can be retrieved for analysis using the Web links and related tools listed in Table 9.1 (Wheeler et al., 2002; Dufresne et al., 2002). The size and number of records in these databases are growing exponentially. As of March 2003, GenBank contained more than 20 billion bases of nucleic acid sequence data in more than 20 million records.

Protein databases include the following:

- The Swiss Protein Databank (SWISS-PROT)
- The Translation of DNA Sequences on EMBL (TrEMBL)
- The Protein Information Resource (PIR)
- The Munich Information Center for Proteins (MIPS)
- The Protein Data Bank (PDB)

TABLE 9.1
List of Genomic and Proteomic Databases and Web Links

Name	Content	Organization	Web Link
GenBank	DNA, protein	National Center for Biotechnology Information (NCBI)	http://ncbi.nlm.nih.gov/Genbank/ or http://www.ncbi.n.m.nih.gov/entrez/
EMBL	DNA	EBI, European Molecular Biology Laboratory (EMBL)	http://www.ebi.ac.uk/Tools/ or http://www.embl-heidelberg.de/Services/index.html
DDBJ	DNA	Japan National Institute of Genetics, Center for Information Biology	http://www.ddbj.nig.ac.jp/searches-e.html
SWISS-PROT	Protein	ExPASy	http://www.expasy.ch/
TrEMBL	Protein	ExPASy, EMBL	http://www.expasy.ch/sprot
PIR	Protein	National Biomedical Research Foundation/Georgetown University Medical Center	http://www-nbrf.georgetown.edu/pirwww/pirhome.shtml
MIPS	Protein	GSF National Research Center for Environment and Health	http://www.mips.biochem.mpg.de/
PDB	Protein, 3-D structures	Protein Data Bank	http://www.rcsb.org/pdb/
EST	ESTs	NCBI	http://www.ncbi.nlm.nih.gov/dbEST/index.html
UniGene	EST sets	NCBI	http://www.ncbi.nlm.nih.gov/entrez/query.fcgi?db=unigene
EPD	Promoters	Swiss Institute for Experimental Cancer Research	http://www.epd.isb-sib.ch/

The recently improved Simple Modular Architecture Research Tool (SMART) database from EMBL (http://smart.embl-heidelberg.de) also allows search and annotation of intrinsic protein domain features, including transmembrane regions, coiled-coils, signal peptides, and internal repeats (Letunic et al., 2002).

In addition to sequences of well-defined and characterized genes and proteins, hundreds of thousands of expressed sequence tags (ESTs) have been deposited. ESTs are short pieces of DNA (cDNA) derived from mRNA fragments via reverse transcription. These sequences may be accessed through databases, such as the EST database of GenBank and UniGene of NCBI, which specifically collect ESTs of different organisms. EST databases represent a valuable resource for gene discovery. However, when using EST databases, one should keep in mind that many ESTs or EST sets may be derived from 5' and 3' incomplete reads from the same cDNA clone without an overlapping region. Others may represent different splicing variants of a gene. Gene intron sequence information is not typically available from EST sequence data. The Eukaryotic Promoter Database (EPD) contains a collection of functional and nonredundant eukaryotic Pol II promoters and promoter elements. Sequence data in this EPD database are an especially valuable resource for promoter discovery and comparative sequence analysis to determine the functionality of promoter motifs.

National and multinational DNA sequencing projects over the past decade have resulted in the sequencing of entire genomes of many organisms, ranging from viruses to bacteria, insects, plants, and mammals, including *Homo sapiens*. These achievements led to establishment of several genome databases, including the following:

- The Institute of Genome Research (TIGR; http://www.tiger.org)
- The University of California/U.S. Department of Energy Joint Genome Institute (JGI; http://jgi.doe.gov)
- The Sanger Center (Wellcome Trust/Medical Research Council, UK; http://www.sanger.ac.uk)
- The National Center for Genome Research (NCGR, Sante Fe, NM; http://www.ncgr.org)
- The Berkeley Drosophila Genome Project (BDG; http://www.fruitfly.org)
- The Whitehead Institute Center for Genome Research (WICGR; http://broad.mit.edu)
- The National Institute of Agrobiological Sciences (RiceGAAS, Japan; http://RiceGAAS.dna.affrc.go.jp)

These databases contain a wealth of genomic DNA sequence information for gene and promoter discovery, genome organization, and gene functionality and interactions.

Plant DNA and protein sequences can be searched and retrieved from many of the above-mentioned databases. Additionally, several databases have been developed specifically to store plant-related sequence data. For instance, the Arabidopsis Information Resource (TAIR; http://www.arabidopsis.org) is an excellent integrated *Arabidopsis* genomic and proteomics resource developed by the Carnegie Institution of Washington and the Department of Plant Biology, Stanford University in collaboration with NCGR. PLANT-Pis (http://bighost.area.ba.cnr.it/PLANT-Pis) is a database of plant protease inhibitors and their genes (De Leo et al., 2002). The database of Plant Cis-acting Regulatory DNA Elements (PLACE, http://www.dna.affrc.go.jp/htdocs/PLACE/) was developed for the storage of sequence data of cis-acting regulatory DNA elements from vascular plant species (Higo et al., 1998). The database of Plant Mitochondrial tRNA (PLMItRNA, http://bighost.area.ba.cnr.it:8000/srs/) collects and manages higher plant mitochondrial tRNA and tRNA gene sequence data (Ceci et al., 1999).

SEQUENCE ANALYSIS SOFTWARE PROGRAMS

EARLY SOFTWARE AND SOFTWARE OF LIMITED FUNCTION

Specialized computer software programs that enable scientists to compile, edit, analyze and convert text sequence data into graphical, functional, and structural data are vital to research in genetic engineering involving gene discovery, DNA analyses and manipulation, and plasmid vector construction. In the early 1980s, many stand-alone DNA software programs were developed to perform simple analyses and conversion functions. For example, the CloneMap program released by CGC Scientific, Inc., Ballwin, MO (http://www.cgcsci.com), was designed as a low-cost package to facilitate simple assembling and editing of DNA sequences, annotation of sequence features in plasmid graphics, and analysis of restriction enzyme sites, PCR primers, and translatable sequences. Many of these simple DNA and protein analysis features can now be performed using publicly available freeware, shareware, or add-ins that can be downloaded from the internet. An example of such an internet resource is the Software Library from the Biotechniques website (http://www.bio-techniques.com). Among the software programs listed there, the Molecular BioComputing Suite (MBCS) is particularly useful. The MBCS was implemented in Microsoft Visual Basic for Applications, allowing software to be operated as an add-in for Microsoft Word. Detailed applications of this software can be found in an article by Muller et al. (2001). The freeware program pDRAW32 has been made available by Kjeld Olesen of AcaClone Software (http://home.get2net.dk/acaclone/). This software accepts various DNA file formats including GCG, EMBL, and FASTA. Its restriction enzyme site database can be updated by importing source files from the GCG website. DNA sequences can also be assembled and edited into new or existing files. However, its graphic display and annotation features lack the versatility of more powerful commercial software programs. Nonetheless, pDRAW32 is very useful for students and small laboratories for the development of plasmid vector maps and for performing simple DNA cloning analyses.

Desktop Computer Software Packages

Three powerful and highly integrated desktop software packages designed for DNA and protein analysis are as follows: Lasergene from DNAStar, Madison, WI; Discovery Studio Gene from Accelrys, Inc., San Diego, CA, USA; and Vector NTI Advance Suite from InforMax, Inc., Frederick, MD.

The Lasergene sequence analysis software package was initially developed for use on graphics-oriented Macintosh computer platforms, and subsequently became available for use on DOS-based PCs running Microsoft Windows. This software is packaged as seven independently operable modules, including EditSeq, GeneQuest, MapDraw, MegAlign, PrimerSelect, Protean, and SeqMan. The modular construction allows for addition of new functional and analytical tools as required. Each module is designed to perform a specific function associated with the analysis of DNA or protein sequence data, such as entering and assembling sequence data, restriction enzyme site analysis, gene discovery, annotation, gene product analysis, sequence homology and similarity searches, sequence alignment, polygenetic analysis, PCR primer design, cloning strategies, and submission of analysis results to public databases. In addition, each module includes built-in internet connectivity, facilitating access to all popular sequence formats and the sharing of results with others. However, the modular organization of the package detracts from its integral functionality, and repeated launching of specific modules and reopening of data sets may be required to perform different analysis tasks. Individual modules of Lasergene can be purchased separately or as a complete package to suit an individual's budget and needs. A trial version of Lasergene is freely available for download (http://www.dnastar.com).

Discovery Studio Gene or DS Gene is one of several products on the Discovery Studio platform developed by Accelrys, Inc. It is the replacement software package for an earlier version, Omiga. The Omiga program was developed by the same company as a DNA and protein sequence analysis package for Macintosh computers. The improved DS Gene supports a wide range of commonly used analysis functions and offers enhanced interface access with other network-based databases and sequence analysis software, such as GCG Wisconsin Package, in order to provide for enhanced sequence searching, data management, and data mining. The Discovery Studio platform also allows the integration of other applications, such as ViewerPro for molecular modeling and visualization and MedChem Explorer for pharmacophore modeling and design. Additional information and demo versions of DS Gene and related products are available (http://www.accelrys.com/desktop/).

Vector NTI Advance suite is the latest version of the Vector NTI package to be released by InforMax, Inc. Vector NTI Advance package is also modular in design and contains five modules within the Advance platform, including Vector NTI, AlignX, BioAnnotator, ContigExpress, and GenomBench. The Vector NTI Advance modules are powerful and versatile programs with seamless integration with a variety of data and analytical tools. Details of the functional capacity provided by each module and related review articles on this software package can be viewed at the InforMax website (http://www.informaxinc.com/solutions/vectorntia/).

Because space does not permit instruction in use of all programs, I have chosen the Vector NTI Advance package to demonstrate use of such software for DNA editing, sequence analysis and alignment, plasmid vector construction, and access to common nucleic acid and protein databases. A demonstration version of this software package can be obtained by contacting the company. Although the demo version is not fully functional, it may be helpful in following some of the examples provided below.

Major Applications of Vector NTI Advance Package

Database Explorer

The Vector NTI database stores five different types of Molecule files, including those for nucleic acids (DNA and RNA); proteins; restriction enzyme sites; oligonucleotides for PCR, sequencing primers, and hybridization probes; and gel markers. All of these files are organized under the program Database Explorer, which is automatically launched when Vector NTI is opened. In addition, user-defined sub-bases can be created within the database, based on parameters such as molecular function, sequence similarity, creation date, authors, and so on. Each molecule file can be loaded with source information, including commercial source, freezer position, organism, and WWW database source, for user field definition management. These sub-bases facilitate storage, search, and organization of information related to specific cloning and analysis projects. Through Database Explorer molecular data can be imported from other databases such as GenBank/GenPept and EMBL/SWISS-PROT and exported to other computers for information sharing. Database Explorer also permits users to perform tasks such as data backup, restoration, and cleaning for database file management.

Vector NTI

Vector NTI is the cornerstone module of the Vector NTI Advance package. It provides a range of functions from database creation and management, to input and editing of DNA sequences, translation and ORF prediction, analysis of restriction enzyme and methylation sites, sequence similarity and BLAST search, design and alignment of PCR primers, construction of plasmid vector, and compilation and management of gel markers. All these features are performed in an integrated Windows display environment.

Molecule Display

When using Vector NTI, a Molecule file can be launched from either Database Explorer or the Vector NTI program. A Molecule Display Window will be opened to display all information associated with the chosen file. As illustrated in a Molecule Display Window of pUC18 (Figure 9.1), there are four main sections or panes within the display window:

1. A Tool Pane on top, containing main menu, main toolbars, and active pane selection toolbars
2. A Text Pane on the left side, containing folders harboring text descriptions and notations, and analyses results
3. A Graphics Pane on the right side, illustrating a graphical representation of a molecule with annotated features
4. A Sequence Pane on the bottom, showing a nucleotide or amino acid sequence of a molecule

The size and format of these display panes can be custom-adjusted or rearranged. Within the Tool Pane there are a wide range of functional keys for manipulation of sequence, format, and annotation of the molecule. Clicking on the drop-down menu from a desired functional key will access specific functions. In addition, the Text Pane allows for addition of project description information and notes for the creation of related molecules.

Creating a Molecule File

There are three ways to create a new Molecule file using Vector NTI. First, molecule data in text format can be imported from outside sources. Vector NTI automatically converts the imported text files in GenBank, EMBL, and FASTA formats to its own Molecule display format with fully annotated features as shown in Figure 9.1. The second method of creating a Molecule file is to enter sequence data manually or to paste text passages containing sequence data from a clipboard

Tool Pane

Text Pane

Graphics Pane

Sequence Pane

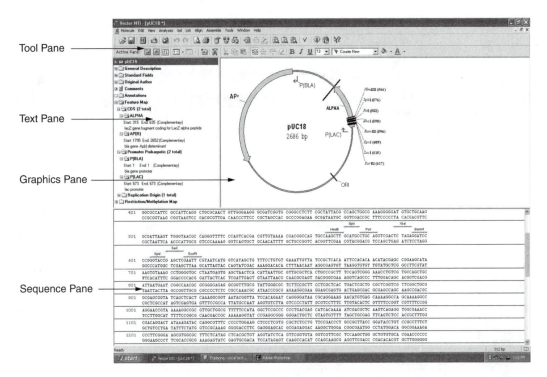

FIGURE 9.1 Molecule display window of pUC18. (Copyright 2004 Invitrogen Corporation. Used with permission.)

or other text program, such as Notepad. Unlike the Lasergene package, wherein sequence data can be read back by an animated computer sound to verify input accuracy, Vector NTI does not offer a read-proof function for sequence entry. Most sequencing data are obtained in the form of text files. Thus, data entry by cut-and-paste is primarily used to create Molecule files using Vector NTI. Sequence fragments also can be copied from other existing donor Molecule files and inserted into a receptor molecule at a desired position to create a new Molecule file. The third method of creating a new Molecule file is to let Vector NTI design new molecules. A list of candidate fragments can be compiled, and the program will choose appropriate restriction sites, linkers, adapters, or terminal modifications to assemble the fragments into a new molecule. This molecule design function is particularly useful for identifying the most effective method for creating recombinant molecules.

The following is a step-by-step procedure for the creation of a new molecule pBR322-UC19MCS. This molecule contains a multiple cloning site (MCS) covering *Eco*RI to *Hind*III from pUC19, and a plasmid backbone from pBR322. Additional details on how to create similar molecules are available in the *Vector NTI Advance User's Manual* provided by Informax, Inc. (http://www.informaxinc.com/support/tutorials.html). This procedure is easy to follow and can be applied to the construction of any plasmid vectors using a variety of DNA molecules and donor fragments.

1. Launch two donor Molecule files from the Database Explorer by clicking on pUC19 and pBR322 files.
2. On the pUC19 display window, choose the **Add Fragment to Molecule Goal List** button within the Active Pane to open the Fragment Wizard dialog box.
3. Select the **Construction Fragment** option and choose **Next**.
4. Select the **Set a Restriction Site** function and then highlight a desired restriction site, *Eco*RI, on the molecule within the Graphics Pane to define the 5' terminus of the donor fragment at nucleotide position 397 (Figure 9.2).

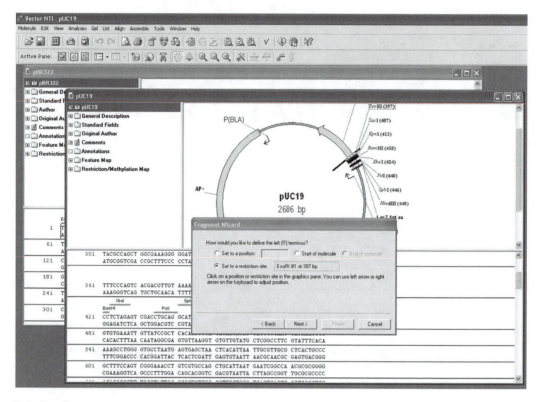

FIGURE 9.2 Defining 5' terminus of a donor fragment from pUC19. (Copyright 2004 Invitrogen Corporation. Used with permission.)

5. Choose **Next** to repeat above step to define the 3' terminus of the donor fragment by holding the shift key when selecting the *Hind*III site on the molecule.
6. Select **Finish**. A new box will appear to show information about the selected donor fragment, i.e., the MCS region of pUC19 spanning *Eco*RI to *Hind*III sites. Click **Add to List** (Figure 9.3).
7. Repeat similar steps with pBR322, except that the *Hind*III site is used as the 5' terminus and the *Eco*RI site as the 3' terminus (Figure 9.4).
8. Choose **Open Goal List** on the Main Toolbar. This opens the Lists dialog box, which shows selected donor fragments and also allows the alternation and editing of name, creation date, molecule type, and other general information of the new molecule. Selected donor fragments can also be edited for sequence modification and addition of linkers and adaptors.
9. Click **Run** to open the Construct Molecule box. Type a new name for this molecule, pBR322-UC19MCS. Select **Construct** (Figure 9.5).
10. A destination database box will appear next. Save the new Molecule file to a desired sub-base. A new Molecule Display Window opens with the newly constructed plasmid vector (Figure 9.6).

Sequence Analysis and Primer Design

The Vector NTI program provides a variety of sequence analysis tools using its own stored data or data from public databases. By using these tools, one can identify and optimize the physical parameters of PCR primers, sequencing primers, and hybridization probes. These tools also permit search for open reading frame (ORF) and translation analysis (Figure 9.7), prediction of functional motifs within promoter regions (Figure 9.8), restriction fragments, and sequence mutagenesis for

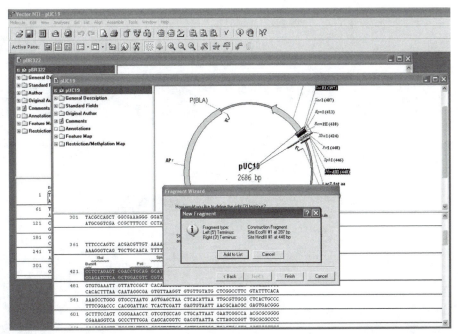

FIGURE 9.3 Defining 3′ terminus of a donor fragment from pUC19. (Copyright 2004 Invitrogen Corporation. Used with permission.)

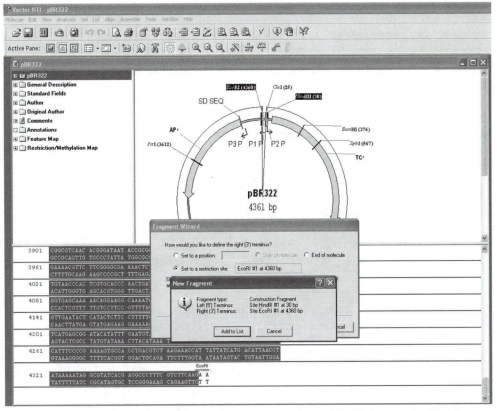

FIGURE 9.4 Defining 5′ and 3′ termini of a donor fragment from pBR322. (Copyright 2004 Invitrogen Corporation. Used with permission.)

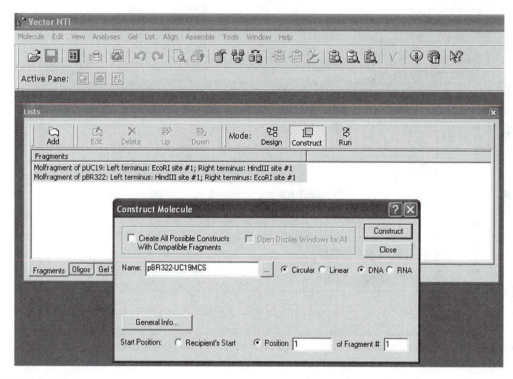

FIGURE 9.5 Construction of a new molecule using two donor fragments. (Copyright 2004 Invitrogen Corporation. Used with permission.)

removal of unwanted restriction sites. Furthermore, many internet links are provided to web-based programs that can be utilized to determine sequence properties ranging from gene features to protein hydrophobicity, peptide mass, and the identification of transmembrane regions.

The homology search and sequence comparison functions of Vector NTI are important tools for the identification of novel genes and genetic elements. Sequence homology searches can be performed against stored sequence data in local databases as well as those in public databases such as GenBank and SWISS-PROT. BLAST searches are often used for sequence homology analysis against known DNA or protein molecules. BLAST searches can be launched from any Molecule Display Window by selecting the **Tools** drop-down menu from the Main Toolbar. The following is an example of a BLAST search for DNA molecules homologous to the ampicillin resistance gene region in pUC18 from the GenBank database.

1. Open the pUC18 Molecule file from the Database Explorer.
2. Select the entire ampicillin resistance gene sequence by clicking on the **AP^r** block on the plasmid molecule in the Graphics Pane. Corresponding DNA sequences will be highlighted in the Sequence Pane.
3. From the Main Toolbar, select **Tools > BLAST Search**. A Sequence Data box will appear. Specify **Selection Only** under **Range** and choose **Direct Strand** for the search.
4. Select an appropriate BLAST server, such as the **NCBI BLAST** server for GenBank database, in the next BLAST Search box. Choose **OK** to proceed.
5. A new BLAST search window will appear and will contain information and the selected sequences. Search options, including the number of retrieved hits, and proxy configuration can also be manipulated via the drop-down menu from the View toolbar.

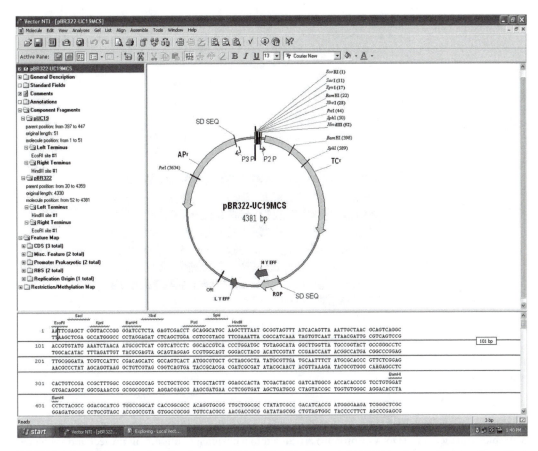

FIGURE 9.6 Newly created molecule pBR322-UC19MCS. (Copyright 2004 Invitrogen Corporation. Used with permission.)

6. Once the automatic search retrieval has finished, results can be viewed by left clicking on the text line on the Search Status Pane and then selecting the **Open Viewer** from the drop-down menu, or by simply double clicking on the text line.

7. A BLAST Results Viewer window will appear. It contains a toolbar menu, a Text Pane, a Graphics Pane, and a Sequence Pane (Figure 9.9). Any hit files containing sequences homologous to the selected ampicillin resistance gene region from pUC18 can be retrieved from the source database and opened in Vector NTI by clicking on the respective database record file from the Text Pane for further analysis.

The design of primers for PCR amplification of DNA fragments of interest, using the Vector NTI primer design tool, has become a relatively simple operation involving the following steps.

1. Launch a target Molecule file and select a fragment for PCR amplification on either the Sequence Pane or the Graphics Pane.

2. Select **Analyses > Primer Design > Find PCR Primers** on the menu bar.

3. A Find PCR Primers dialog box appears. A variety of primer parameters, including primer length, orientation, sequence specificity, structure features, and Tm and %GC can be specified.

4. Press **OK** to initiate PCR analysis based on entered values.

5. All possible sets of PCR primers will be displayed in a newly created subfolder on the Text Pane of the Molecule file. Inspect and select ideal primers for PCR operation.

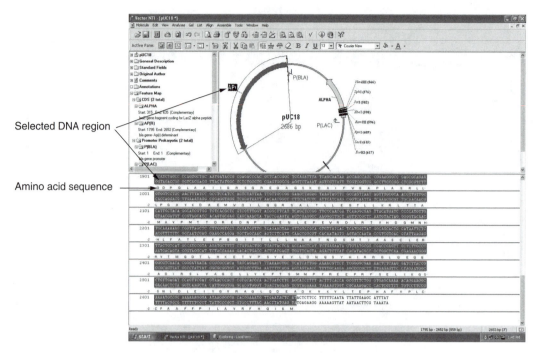

Selected DNA region

Amino acid sequence

FIGURE 9.7 Translation analysis of the AP^r region of pUC18. (Copyright 2004 Invitrogen Corporation. Used with permission.)

AlignX

Analysis of sequence homology is performed by using the highly integrated AlignX program. Multiple sequence files (DNA or protein) are added to a designated alignment project and subjected to analysis. Results are displayed in a four-pane window. Two text panes display file names and a phylogenetic tree; the Graphics Pane indicates the similarity analysis between the sequences of the compared molecules and the complexity indexes of a reference molecule. The Sequence Pane displays the aligned sequences with color-highlighted letters indicating the level of similarity and a consensus sequence. Figure 9.10 illustrates the results of a sequence alignment analysis of six plasmid vector molecules.

BioAnnotator

BioAnnotator is a comprehensive program for DNA and protein sequence analysis. Many predefined nucleic acid and protein analysis scales, such as sequence composition and free energy transfer, are available for use, and the analysis results are displayed as linear graphics for convenient visualization of the physiochemical characteristics of a molecule. For example, there are over 50 predefined analysis scales for protein molecules, including hydrophobicity, hydrophilicity, and many secondary structure parameters. Analysis scales for nucleic acids include GC content, melting temperature, and nucleic acid complexity. To demonstrate the use of BioAnnotator, a protein molecule corresponding to the dihydrofolate reductase of *E. coli* from the InforMax-provided database is analyzed, using the following steps.

1. Open the DYR4 ECOLI protein file containing the peptide sequence of the dihydrofolate reductase of *E. coli* from the Database Explorer window.
2. From the Main Toolbar, select **Analyses > BioAnnotator > Analyze Selected Molecule** (Figure 9.11).

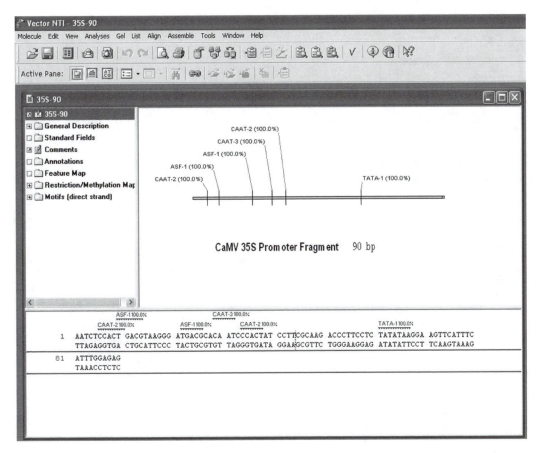

FIGURE 9.8 Sequence analysis of a CaMV 35S promoter fragment. (Copyright 2004 Invitrogen Corporation. Used with permission.)

3. A new BioAnnotator window will appear. From the Main Toolbar, select **Analyses** to activate **BioAnnotator Setup**. Double click the desired predefined analysis procedure or the protein scales from the Available analyses window of the setup box (Figure 9.12). Choose **OK** to proceed.

4. Analysis results will be displayed in three panes: a Text Pane indicating the general physiochemical information of the molecule; a Graphics Pane showing results of selected analyses or scales as multiple plots; and a Sequence Pane providing sequence positions on the molecule correlated with graphic display (Figure 9.13).

CONCLUSION

Advances in techniques of molecular biology have resulted in an explosion of DNA, RNA, and protein sequence data. Accessing such large amounts of data in a useful manner requires computer analysis programs. Fortunately, such programs are available and are accelerating the pace of progress and our understanding of genetic mechanisms.

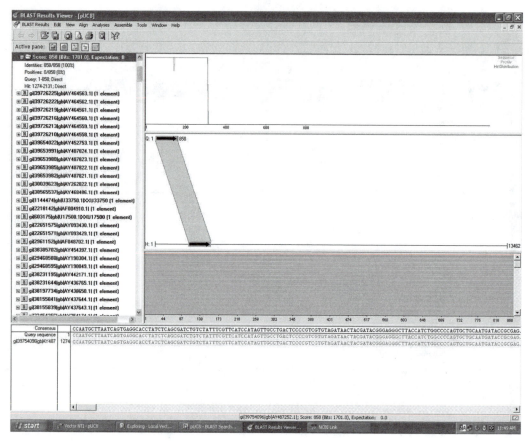

FIGURE 9.9 Results of BLAST search using Vector NTI Advance. (Copyright 2004 Invitrogen Corporation. Used with permission.)

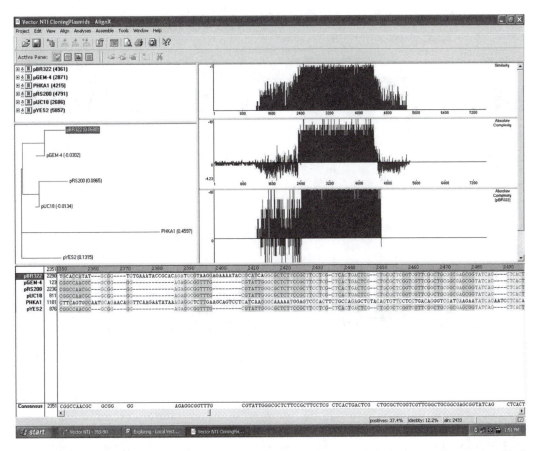

FIGURE 9.10 Alignment analysis of several plasmid vectors. (Copyright 2004 Invitrogen Corporation. Used with permission.)

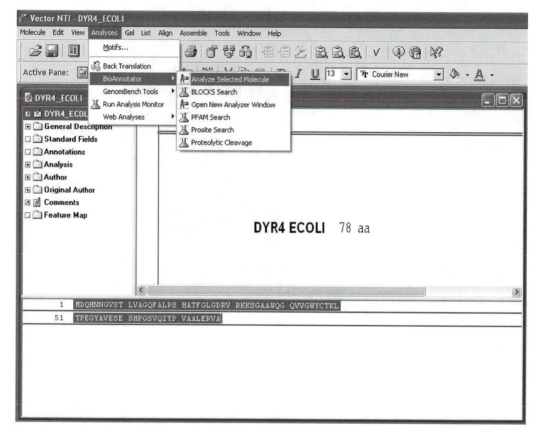

FIGURE 9.11 Alignment analysis of several plasmid vectors. (Copyright 2004 Invitrogen Corporation. Used with permission.)

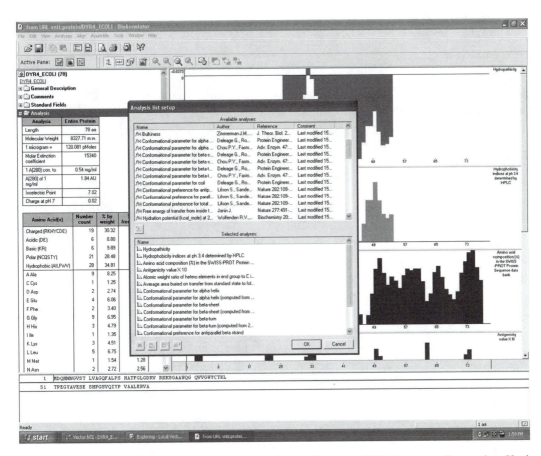

FIGURE 9.12 Selection for various BioAnnotator analyses. (Copyright 2004 Invitrogen Corporation. Used with permission.)

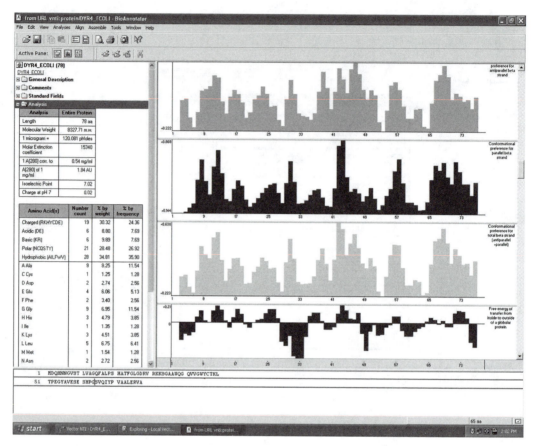

FIGURE 9.13 Results of BioAnnotator analysis of the DYR4 ECOLI protein molecule. (Copyright 2004 Invitrogen Corporation. Used with permission.)

LITERATURE CITED

Ceci, L.R. et al. 1999. PLMtRNA, a database for higher plant mitochondrial tRNA and tRNA genes. *Nucl. Acids Res.* 27:156–157.

De Leo, F. et al. 2002. PLANT-Pis: A database for plant protease inhibitors and their genes. *Nucl. Acids Res.* 30:347–348.

Dufresne, G. et al. 2002. Patent searches for genetic sequences: How to retrieve relevant records from patented sequence databases. *Nat. Biolotech.* 20:1269–1271.

Higo, K. et al. 1998. PLACE: A database of plant cis-acting regulatory DNA elements. *Nucl. Acids Res.* 26:358–359.

Letunic, I. et al. 2002. Recent improvements to the SMART domain-based sequence annotation resource. *Nucl. Acids Res.* 30:242–244.

Muller, P.Y., E. Studerand, and A.R. Miserez. 2001. Molecular BioComputing Suite: A word processor add-in for the analysis and manipulation of nucleic acid and protein sequence data. *Biotechniques* 31:1306–1313.

Wheeler, D.L. et al. 2002. Databases resources of the National Center for Biotechnology Information: 2002 update. *Nucl. Acids Res.* 30:13–16.

10 Molecular Approaches to the Study of Plant Development

Albrecht G. von Arnim

CHAPTER 10 CONCEPTS

- Many complex and long-standing problems in plant developmental biology are becoming understood through a combination of powerful experimental tools. These include molecular cloning and gene expression analysis, genetic reagents such as mutants, and cell biological and biochemical approaches.

- Forward genetics, the classical direction of molecular genetic investigation, dissects a biological process beginning with mutational analysis, followed by molecular cloning of the genes involved and biochemical investigation of gene function. Reverse genetics, in contrast, begins with the gene as a molecular tool and seeks to identify a biological activity for this gene using transgenic approaches.

- The availability of the complete genome sequence and of shared resources has accelerated the developmental biology of *Arabidopsis* by facilitating gene cloning and gene knockout studies. Similar effects are imminent in other species whose genome sequences are about to become available.

INTRODUCTION

As we interact with plants on a daily basis, perhaps enjoying our backyard or a landscaped park, hiking across an alpine meadow, or harvesting the fruits of agricultural labor, we might ask ourselves: why does a daffodil flower bloom in the spring, and a dahlia in late summer? Why are the leaves on a tobacco plant undivided (whole), while those on the related solanaceous species, tomato, are divided (compound)? What prevents the premature germination of corn kernels before they have been shed by the mother plant? These are just a few examples of the questions that have long inspired and fascinated plant developmental biologists. Each of them can be boiled down to one or more underlying conceptual events. For example, the induction of flowering by photoperiod and the induction of germination by stratification are both examples of developmental phase transitions. This and other concepts in plant development are summarized in Table 10.1.

Development is associated with an increase in biological complexity over time. Evidently, the competence for a specific developmental program is laid down in the DNA sequence, the genome of a fertilized egg cell (zygote). However, how the genetic information is realized is a far more complicated question. In the search for the mechanistic basis of developmental events, certain underlying themes emerge again and again as follows:

TABLE 10.1
Conceptual Events in Plant Development, with Examples

Concept	Examples
Developmental phase transitions	Seed germination, flowering, transition from juvenile to adult
Pattern formation	Phyllotaxis, flower structure, arrangement of hairs on leaf surface
Histogenesis, including cell fate determination and cell differentiation	Formation of vasculature and wood, hairs, pericarp
Organogenesis	Emergence of leaves from the apical meristem, emergence of lateral roots from the pericycle, embryogenesis
Morphogenesis	Shapes of leaves and fruits

- Are there environmental factors that modulate a given genetically programmed event? If so, how do they work?
- From where does a developing cell receive its cues? Is a particular developmental event informed by cell lineage information — information handed down to the cells in question from their progenitors — or by positional information — information transmitted to the cells from their current neighbors?

In summary, although this chapter focuses primarily on the molecular toolkit of plant developmental biology, it is important to keep in mind the big picture; that is, the types of questions that are supposed to be answered using these tools.

ARABIDOPSIS

One critical event in plant developmental biology was the adoption of the small crucifer *Arabidopsis thaliana* as an experimental platform. *Arabidopsis*, a herbaceous (winter) annual with a generation time of about 3 months, is easily grown in the laboratory in either soil or defined growth media. It is self-compatible and produces an abundance of seeds. Its small genome (~125 million base pairs) is easily mutagenized. *Arabidopsis* is also very amenable to the addition of new genes (transformation) using the bacterium *Agrobacterium tumefaciens* as a vehicle. The bacteria are conveniently delivered to the *Arabidopsis* flowers by infiltration in the presence of detergent.

Perhaps the main advantage of *Arabidopsis* lies in the fact that it has been adopted by hundreds of research groups around the world. The combined thrust has resulted in the DNA sequencing of essentially the entire genome, and has led to numerous associated resources, including genomic libraries, full-length cDNA collections, insertional mutagenesis programs, and the collegial sharing of reagents and tricks of the trade.

SCOPE OF THIS CHAPTER

This chapter is meant to deliver a concise introduction to the toolbox of molecular genetics as applied to the investigations of plant development. The emphasis is on outlining the capabilities, limitations, and applications of these molecular approaches in the experimental model organism *A. thaliana* (Brassicaceae). As background, a basic understanding of processes such as molecular cloning, gene structure and function, DNA sequencing, the polymerase chain reaction (PCR), genetic mapping, and Mendelian genetics can be obtained from introductory genetics textbooks. Informative examples are given whenever possible. Unfortunately, it has proven impossible to acknowledge all of the contributions of individual investigators. When practical, the names of key investigators have been associated with specific examples.

GENE CHARACTERIZATION USING MUTANTS

OVERVIEW

Many aspects of plant development can be understood as a chain of events between signals (e.g., the hormone auxin), that are perceived via signal intermediates (e.g., auxin responsive genes) and resulting responses (e.g., growth regulation in axillary buds). In turn, the response often serves as a signal for a subsequent response. Signals, signal intermediates, and responses are either the products of genes or are dependent on gene products for their synthesis or perception. Therefore, much insightful information can be gained from carefully observing the phenotypic effects that result when the function of the underlying gene is altered by mutations.

Mutational analysis underlies many of the experiments in developmental biology. Mutations may either abolish gene function (null allele), compromise gene function (partial loss-of-function allele), or artificially enhance gene function (gain-of-function allele). Briefly, for a mutant screen, a family of plants whose genome has been exposed to a mutagen is screened for defects in a specific aspect of development, for example pigmentation (Figure 10.1A), or the response to the hormone ethylene. Phenotypically abnormal plants (putative mutants) are identified (Figure 10.1B) and strains with single genetic defects are isolated via backcrossing (Figure 10.1D).

Independent mutations are tested for allelism to distinguish mutations that are allelic (i.e., affecting the same gene) from those affecting different genes (Figure 10.1C). As a result, one can arrive at a minimal estimate of the number of genes involved in the developmental event under study. The precise phenotypic defects are examined at the physiological and anatomical levels, which often distinguish pleiotropic mutations (those affecting many characters) from more specific ones (Figure 10.1E). Recessive mutations, usually suggesting reduced function of a gene, are distinguished from dominant ones, which may be due to gain of function, haploinsufficiency, or dominant negative loss of function. Much can be learned about the operational order of the genes involved by examining the phenotypes of plants with mutations in two genes at once (double mutants) for epistatic relationships between the alleles.

The resulting data are often summarized in the form of a "genetic pathway," which is nothing more than a set of hypotheses that propose a hierarchy in which the gene products contribute to their overall function. For example, Figure 10.2A shows the famous ABC model first proposed by Coen and Meyerowitz (1991), which addresses the specification of the four different whorls of a dicot flower as sepals, petals, stamens, and carpels. The ABC model explains the determination of organ types on the basis of spatially overlapping activities of no more than three classes of genes. It was derived primarily from the phenotypic analysis of flowering genes, and their mutations that caused characteristic transformations from one organ type to another (homeotic mutations).

MUTAGENS

Mutagens come in three types: chemical, physical, and biological. They differ in the spectrum of mutations caused, which has implications for the severity of phenotypes that can be expected as well as for the prospect of an eventual molecular cloning of the underlying gene (Table 10.2).

Chemical Mutagens

The methylating chemical ethyl methane sulfonate (EMS) is widely used due to its moderate toxicity and high effectiveness in inducing multiple mutations per genome. Moreover, the mutations are usually single-base substitutions. Thus, EMS is the mutagen of choice when mild, partial loss-of-function alleles are desired, although null alleles ("knockouts") and gain-of-function alleles will also be produced. Most other mutagens are more likely to cause complete knockouts.

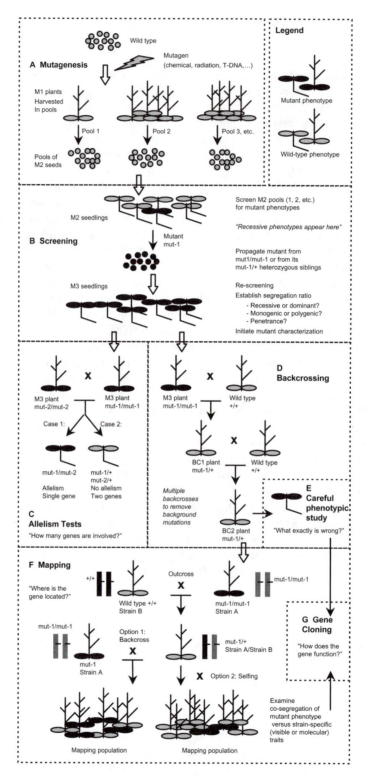

FIGURE 10.1 Mutagenesis. This figure provides a bird's-eye view of the genetic procedures and analyses conducted in order to identify mutations in a specific developmental program. For ease of illustration, the mutant phenotype shown here is excess pigmentation. For details, see text.

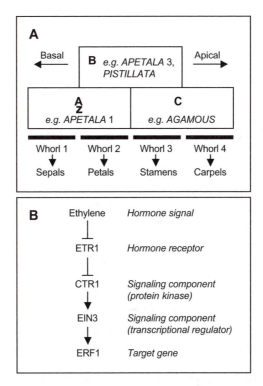

FIGURE 10.2 Genetic pathways. (A) A simplified version of the ABC model for the specification of floral whorls. A, B, and C are three distinct gene activities that are expressed in a spatially overlapping fashion in basal, medial, and apical domains of the flower meristem. Each activity is provided by several genes, only a few of which are exemplified here. Thus, the combined presence of A and B activities specifies the formation of petals in whorl 2, etc. (B) The genetic pathway in (B) is a simplified model for the mode of action of the hormone ethylene. The pathway was derived in large part from mutational analysis in *Arabidopsis* through the efforts of Joe Ecker (Salk Institute), his co-workers, and many other groups. Ethylene is perceived by the ethylene receptor ETR1, which causes changes in gene expression through downstream signaling components that include the protein kinase CTR1 and the transcriptional regulator EIN3, which binds to DNA sequences in target genes such as the *ERF1* gene. Blocked arrows indicate repression and regular arrows symbolize activation.

Physical Mutagens

Among the physical mutagens (radiation), fast neutrons and x-rays cause deletions, which are often knockouts. Deletion alleles can be very useful during map-based cloning of a mutated gene (see below), when the task at hand consists of locating the gene within a larger chromosomal region containing, perhaps, several dozen genes. A deletion will cause a restriction fragment length polymorphism (RFLP), whereas base substitutions usually do not. Probing for the mutated gene by means of Southern blotting will reveal a difference in the banding pattern between wild type and mutant plants. This is a powerful way of distinguishing the gene of interest from surrounding genes.

Biological Mutagens

Biological mutagens are fragments of DNA that insert themselves into the plant genome. They take the form of transposons or of Agrobacterium T-DNA. Their insertion tends to cause a knockout of the affected gene (Figure 10.3A), although an insertion into a gene's regulatory region may result in a weak allele or even a gain-of-function allele. Although time-consuming to produce, insertional

TABLE 10.2
Mutagens, Their Effects, and Implications for Molecular Analysis

| Mutagen | Predominant Molecular Lesion | Implication for... | |
		Typical Genetic Defect	Molecular Analysis
Ethyl methane sulfonate (EMS)	Base substitution (transition)	Reduced function, loss of function; efficient mutagen	—
Fast neutron	Deletion (several kb)	Null allele (knockout)	Molecular lesion usually reflected in RFLP
X-ray	Deletion, chromosome break	Broad spectrum	—
Transposon (stable)	Insertion	Null allele	Sequence tag
Transposon (unstable)	Small insertion (footprint)	Broad spectrum, including frameshift	—
T-DNA (standard)	Insertion	Null allele	Sequence tag
T-DNA (activation tagging)	Insertion	Gain of function; overexpression, ectopic expression	Sequence tag

mutants have one powerful advantage: the DNA sequence of the insertion element is known. The insertion element thus provides a "tag," which can be used to obtain the unknown DNA sequence of the gene.

One derivative of the T-DNA tagging strategy is "activation tagging." Here, the T-DNA contains a strong transcriptional enhancer sequence, which may lead to a dominant gain-of-function allele of the gene flanking the T-DNA, due to its overexpression expression in the wrong cell type (Figure 10.3B).

Transposable elements are more efficient mutagens than T-DNA if the transposon catalyzes its own excision and reintegration. However, certain inconveniences result. First, a transposon that re-excises from a tagged gene may cause a mutation by leaving behind no more than a small insertion ("footprint"), which is not sufficient to serve as a tag for molecular cloning. Second, if the transposon accumulates to a high copy number in the genome, it can be difficult to track down which specific copy of the transposon caused the mutant phenotype.

BUILDING A GENETIC PATHWAY FROM MUTANTS AND THEIR GENETIC INTERACTIONS

A successful mutant screen will usually yield a series of mutants in a number of genes that affect the process of interest. Once a set of mutants have been isolated, one can attempt to organize the underlying genes into a genetic hierarchy ("genetic pathway"), as exemplified by the ethylene signaling pathway illustrated in Figure 10.2B (Alonso and Ecker, 2001). There are several genetic tools to this end.

Pleiotropic versus Specific Effects

The first order of the day is the careful observation of the various mutant phenotypes. Certain mutations will lead to very specific defects, whereas others may have broader consequences, referred to as pleiotropic effects. For example, mutations causing defects in the response to ethylene may display a defect in just one ethylene response or all ethylene responses. Mutations causing defects in root growth may affect the entire root or may affect just a single cell layer. Often, the more pleiotropic mutation marks a gene that functions high up in the hierarchy. This is often referred to as "upstream in the genetic pathway," whereas the mutation with a more specific defect is said to mark a "downstream" gene.

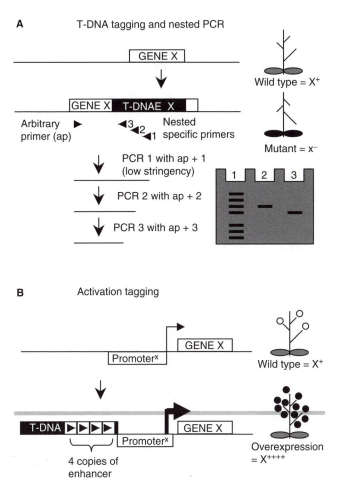

FIGURE 10.3 T-DNA tagging. (A) A T-DNA insertion into a postulated "Gene X" causes aberrant branching and pigmentation. Gene X can be identified through a stepwise PCR procedure using nested PCR primers. While round one results in numerous unspecific products, round two and round three amplify the desired product more specifically, as seen by the shift in the product size evident in the gel electrophoresis pattern. (B) A T-DNA carrying a transcriptional enhancer can boost the expression of a neighboring gene, causing an overexpression phenotype such as profuse flowering.

Activators versus Repressors

An important characteristic is whether the gene defined by mutation affects the process under study in a positive fashion, as an activator, or in a negative fashion, as a repressor (Figure 10.2B). For instance, the long hypocotyl (*hy*) mutations of *Arabidopsis* mark positive regulators in the response of seedlings to light, given that an abnormally long hypocotyl is evidence of reduced sensitivity to light in the *hy* mutant. Not surprisingly, by the way, several *hy* mutations are due to lesions in phytochrome or cryptochrome photoreceptor genes. In contrast, constitutive photomorphogenesis (*cop*) mutants mark negative regulators of the light response. The *cop* mutants appear like light-grown plants when germinated in darkness, due to a failure to repress the default light response pathway in the darkness.

When we make an inference about a wild-type gene based on its mutant phenotype, we must be certain that the mutation is showing us the loss-of-function phenotype of the gene in question. In the case of recessive mutant alleles, this assumption is usually justified.

Allelic Series

For certain genes, the mutant phenotypes of individual alleles do not all look alike, but instead form an allelic series. An allelic series can be very informative when it comes to defining the function of a specific gene at different developmental stages.

The *COP1* gene of *Arabidopsis* is a good example. Its severe alleles are lethal at an early stage of seedling development, whereas milder alleles have defects more specifically in the light regulation of seedling and vegetative development, suggesting that the *COP1* gene functions in different roles at different stages of development.

Double Mutant Analysis

One of the more powerful tools for defining the hierarchy of multiple genes in development is double mutant analysis. Let us assume that we possess two loss-of-function mutants with opposite phenotypes. One mutant fails to carry out a specific response even when the appropriate stimulus is present (e.g., the wild-type allele is a positively acting gene); the second mutant carries out the specific response constitutively (e.g., marking a repressor gene). Suitable examples are the already familiar *hy* and *cop* mutants, respectively. We can answer the question whether the *HY* or the *COP* gene functions higher in the pathway by examining the phenotype of a double mutant, for instance between *hy3* and *cop1*, which is generated by genetic crossing. If the *hy3/cop1* double mutant were to display the *hy* phenotype, then the *HY3* gene would be placed below the *COP1* gene in the hierarchy. In fact, however, the *hy3/cop1* double mutant displays the *cop1* phenotype, regardless of light conditions, thus placing the *COP1* gene below ("downstream of") the *HY3* gene in the light signaling pathway. Consistent with its position upstream in the light signaling pathway, the *HY3* gene encodes the phytochrome B photoreceptor. Similar studies have contributed to the genetic models in Figure 10.2.

Clonal Sector Analysis

Another elegant tool for defining the action of developmental regulator genes is clonal sector analysis. Let us assume that we possess a mutant with only rudimentary hairs on the leaf surface (glabrous). We are interested in determining whether the underlying *GLABRA2* gene functions within the hair precursor cell or in neighboring cells to induce the outgrowth of hairs. The Kilby group at the University of Cambridge was able to generate plants in which the leaves consist of marked (labeled) clonal sectors of mutant tissue (rudimentary hairs) and wild-type tissue (regular hairs). By carefully observing the hair growth near the border between mutant and wild-type sectors, researchers were able to discern that the *GLABRA2* gene functions in a cell-autonomous fashion because hairs were confined exclusively to the wild-type sector (Figure 10.4). If the gene were not cell-autonomous, then at least the edge of the mutant sector would have carried regular trichomes. Clonal sector analysis can sometimes distinguish the functions of genes whose mutations are otherwise completely identical.

Although careful mutant characterization can in many cases define the relative position and functional interactions of distinct developmental regulator genes, there will often be genes that cannot be distinguished functionally. For example, the *CLAVATA1* and *CLAVATA3* genes both serve to restrict stem cell growth in the *Arabidopsis* shoot apical meristem (Sharma and Fletcher, 2002); thus, both *clv1* and *clv3* mutants share the phenotype of an enlarged shoot apex. In this case, only molecular analysis of the genes can distinguish their precise activities. In fact, as we will see, not only are the two *CLAVATA* genes expressed in different regions of the apical meristem, they also encode very different types of proteins. However, these two proteins function together as a cell-surface receptor (CLV1) and the cognate ligand molecule (CLV3), which explains the identical phenotypes of their mutants.

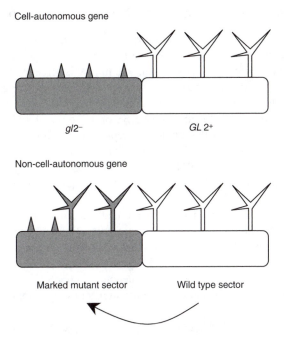

Cell-autonomous gene

gl2⁻   *GL 2⁺*

Non-cell-autonomous gene

Marked mutant sector   Wild type sector

FIGURE 10.4 Principle of clonal sector analysis. In the case of the cell-autonomous gene *GL2*, mutant and wild-type phenotypes (hairs) are confined precisely to genetically mutant and wild-type sectors of tissue, respectively (shading). In the case of a hypothetical non-cell-autonomous gene, gene activity in a wild-type sector rescues the defect in a nearby mutant sector.

GENE ISOLATION: METHODS OF CLONING GENES

Selecting the best among numerous strategies for cloning genes will depend on two things: What is our long-term objective, and which tools and reagents do we have at our disposal?

1. Are we looking for the one-and-only gene that is defective in a specific, developmentally dysfunctional mutant? In this case, a map-based ("positional") cloning strategy may be called for.
2. Are we looking for a gene that is responsible for a particular enzymatic reaction? This question might lend itself to functional complementation of a corresponding mutation in yeast.
3. Are we generally interested in genes that are expressed in response to a particular stimulus, for example cytokinin treatment? Subtractive hybridization may lead to the goal in this case.
4. Finally, we might be interested in additional gene family members of known genes; for example, the homologue in tomato of an *Arabidopsis* flowering gene. In this case, a screen of a cDNA library by hybridization at low stringency may be the way to go.

In principle, the aforementioned examples are representative of four general cases, the selection of genes based on:

1. Mutant phenotype
2. Biochemical activity of the gene product
3. Expression pattern
4. DNA sequence

CLONING GENES BASED ON DNA SEQUENCE IS DONE BY HYBRIDIZATION

Molecular cloning based on DNA sequence is a straightforward method for isolating specific genes. The principle is simple: the genome of interest is converted into a library of clones using a suitable cloning vector, usually bacteriophage lambda, and the library is screened by hybridization with an available fragment of DNA (the "probe") to identify a new gene that resembles the probe in its DNA sequence. Examples of applications include identifying a rice gene for starch synthesis that corresponds to a gene previously cloned from maize, or using one *Arabidopsis* ethylene receptor gene to find additional members of the ethylene receptor gene family. Depending on the circumstances, the library is either a genomic library, representing chromosomal DNA, or a cDNA library, representing transcribed sequences (mRNAs). One can adjust the stringency of hybridization to select genes that are nearly identical to the probe in their sequence or genes with only moderate sequence similarity.

In the case of reverse translation, let us assume that the amino acid sequence of a protein of interest, for example a DNA binding protein, has been partially determined by biochemical means. Using the genetic code, we then reverse translate a short amino acid sequence into a set of corresponding DNA sequences. After chemically synthesizing the DNA, we proceed to isolate the gene of interest by hybridization. As protein-sequencing technology by mass spectrometry has improved dramatically in sensitivity and accessibility in recent years, cloning by reverse translation has become very popular.

Library screening based on DNA sequence similarity is becoming all but obsolete in plants with a fully sequenced genome — currently *Arabidopsis* and rice — once molecular clones for individual genes have been catalogued and made available for distribution by genetic stock centers.

CLONING GENES BASED ON THE BIOCHEMICAL ACTIVITY OF THE GENE PRODUCT

The cloning strategies in this section, in common, all rely on some form of biochemical assay for the gene product in question. In principle, a cDNA library is constructed in a gene expression vector. Clones are selected from the library based on the biochemical activity of the expressed gene product. The activity assay is preferably conducted *in vivo*, and in a microorganism such as yeast or *E. coli*.

For example, the G-box is a DNA sequence motif with a regulatory role in many plant promoters regulated by light as well as abscisic acid. The group of Anthony Cashmore at the University of Pennsylvania selected genes for G-box DNA binding proteins by expressing a plant cDNA library in *E. coli* and probing this library with a labeled version of the G-box (Schindler et al., 1992).

One elegant, powerful, and very popular cloning strategy in the category of activity-based methods is the yeast two-hybrid system. In order to dissect a developmental pathway at the molecular level, we often wish to identify gene products that physically interact with a known protein. In principle, the yeast two-hybrid system consists of two expression plasmids and a reporter plasmid (Figure 10.5A). A physical interaction between two proteins X and Y is visualized by expressing protein X as a translational fusion protein with a specific DNA binding domain — "tagging" X with DBD — and tagging protein Y with a transcriptional activation domain (AD). X-DBD and Y-AD are then coexpressed in the same yeast cell. This yeast cell also possesses an inducible reporter gene, characterized by a promoter that contains a binding site for DBD. If X and Y interact by heterodimerization, then Y-AD will activate transcription of the reporter gene because a transcriptional activation domain is recruited to the promoter by the DBD-X partner protein. Two reporter genes are widely used: the yeast *LEU2* selectable marker gene, which confers the ability to synthesize the amino acid leucine (leucine auxotrophy), and a colorimetric marker gene, such as *lacZ* (beta-galactosidase). To select new genes that interact with Y, a cDNA library is made in the vector contributing the DNA binding domain and candidate interactors of Y are selected by looking for growth of yeast clones on medium lacking leucine.

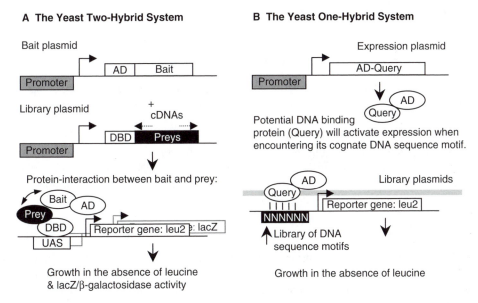

A The Yeast Two-Hybrid System

Bait plasmid

Promoter | AD | Bait

Library plasmid
+
cDNAs

Promoter | DBD | Preys

Protein-interaction between bait and prey:

Bait
AD
Prey
DBD
UAS | Reporter gene: leu2 e: lacZ

Growth in the absence of leucine
& lacZ/β-galactosidase activity

B The Yeast One-Hybrid System

Expression plasmid

Promoter | AD-Query

AD
Query

Potential DNA binding
protein (Query) will activate expression when
encountering its cognate DNA sequence motif.

AD   Library plasmids
Query
| | | | | | Reporter gene: leu2
NNNNNN
Library of DNA
sequence motifs

Growth in the absence of leucine

FIGURE 10.5 Library screening in yeast. (A) The yeast two-hybrid system is a tool to identify proteins that interact with a given starter protein, the bait. The bait protein is expressed as a fusion to a transcriptional activation domain (AD). Plant cDNAs are selected from a library (black) constructed as a fusion to a DNA binding domain (DBD). Prey cDNAs are selected based on their ability to interact with the bait (double-headed arrow) as evidenced by reporter gene activation. For details, see text. (B) The yeast one-hybrid system is a strategy to identify the DNA binding sites of a given query protein from a library of DNA sequence elements (black) in a reporter plasmid. The query protein is expressed as a fusion to a transcriptional activation domain (AD). Only cells containing the cognate DNA sequence motif will show *leu2* reporter gene activation and thus growth in the absence of leucine. (UAS, upstream activating sequence.)

The yeast two-hybrid system is just a specialized case of a more general strategy, cloning by mutant rescue in yeast. Let us assume our charge consists of identifying a plant gene for a specific enzymatic step in amino acid biosynthesis. If a yeast loss-of-function mutation is available in the corresponding gene of yeast, then it may be possible to isolate the plant version of this gene as follows: A plant cDNA library is constructed in a yeast expression vector and is screened for cDNAs that rescue the yeast mutant defect. However, few if any plant developmental genes lend themselves to this approach because such genes are unlikely to be functionally conserved in yeast.

CLONING BASED ON GENE EXPRESSION CHARACTERISTICS

In order to identify genes based on their expression characteristics, such as developmental timing, spatial pattern, or inducibility of expression, there are, technically speaking, four categories of approach:

1. Differential (subtractive) cloning or differential hybridization
2. Differential PCR amplification
3. Microarrays
4. Promoter trapping

The underlying assumption is that plant development is regulated by changes in the abundance of specific transcripts. Vice versa, a gene whose expression changes in response to a developmental stimulus might lead us to better understand developmental regulation. For instance, the *CONSTANS* gene, which is regulated in response to day length in *Arabidopsis*, is a key player in the induction of flowering, a day-length-dependent process (Yanovsky and Kay, 2002).

Differential (Subtractive) Cloning or Differential Hybridization

The first set of methods begins by isolating mRNA transcripts from the two tissue samples to be compared; for example, shoot apices that have received a stimulus for flowering versus a control. During subtractive cloning, the population of transcripts is biochemically enriched for those that are present primarily in one sample and absent in the other, followed by selective cloning of the unique cDNAs. For comparison, during differential hybridization, the library to be screened is left unselected. Instead, each set of mRNAs is converted to a set of cDNAs and each set of cDNAs is used to probe the same library; for example, the library made from apices induced to flower. Clones are selected that hybridize preferentially to one of the probes but not to the other.

Differential PCR Amplification

The two methods described in the previous section are not only technically demanding, but also suffer from the disadvantage that they tend to select for highly expressed genes, even though the most interesting developmental regulator genes are often expressed at low levels. Differential PCR amplification takes a stab at overcoming those problems.

Again, two different mRNA samples to be compared are isolated and converted into cDNA. In principle, two PCR primers of essentially arbitrary sequence are designed and the cDNA then serves as the template for PCR. Because the primers are short and degenerate, they amplify multiple cDNAs from each cDNA collection. Consequently, it is plausible that the resulting two sets of PCR products might differ in the abundance of a subset of products, which would reflect differentially expressed genes. The amplified fragment length polymorphism (cDNA-AFLP) method puts this idea into practice. cDNAs from two different samples are first digested into smaller fragments using a pair of restriction enzymes, then ligated to a short synthetic adapter of arbitrary DNA sequence. Because the base pair sequence of the adapter is known, the cDNA fragments are now PCR amplifiable with primers matching the adapters. Bulk PCR products are separated and distinguished according to their length by high-resolution gel electrophoresis. Again, products that accumulate to different levels when the two cDNA samples are compared may reflect differentially expressed genes. For practical reasons the adapter-modified cDNAs must be preselected during PCR to keep the total number of amplified fragments manageable. To this end, the primers contain an arbitrary 2-base extension at their 3' end, which will allow amplification of only a subset (~1/16) of cDNA fragments.

Microarrays

With the advent of microarray-based expression profiling, a new way of identifying differentially expressed genes has arisen. A microarray ("DNA chip") is a device that permits us to measure the expression level of, in principle, all the genes in the genome of a plant. The limiting factor is the production of the microarray itself, which goes beyond the scope of this chapter. Suffice it to say that microarrays are becoming commercially available for several plant species. By probing the microarray with cDNA derived from two developmentally distinct plant samples, it is possible to identify all the genes that are induced or repressed in one sample over the other.

Promoter Trapping

In contrast to the transcript-profiling methods covered above, promoter trapping is an insertional mutagenesis method designed to reveal the expression patterns of specific chromosomal genes

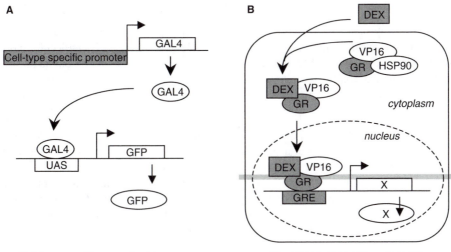

FIGURE 10.6 Inducible gene expression systems. (A) The cell-type specific expression of the GAL4 tran-scriptional activator protein drives the cell-type specific accumulation of the green fluorescent protein (GFP) gene due to a GAL4 binding site (UAS) in the *GFP* promoter. (B) The gene (X) is inducible by the synthetic steroid, dexamethasone (DEX). In the absence of DEX, the glucocorticoid receptor (GR) is sequestered into an inactive form by the HSP90 protein. In the presence of DEX, the GR binds to a glucocorticoid response element (GRE) in the promoter of X. VP16 is a transcriptional activation domain that has been fused to the glucocorticoid receptor.

in situ. To this end, a promoterless, or "silent," reporter gene, for example β-*glucuronidase* (*GUS*), is integrated as a transgene into the plant genome. A collection of individual transgenic lines is created, each containing the reporter gene at exactly one random chromosomal location. Crucial to promoter trapping is the fact that by inserting into or next to an endogenous gene, the reporter gene may become expressed in a pattern similar to that of the endogenous gene, which is then said to be trapped. For example, Qing Gu and Robert Martienssen at the Cold Spring Harbor Laboratory discovered a promoter trap line in the *Arabidopsis FRUITFULL* gene by virtue of a characteristic *GUS* expression pattern in the floral meristem and the developing fruit (Gu et al., 1998). The *FRUITFULL* gene plays a role in coordinating fruit and seed development in *Arabidopsis*.

Two different reporters have proven to be most useful for promoter trapping. *GUS* is easily visualized by histochemical staining *in situ*. By comparison, green fluorescent protein (GFP) has the benefit of being visible in a live plant. Because GFP itself is not as sensitive a reporter as *GUS*, gene trapping with GFP is sometimes carried out with an amplification step (Figure 10.6A). The promoter-trapping cassette contains a transcriptional activator protein with sequence specific DNA binding activity, the yeast GAL4 protein. The expression pattern of GAL4 is visualized indirectly by operating in a plant line that harbors a GFP transgene with a GAL4 dependent promoter.

A promoter–reporter fusion with a defined expression pattern may serve as a starting point for subsequent mutagenesis to identify other genes that regulate its expression, thus moving one step forward in a developmental pathway. This particular application has made use of yet another reporter enzyme, the light emitting firefly luciferase. For example, the group of Steve Kay and others has discovered genes regulating the circadian control of gene expression in *Arabidopsis* (Millar et al., 1995). Although visualizing luciferase bioluminescence *in situ* requires a sensitive camera system, this disadvantage is outweighed by the advantages of real time *in vivo* and quantitative data acquisition.

CLONING GENES BASED ON WHOLE-PLANT MUTANT PHENOTYPE

The most direct way of uncovering novel developmental regulator genes is to identify a mutant plant with a phenotypically informative defect in a single gene followed by molecular cloning of the corresponding gene. Two rather different approaches to this problem are gene tagging (insertional mutagenesis) and map-based cloning.

Gene Tagging

The key element of gene tagging is that the mutation of interest must be generated by insertion of a DNA element of known sequence, usually a transposon or a T-DNA (see earlier). Once this has been accomplished and established by confirming genetic linkage (cosegregation) between the mutant phenotype and a suitable marker gene on the insertional element, such as antibiotic resistance, the identification of the corresponding gene is often straightforward. Cloning a tagged gene hinges on isolating and sequencing the chromosomal DNA flanking the insertion element. One of several methods to this end is nested PCR amplification (Figure 10.3A).

Map-Based Cloning

Alas, many mutations with interesting developmental phenotypes are not associated with a DNA sequence tag. In this case, the corresponding gene of interest may be identified by map-based cloning or positional cloning (Figure 10.7). In the first stage, the mutation is mapped to smaller and smaller genetic intervals, with the help of molecular markers (Figure 10.8). The smallest genetic interval that contains, among other genes, the gene of interest, is cloned in its entirety (Figure 10.9), and the gene of interest is identified from within the interval by preferably two independent means. First, cloned fragments are tested for their ability to rescue the mutant defect when reintroduced as a transgene into the mutant background (mutant rescue or complementation). Second, independent mutant alleles of the gene of interest are sequenced in order to demonstrate that each mutant allele contains a specific sequence alteration that is a plausible basis for the mutant defect (sequencing of an allelic series).

Map-based cloning relies on a dense map of genetic markers, most of which should be molecular markers, such as restriction fragment length polymorphisms (RFLPs) and short sequence length polymorphisms (SSLPs), which can be scored painlessly using PCR in the large families necessary to establish tight linkage. Map-based cloning is also facilitated tremendously by shared genomic resources, such as a completed genome sequence, a dense map of available markers, and large-insert genomic libraries. The portal for these resources for *Arabidopsis* is found at http://www.Arabidopsis.org/.

METHODS FOR ANALYZING GENE FUNCTION

Once a developmentally important gene has been cloned, the next question is, How does the gene function? Experimentally, this question can be approached in a number of ways:

1. Domain structure: Does the predicted protein sequence contain clues?
2. Gene expression: Where and when is the gene expressed?
3. Biochemical activities: What are its enzymatic activities or partner proteins?
4. Protein localization: Where in the cell is the protein found?
5. Reverse genetics: What happens when the expression of the protein is modified?

PROTEIN DOMAIN STRUCTURE

Comparing the encoded protein sequence of the gene against other sequences deposited in the Genbank database often leads to a plausible hypothesis. For example, the gene product might appear

Cloning a Gene based on its Map Position

Define the Gene of Interest (GOI)

Mutant phenotype
Mode of inheritance
Allelic series

Mapping

Identify flanking molecular markers
Coarse and fine mapping

Map interval for GOI

Chromosome walk

Isolate molecular clones
spanning the GOI

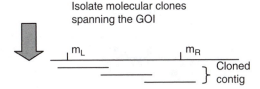

Select the Gene of Interest

Complementation
DNA sequence differences between alleles

DNA sequencing

cDNA and/or gnomic DNA

Functional Analysis

Gene expression
Biochemical activity
Structure-function studies

FIGURE 10.7 Cloning a gene based on its map position (positional cloning). The flowchart demonstrates the steps involved. For details, see Figures 10.8 and 10.9 and the text. Bars indicate chromosomal regions with markers (m) interspersed along them.

to be a biosynthetic enzyme, a protein kinase, a transmembrane channel, a DNA binding protein with or without transcription activation domains, or a protein with interaction surfaces for other proteins. Does the protein consist of single or multiple domains, and have other proteins with similar domain combinations been identified? Are there known orthologs of the protein in other species, and is their function already understood? Answers to these questions can speed the subsequent analysis tremendously. Nonetheless, we have to bear in mind, first, that any suggestion derived from sequence similarity is merely a hypothesis seeking experimental validation; and second, that the protein sequence search may lead to no useful clues whatsoever.

Short sequence length polymorphism (SSLP) or microsatellite

A

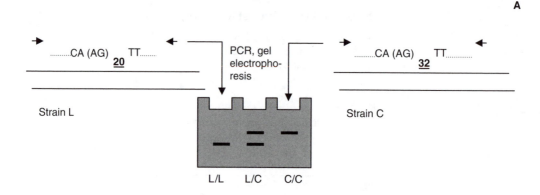

Restriction fragment length polymorphism (RFLP)

B

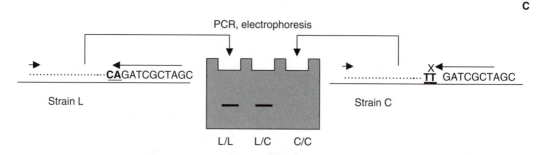

Randomly amplified polymorphic DNA (RAPD)

C

FIGURE 10.8 Types of molecular markers used in gene mapping. Each panel shows the DNA sequence difference (polymorphism) underlying the molecular marker for each of two strains of plants, L and C. The sequence difference is in boldface. Note that strains L and C are diploid, as indicated by the duplicated sequence in the SSLP panel. For simplicity, only one allele is shown for RFLPs and RAPDs. Short arrows symbolize PCR primers. The combed box represents an electrophoresis gel used to score the markers. L/L and C/C: marker scored as homozygous; L/C: marker scored as heterozygous. SSLPs and RFLPs are scored with two different PCR primers that perfectly match the genomic sequence around the polymorphism. In contrast, RAPDs are scored with a single short primer. Small sequence differences between strains can interfere with efficient annealing of the primer, as shown here for strain C.

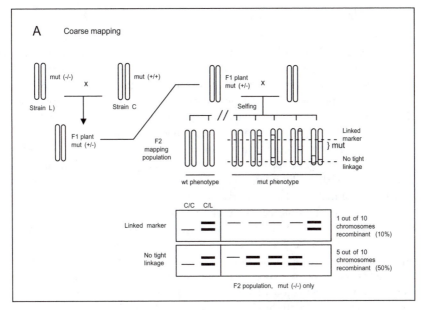

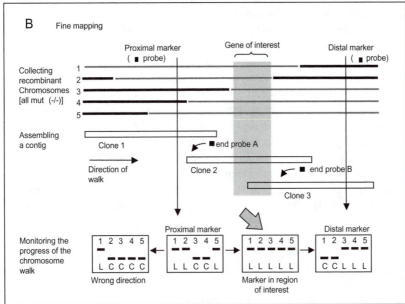

FIGURE 10.9 Mapping a gene with molecular markers. (A) Coarse mapping. The panel shows the three-generation crossing scheme to derive a mapping population (here: F2). A subset of F2 plants is shown to illustrate the principle of detecting linkage between the gene (*mut*) and two SSLP markers. The squared boxes represent gel electrophoresis patterns. Only one of the two markers shows clear linkage. Note that one of the ten chromosomes scored with the linked marker has a rare recombination event between *mut* (from strain L) and the marker (strain C). (B) Fine mapping and chromosome walk. After collecting a series of plants with recombinant chromosomes due to recombination events on either side of the gene of interest, the molecular markers are used to identify DNA clones near the region of interest (Clone 1 and Clone 3). Additional clones spanning the region of interest are derived, starting with probe A from Clone 1, by chromosome walking (Clone 2). The resulting set of clones is called a contig. Newly developed molecular markers (thick arrow) can be used to define the region of interest to a small interval, which must contain the gene of interest. Note that in a real map-based cloning project, a much greater number of plants, chromosomes, and markers must be scored than are shown here.

GENE EXPRESSION PATTERN

Much about gene function can be learned from the gene's spatial expression pattern and its dependence on experimental conditions. Genes involved in pattern specification, organogenesis, and cell differentiation are often expressed in a very specific cell type or in a restricted region of the plant. Conversely, genes involved in developmental phase transitions may follow a temporal pattern of induction or repression, whereas genes involved in a hormone response pathway may themselves be regulated by the same hormone. To visualize the gene expression pattern, the mRNA level of the gene is examined using RNA gel electrophoresis, followed by RNA hybridization ("northern blotting"). Superior spatial resolution of expression is provided by *in situ* hybridization techniques. For example, the gene for the transcription factor *APETALA3*, which helps to specify the petals and stamen whorls in the flower (see Figure 10.2A), is indeed expressed in a region of the early floral meristem that will later give rise to petals and stamens.

Another *in situ* gene expression technique makes use of fusing the 5' upstream region of the gene (assumed to be the promoter) to a reporter gene, such as *GUS*. This recombinant promoter-*GUS* fusion is introduced into the genome as a transgene and can be used to trace the expression of the corresponding gene by histochemical *GUS* staining. By crossing the transgene into specific mutant genetic backgrounds or by exposing it to specific environmental conditions, a researcher can gain insights into the regulation of the gene in question. For example, the group of Tom Guilfoyle created a *GUS* fusion to the auxin-responsive DNA sequence motif known as DR5, which has become popular for reporting on endogenous auxin levels in plant tissues.

Determining whether and how the expression of a gene depends on other genes or environmental parameters can help the researcher establish a functional network around the gene of interest. Often this is referred to as a "genetic pathway," with "upstream" events that regulate the gene and "downstream" events by which the gene exerts its effects on plant development. These results should be reconciled with similar "pathway" results from double mutant analysis (see earlier).

Although the mRNA expression pattern is usually a reliable indicator for where the gene functions, this is not always true. Exceptions to the rule are among the most exciting recent discoveries because they suggest how genes can function over a physical distance of one or more cell layers. Plant cell differentiation is exquisitely sensitive to position effects, that is, specific signals communicated by neighboring cells, but the underlying basis for such position effects has long been unclear. Philip Benfey and co-workers at New York University have investigated the establishment of the radial pattern of cell types in the *Arabidopsis* root. They found that the mRNA of the regulatory gene, *SHORTROOT*, is found in the central stele and pericycle cells of the root (Nakajima et al., 2001). This result was surprising because the most striking phenotypic defects in the *shortroot* mutant are observed in the more peripheral endodermal cell layer. It turns out that the SHORTROOT protein product is capable of migrating from the more central pericycle cells into neighboring endodermis cells, presumably via symplastic intercellular connections, the plasmodesmata.

Similar examples of mobile proteins have been discovered elsewhere. In the maize shoot apex, for instance, the Knotted1 protein is transported from the subepidermal layer, where its mRNA level is high, into the epidermal layer, where its mRNA level is negligible. Likewise, the CLAVATA3 protein appears to migrate from its point of synthesis in the central shoot apical meristem to more basal regions of the shoot apex where it interacts with a cell surface receptor, CLAVATA1. CLAVATA3 is a small peptide and is thought to migrate through the extracellular cell wall space (Sharma and Fletcher, 2002).

BIOCHEMICAL ACTIVITIES, INCLUDING PROTEIN–DNA AND PROTEIN–PROTEIN INTERACTIONS

A protein with a role in plant development can have any conceivable biochemical activity; it might function as a structural cell wall protein, a protein kinase, or a sequence-specific RNA binding protein, to give just a few examples. Although the quest for the biochemical activity of a protein is key to understanding its mode of action, there exists no systematic path that will guarantee a solution. Rather, defining this activity resembles a random walk, or searching for a needle in a haystack, and involves some trial and error. Nonetheless, two standard types of assays are applied routinely, assays for protein–DNA interactions and assays for protein–protein interactions.

Protein–DNA Interactions

If there is reason to believe that the protein might bind DNA, this can be tested by asking whether the protein retards the electrophoretic mobility of DNA in a gel (gel-retardation assay).

Assays to define the specific DNA target sequence of the protein include PCR based selection processes, gel retardation assays, and DNAse footprinting assays. An alternative method is first to express the protein in yeast as a fusion to a transcriptional activation domain. Subsequently, one screens for the DNA sequence motif recognized by the protein using a randomized library of promoter DNA sequence elements that are linked to a suitable reporter gene (the yeast one-hybrid system; Figure 10.5B). One increasingly popular method for finding specific target genes for proteins that are associated with chromatin is to enrich the protein, together with its bound target DNA, from a nuclear extract using an antibody against the protein under scrutiny. The enriched fraction is then tested by hybridization for enrichment of individual plausible target genes (chromatin immunoprecipitation or ChIP).

Protein–Protein Interactions

Once a model for a genetic pathway is established it is time to test it experimentally, for example by addressing whether a genetic interaction between two genes is reflected in a direct biochemical interaction between their gene products. For proteins that appear to function by binding to other proteins, the yeast two-hybrid cloning strategy outlined above is a powerful way to close in on the relevant cellular partners. Alternatively, one can develop a biochemical purification scheme for the protein at hand and determine the identity of any copurifying proteins. As hinted at earlier, identifying even small amounts (less than 1 µg) of copurifying proteins has become quite straightforward in recent years, due to advances in mass spectrometric peptide sequencing. If a full genome sequence is also available, one can often identify the correct gene corresponding to the peptide sequence *in silico*, i.e., by a search of the genome sequence database.

For example, the *Arabidopsis* repressor of light signaling, COP9, did not reveal any clues to its mode of action when first sequenced. However, a purification scheme developed by Daniel Chamovitz and Xing-Wang Deng at Yale University resulted in the copurification of an entire protein complex of eight subunits, the COP9 signalosome (Chamovitz et al., 1996). Because the sequence of additional subunits resembled subunits of the proteasome proteolytic complex, it was hypothesized, and eventually confirmed, that COP9 and its partners are involved in regulating the degradation of light-regulatory transcription factors.

Coimmunoprecipitation is another popular biochemical technique used to examine whether two functionally related developmental regulators do in fact interact physically. Protein-protein interaction assays are, unfortunately, notorious for their false-positive results. Results derived from one method must therefore be reproduced with an independent technique.

In response to the potential for false-positive results under *in vitro* conditions, one recent trend is to study protein interactions *in vivo*. To this end, specific optical probes, such as fluorescent proteins or luciferase, are attached to the two suspected interaction partners using recombinant techniques. An exquisitely distance-dependent physical phenomenon, fluorescence (or bioluminescence) resonance energy transfer (FRET/BRET) can be exploited to address whether two cellular proteins do in fact interact physically in a live cell.

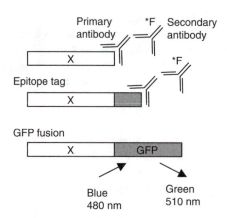

FIGURE 10.10 Determining the subcellular localization of a protein. Shown here are three approaches for visualizing the subcellular localization of protein X. Top: Indirect immunofluorescence using a primary antibody and a fluorescently labeled secondary antibody. Middle: A recombinant epitope tag (antibody binding site) is introduced into the protein to be detected. Bottom: GFP-fusion proteins are detected by exciting their fluorescence at 480 nm and detecting fluorescence emission at 510 nm.

PROTEIN LOCALIZATION

Many proteins contain one or more subcellular targeting signals that restrict where in the cell the protein will accumulate. Apart from their default location in the cytosol, proteins may be targeted to the chloroplast, to mitochondria, to the nucleus, to the endomembrane system, and to numerous more specific subcellular addresses. Experiments in visualizing the subcellular localization of a developmentally important gene product are informative for multiple reasons. First, the result will bolster or exclude certain hypotheses regarding its biological activity. For example, most proteins thought to bind DNA would be expected to reside in the nucleus. A protein thought to bind actin would be expected to be cytoplasmic. Second, the activity of many proteins is regulated at the level of subcellular localization. For instance, the *Arabidopsis* light regulatory protein, COP1, appears to be excluded from its site of action in the nucleus in response to light or developmental signals; meanwhile, the photoreceptor phytochrome is imported into the nucleus in a light-dependent fashion.

The cellular location of a gene product may be determined with the help of an antibody against the protein in conjunction with epifluorescence microscopy (Figure 10.10) or immuno-electron microscopy. Alternatively, especially when no specific antibody is available, the protein may be expressed transgenically as a fusion to a small peptide for which a specific antibody is commercially available (epitope tag). Antibodies are usually detected using a commercially available secondary antibody that carries a fluorescent or similar tag. Another powerful tag for studying the localization of proteins and their regulation is the green fluorescent protein (GFP) or its yellow, cyan, blue, and red siblings. For instance, Tai-Ping Sun and co-workers at Duke University demonstrated that the RGA protein, a negative regulator of gibberellic acid signaling, was excluded from the nuclei of *Arabidopsis* root cells and degraded in response to gibberellic acid (Dill et al., 2001). GFP can be visualized in live cells, thus opening the door to elegant real-time analysis of changes in protein localization in response to experimental treatments.

REVERSE GENETICS

Groundbreaking progress in developmental biology has been made based on the "forward genetic" paradigm. That is, phenotypic characterization of genetic lesions is followed by molecular cloning of the responsible gene and biochemical dissection of its activity. Reverse genetics makes use of the same tools but in the opposite order. That is, reverse genetics seeks to ascribe a phenotype and thus a biological role to a gene that has been defined initially at the molecular level.

The typical starting point for this effort is as follows: Standard forward genetics may have identified a novel gene. A thorough search of the plant's genome then reveals that the novel gene is simply the first known member of a larger gene family of structurally and evolutionarily related genes. The question arises, then: What are the biological roles of the remaining gene family members? Do they function in the same developmental pathway or not?

In exceptional cases, the protein sequence per se is informative. For example, once two physiologically defined phytochrome photoreceptor genes, *PHYA* and *PHYB*, had been cloned through the efforts of Robert Sharrock and Peter Quail, it became apparent that the *Arabidopsis* genome contains an additional three phytochrome-like genes, *PHYC, D*, and *E*. The notion that the *PHYC, D*, and *E* gene products are indeed photobiologically active phytochromes with special roles has since been confirmed.

However, more typically, protein sequence analysis will provide little if any clues with respect to the biological activity of otherwise anonymous gene family members. To illustrate this point, the *Arabidopsis* genome encodes a family of more than 500 proteins that share the "F-box" sequence motif. All that is known is that the F-box implicates the proteins in the specification of protein turnover events. However, it is impossible to predict from sequence data alone which biological process might be affected by a given F-box protein, or which specific proteins are the targets of a given F-box protein. To make progress, it is essential to generate or identify strains of plants in which the expression of the gene in question is altered, whether increased, decreased, abolished, or altered qualitatively. These strains can then be examined phenotypically for developmental abnormalities and can be subject to the molecular and biochemical gauntlets described earlier. Approaches include the following:

1. Overexpression and ectopic expression
2. Synthetic gain-of-function or loss-of-function alleles
3. Loss of function by RNA interference (RNA silencing)
4. Assembly of an allelic series by TILLING
5. Gene knockouts

Overexpression and Ectopic Expression

Simple overexpression of a gene involves inserting additional copies of the gene into the genome, with the reasonable assumption that the additional copies are regulated in the same fashion as the original copy. By comparison, ectopic expression refers to the artificial (over)expression of the gene in cell types where the gene is normally silent. Ectopic expression can have major consequences if the gene is involved in pattern formation. For example, when proteins of the *Knotted* family of DNA binding proteins, which are normally expressed predominantly in the shoot apex and the stem, are ectopically expressed in the leaf, striking abnormalities result. In maize, the result is outgrowth of extraneous tissue ("knots") on the leaf surface, whereas in tomato the leaves become excessively branched (more "compound"). These results provide evidence that the spatial restriction in the expression of *Knotted* in the stem contributes to cell fate specification in the stem versus the leaves.

Synthetic Alleles

Certain proteins lend themselves to the construction of synthetic gain- or loss-of-function alleles. As an example, the ethylene receptor, ETR1, is a member of a small gene family that includes the ethylene response sensor proteins ERS1 and ERS2. The functionality of the ERS proteins was explored in the Meyerowitz lab at Caltech by introducing specific amino acid substitutions into them, substitutions that were known to turn the ETR1 protein into a dominant gain-of-function allele. When introduced into *Arabidopsis*, the synthetic mutant alleles of ERS1 and ERS2 conferred the same ethylene insensitivity phenotype as did homologous changes in ETR1. This result indicates that ERS proteins are also ethylene receptors.

Aside from altering the protein sequence, one can create synthetic alleles by altering the regulation of gene expression. A trivial yet popular way of accomplishing this is by expressing a gene of interest under the control of a constitutive promoter such as the 35S promoter of cauliflower mosaic virus (CaMV 35S). More sophisticated are inducible expression systems. A variety of designs have been described, including heavy metal inducibility, ethanol inducibility, and heat shock inducibility. Perhaps the most widely adopted design is based on the glucocorticoid receptor, which provides inducibility of a target gene by externally added steroid hormones (Figure 10.6B). This strategy was pioneered by Alan Lloyd and Ron Davis at Stanford and has since been improved upon by the Chua laboratory at Rockefeller University.

RNA Interference (RNAi) and Related Gene Silencing Tools

RNAi is an RNA nuclease activity that results in the degradation of specific mRNAs in response to a trigger of double-stranded RNA (dsRNA). RNAi was originally characterized in plants under the terms post-transcriptional gene silencing (PTGS) and cosuppression. The molecular details of the RNAi process, which is thought to represent an antiviral defense mechanism, are still under investigation. Regardless, through RNAi, a specific mRNA sequence can be destroyed reliably. To this end, a transgene is constructed that drives expression of an inverted repeat of the mRNA, with an intron separating the two repeats. Upon transcription, the intron is spliced out and the RNA folds into a dsRNA hairpin structure, which all but abolishes the expression of its target gene.

Application of the RNAi technique is still in its infancy, yet it is gaining popularity rapidly. Cosuppression and RNAi can result in a variety of phenotypic effects, from mild to severe, mimicking the effects of a conventional allelic series. For example, cosuppression of the *Arabidopsis* light regulatory *COP1* gene can result in late phenotypes during the adult stage, which are mild, as well as early effects during the seedling stage, which may be lethal, just as observed with conventional *cop1* null alleles.

Assembly of an Allelic Series by TILLING

In order to collect a traditional allelic series of mutations in the gene of interest using a standard mutagen, such as EMS, Steven Henikoff and co-workers devised the TILLING technique. The acronym stands for *T*argeting *I*nduced *L*ocal *L*esions *IN* *G*enomes. *Arabidopsis* is first mutagenized using EMS, which causes point mutations. Subsets of *Arabidopsis* plants are used for PCR amplification of the gene of interest. The bulk PCR product from individual subsets of plants is examined for point mutations by searching for mismatches with the wild-type allele using a high-throughput analytical biochemistry technique. If a point mutation is present in the subset, iteratively smaller subpools of the mutagenized population are screened for the same mutation until a single plant with this new mutant allele has been identified. The mutant may then be examined for a possible developmentally informative phenotype. The great advantage of the TILLING technique is that mild alleles can be found. Mild alleles are crucial to understanding the late effects of genes that may be required throughout development.

Gene Knockouts

Last but not least, the first, best evidence for the function of an unknown gene comes from its "knockout" phenotype, that is, the phenotype of a plant in which this gene is missing. Other than in the mouse, there does not yet exist a reliable method for eliminating individual genes in the plant genome in a targeted fashion ("gene targeting"). However, at least in *Arabidopsis*, it is relatively easy to generate large collections of T-DNA insertion strains. As discussed for T-DNA tagging above, insertion of a T-DNA into a gene is likely to abolish the gene's function (Figure 10.3A). Such collections may now be screened, using a pooling strategy, for insertions in specific genes. Moreover, research centers affiliated with the Salk Institute in San Diego are in the process of mapping the precise insertion point of individual T-DNAs. The T-DNA locations of large numbers of insertion lines are being deposited in a publicly accessible database (http://www.signal.salk.edu), greatly accelerating reverse genetic analysis. Thus, a researcher looking for a knockout allele of a gene of interest may be able to obtain it with minimal effort on his behalf.

REVERSE GENETICS: CONCLUSION

In summary, reverse genetic tools have brought the goal within reach of characterizing the function of each of the approximately 25,000 genes in the *Arabidopsis* genome, the first plant genome to be sequenced. Many of these genes will have roles associated with the development of *Arabidopsis*. This goal is being pursued under the auspices of the National Science Foundation in its *Arabidopsis* 2010 project.

LITERATURE CITED

Alonso, J.M. and J.R. Ecker. 2001. The ethylene pathway: a paradigm for plant hormone signaling and interaction. *Sci STKE* 2001: RE1.

Chamovitz, D.A. et al. 1996. The COP9 complex, a novel multisubunit nuclear regulator involved in light control of a plant developmental switch. *Cell* 86:115–121.

Coen, E.S. and E.M. Meyerowitz. 1991. The war of the whorls: genetic interactions controlling flower development. *Nature* 353:31–37.

Dill, A., H.S. Jung, and T.P. Sun. 2001. The DELLA motif is essential for gibberellin-induced degradation of RGA. *Proc. Natl. Acad. Sci. U.S.A.* 98:14162–14167.

Gu, Q. et al. 1998. The FRUITFULL MADS-box gene mediates cell differentiation during *Arabidopsis* fruit development. *Development* 125:1509–1517.

Millar, A.J. et al. 1995. Circadian clock mutants in *Arabidopsis* identified by luciferase imaging. *Science* 267:1161–1163.

Nakajima, K. et al. 2001. Intercellular movement of the putative transcription factor SHR in root patterning. *Nature* 413:307–311.

Schindler, U. et al. 1992. Heterodimerization between light-regulated and ubiquitously expressed *Arabidopsis* GBF bZIP proteins. *EMBO J.* 11:1261–1273.

Sharma, V.K. and J.C. Fletcher. 2002. Maintenance of shoot and floral meristem cell proliferation and fate. *Plant Physiol.* 129:31–39.

Yanovsky, M.J. and S.A. Kay. 2002. Molecular basis of seasonal time measurement in *Arabidopsis*. *Nature* 419:308–312.

Section IV

Propagation and Development Concepts

11 Shoot Culture Procedures

Michael E. Kane

CHAPTER 11 CONCEPTS

- Shoot meristem cells retain the embryonic capacity for unlimited division.

- Isolated smaller meristem explants require more complex culture media for survival.

- Meristem and meristem tip culture are methods for disease eradication.

- Shoot culture provides a means to multiply periclinal chimeras.

- Cytokinins disrupt apical dominance and enhance axillary shoot production.

- Increased auxin concentration increases rooting percentage and root number, but decreases root elongation.

- Negative carryover effects of auxins used for Stage III rooting may affect *ex vitro* survival and growth of plantlets.

INTRODUCTION

Micropropagation is defined as the true-to-type propagation of selected genotypes using *in vitro* culture techniques. Four basic methods are used to propagate plants *in vitro*. Depending on the species and cultural conditions, *in vitro* propagation can be achieved by the following:

1. Enhanced axillary shoot proliferation (shoot culture)
2. Node culture
3. *De novo* formation of adventitious shoots through shoot organogenesis (Chapter 12)
4. Nonzygotic embryogenesis (Chapter 14)

Currently, the most frequently used micropropagation method for commercial production utilizes enhanced axillary shoot proliferation from cultured meristems. This method provides genetic stability and is easily attainable for many plant species. Consequently, the shoot culture method has played an important role in development of a worldwide industry that produces more than 350 million plants annually. Besides propagation, shoot meristems are cultured *in vitro* for two other purposes: the production of pathogen-eradicated plants, and the preservation of pathogen-eradicated germplasm (Chapter 28). Concepts related to propagation by shoot and node culture will be discussed in this chapter.

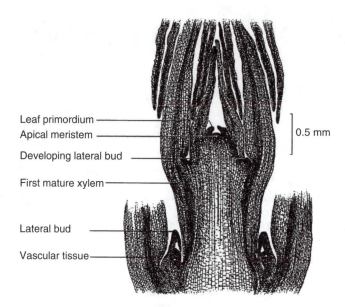

Leaf primordium

Apical meristem

Developing lateral bud

First mature xylem

Lateral bud

Vascular tissue

0.5 mm

FIGURE 11.1 Diagrammatic representation of a dicotyledonous shoot tip. The shoot tip comprises the apical meristem, subtending leaf primordial, and lateral buds.

SHOOT APICAL MERISTEMS

It is important to review briefly the general structure of shoot meristems. Shoot growth in mature plants is restricted to specialized regions that exhibit little differentiation, and in which the cells retain the embryonic capacity for unlimited division. These regions, called apical meristems, are located in the apices of the main and lateral buds of the plant. Cells derived from these apical meristems subsequently undergo differentiation to form the mature tissues of the plant body. Due to their highly organized structure, apical meristems tend to be genetically stable.

There exist significant differences in the shape and size of shoot apices among different taxonomic plant groups (Fahn, 1974). A typical dicotyledonous shoot apical meristem consists of a layered dome of actively dividing cells located at the extreme tip of a shoot, and measures about 0.1 to 0.2 mm in diameter and 0.2 to 0.3 mm in length. The apical meristem has no connection to the vascular system of the stem. Below the apical meristem, localized areas of cell division and elongation represent sites of newly developing leaf primordia (Figure 11.1). Lateral buds, each containing an apical meristem, develop within the axils of the subtending leaves. In the intact plant, outgrowth of the lateral buds is usually inhibited by apical dominance of the terminal shoot tip. Organized shoot growth from the apical meristem of plants is potentially unlimited and is said to be indeterminate. However, shoot apical meristems may become committed to the formation of determinate organs such as flowers (see also Chapters 7 and 13 for meristem structure).

IN VITRO CULTURE OF SHOOT MERISTEMS

The recognized potential for unlimited shoot growth prompted early, but largely unsuccessful, attempts to culture isolated shoot meristems aseptically in the 1920s. By the mid-1940s, sustained growth and maintenance of organization of cultured shoot meristems through repeated subculture was achieved for several species (see Chapter 2). Ball (1946), however, provided the first detailed procedure for the isolation and production of plants from cultured shoot meristem tips and the successful transfer of rooted plantlets into soil. Ball is often called the "Father of Micropropagation" because his shoot tip culture procedure is the most commonly used by commercial micropropagation laboratories today (see Chapter 27). Although these studies demonstrated the feasibility of regenerating shoots from cultured shoot tips, the procedures typically yielded unbranched shoots.

Several important findings facilitated application of *in vitro* culture techniques for large-scale clonal propagation from meristems. The discovery that virus-eradicated plants could be generated from cultured meristems lead to the widespread application of the procedure for routine fungal and bacterial pathogen eradication as well (Morel and Martin, 1952; Styer and Chin, 1984). Demonstration of rapid production of orchids from cultured shoot tips supported the possibility of rapid clonal propagation in other crops (Morel, 1960; 1965). It should be noted that *in vitro* propagation in many orchids does not occur via axillary shoot proliferation, but rather the cultured meristems become disorganized and form spheroid protocorm-like bodies that are actually nonzygotic embryos.

The final discovery was the elucidation of the role of cytokinins in the inhibition of apical dominance (Wickson and Thimann, 1958). This finding was eventually applied to enhance axillary shoot production *in vitro*. Application of this method was expedited by development of improved culture media that supported the propagation of a wide diversity of plant species (Murashige and Skoog, 1962; Lloyd and McCown, 1980).

MERISTEM AND MERISTEM TIP CULTURE

Although not directly used for propagation, meristem and meristem tip culture will be briefly described because these procedures are used to generate pathogen-eradicated shoots that subsequently serve as propagules for *in vitro* propagation. Culture of the apical meristematic dome alone (Figure 11.2) from either terminal or lateral buds, for the purpose of pathogen elimination, is termed meristem culture. In reality, true meristem culture is rarely used because isolated apical meristems of many species exhibit low survival rates and increased chance of genetic variability following callus formation and indirect shoot organogenesis.

Pathogen elimination can often be accomplished by culture of relatively larger (0.2–0.5 mm long) meristem tip explants excised from plants that have undergone thermo- or chemotherapy. The meristem tip is comprised of the apical meristem plus one or two subtending leaf primordia (Figure 11.2). This procedure is therefore termed meristem tip culture. Caution should be taken when interpreting much of the early published literature of successful "meristem" culture because in many instances, meristem tip or even larger shoot tip explants were actually used. The term "meristemming," commonly used in the orchid literature, is equally ambiguous.

SHOOT AND NODE CULTURE

Although not the most efficient procedure, propagation from axillary shoots has proven to be a reliable method for the micropropagation of a large number of species (Kurtz et al., 1991). Depending on the species, two methods, shoot and node culture, are used. Both methods rely on the stimulation of axillary shoot growth from lateral buds following disruption of apical dominance of the shoot apex. Shoot culture (shoot tip culture) refers to the *in vitro* propagation by repeated enhanced formation of axillary shoots from shoot tips or lateral buds cultured on medium supplemented with growth regulators, usually a cytokinin (George, 1993). The axillary shoots produced are either subdivided into shoot tips and nodal segments that serve as secondary explants for further proliferation or are treated as microcuttings for rooting. In some species, modified storage organs such as miniaturized tubers or corms (Figure 11.3) develop under inductive culture conditions from axillary shoots or rhizomes and may serve as the propagule for either direct planting or long-term storage.

When either verified pathogen-free stock plants are used or when pathogen elimination is not a concern, relatively larger (1–20 mm long) shoot tip or lateral bud primary explants (Figure 11.2)

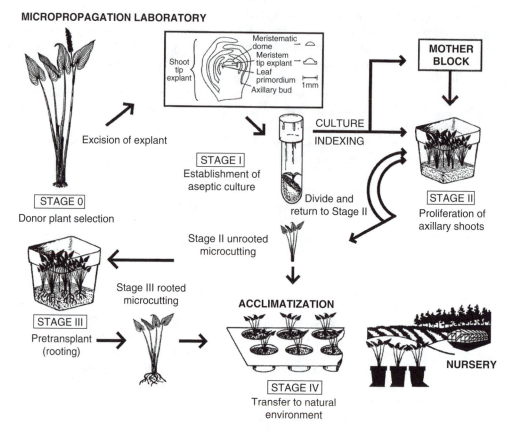

FIGURE 11.2 Micropropagation stages for production by shoot culture.

FIGURE 11.3 In some species, such as *Sagittaria latifolia* L., *in vitro* multiplication may occur through production of shoots (left) or corms (right), depending on culture conditions. Scale bar = 1 cm.

can be used for culture establishment and subsequent shoot culture. Advantages of using larger shoot tips include greater survival, more rapid growth responses, and the presence of more axillary buds. However, these larger explants are more difficult to completely surface sterilize and can potentially harbor undetected latent systemic microbial infection. Compared to other micropropagation methods, shoot cultures have the following characteristics: they provide reliable rates and consistency of multiplication following culture stabilization; they are less susceptible to genetic variation; and they may provide for clonal propagation of periclinal chimeras.

Node culture, a simplified form of shoot culture, is another method for production from preexisting meristems. Numerous plants such as potato (*Solanum tuberosum* L.) do not respond well to the cytokinin stimulation of axillary shoot proliferation observed in the micropropagation of many crops. Axillary shoot growth is promoted by the culture of either intact shoots (from meristem tip cultures) positioned horizontally on the medium (*in vitro* layering) or single or multiple node segments. Typically, single elongated unbranched shoots, comprised of multiple nodes, are rapidly produced. These shoots (microcuttings) are either rooted or acclimatized to *ex vitro* conditions or repeatedly subdivided into nodal cuttings to initiate additional cultures. Although node culture is the simplest method, it is associated with the least genetic variation.

MICROPROPAGATION STAGES

Murashige (1974) originally described three basic stages (I–III) for successful micropropagation. Recognition of the contamination problems often associated with inoculation of primary explants prompted Debergh and Maene (1981) to include a Stage 0. This stage described specific cultural practices, which maintained the hygiene of stock plants and decreased the contamination frequency during initial establishment of primary explants. As a result of our increased information base, it is now agreed that there are five stages (0–IV) critical to successful micropropagation. These stages not only describe the procedural steps in the micropropagation process, but also represent points at which the cultural environment is altered (Miller and Murashige, 1976). This system has been adopted by most commercial and research laboratories as it simplifies production scheduling, accounting, and cost analysis (Kurtz et al., 1991). Requirement for completion of each stage will depend on the plant material and specific method used. Diagrammatic representation of the micropropagation stages for propagation by shoot culture is provided in Figure 11.2.

STAGE 0: DONOR PLANT SELECTION AND PREPARATION

Explant quality and subsequent responsiveness *in vitro* is significantly influenced by the phytosanitary and physiological conditions of the donor plant (Debergh and Maene, 1981; Read, 1988). Prior to culture establishment, careful attention is given to the selection and maintenance of the stock plants used as the source of explants. Stock plants are maintained in clean, controlled conditions that allow active growth but reduce the probability of disease. Maintenance of specific pathogen-tested stock plants under conditions of relatively lower humidity, as well as the use of drip irrigation and antibiotic sprays, have proven effective in reducing the contamination potential of candidate explants. Such practices also permit excision of relatively larger and more responsive explants, often without increased risks of contamination.

Numerous practices are also employed to increase explant responsiveness by modifying the physiological status of the stock plant. These include the following:

1. Trimming to stimulate lateral shoot growth
2. Pretreatment sprays containing cytokinins or gibberellic acid
3. Use of forcing solutions containing 2% sucrose and 200 mg/l 8-hydroxyquinoline citrate for induction of bud break and delivery of growth regulators to target explant tissues (Read, 1988)

Currently, information on the effects of other factors such as stock plant nutrition, light, and temperature treatments on the subsequent *in vitro* performance of meristem explants is lacking.

STAGE I: ESTABLISHMENT OF ASEPTIC CULTURES

Initiation and aseptic establishment of pathogen eradicated and responsive terminal or lateral shoot meristem explants are the goals of this stage. The primary explants obtained from the stock plants may consist of surface-sterilized shoot apical meristems or meristem tips for pathogen elimination or shoot tips from terminal or lateral buds (Figure 11.2).

The presence of microbial contaminants can adversely affect shoot survival, growth, and multiplication rate. Bacteria and fungal contaminants often persist within cultured tissues that visually appear contaminant-free. Consequently, it is essential that Stage I cultures be indexed (screened) for the presence of internal microbial contaminants prior to serving as sources of shoot tip and nodal explants for Stage II multiplication.

The following factors may effect successful Stage I establishment of meristem explants: explantation time; position of the explant on the stem; explant size; and polyphenol oxidation. Time of explantation can significantly affect explant response *in vitro*. In deciduous woody perennials, shoot tips explants collected at various times during the spring growth flush may vary in their ability for shoot proliferation. Shoot tips collected during or at the end of the period of most rapid shoot elongation exhibited weak proliferation potential. Explants collected before or after this period are capable of strong shoot proliferation *in vitro* (Brand, 1993). Conversely, the best results are obtained with herbaceous perennials that form storage organs, such as tubers or corms, when explants are excised at the end of dormancy and after sprouting.

Explants also exhibit different capacities for establishment *in vitro* depending on their location on the donor plant. For example, survival and growth of terminal bud explants are typically greater than lateral bud explants. Often similar lateral meristem explants from the top and bottom of a single shoot may respond differently *in vitro*. In woody plants exhibiting phasic development, juvenile explants typically are more responsive than those obtained from the often unresponsive mature tissues of the same plant. Sources of juvenile explants include the following: root suckers, basal parts of mature trees, stump sprouts, and lateral shoots produced on heavily pruned plants.

The excision of primary explants often promotes the release of polyphenols and stimulates polyphenol oxidase activity within the damaged tissues. The polyphenol oxidation products often blacken the explant tissue and medium. Accumulation of these polyphenol oxidation products can eventually kill the explants. Procedures used to decrease tissue browning include the following: use of liquid medium with frequent transfer; adding antioxidants such as ascorbic acid or polyvinylpyrrolidone (PVP); and culture in reduced light or darkness.

There clearly is no one universal culture medium for establishment of all species. However, modifications to the Murashige and Skoog (MS) basal medium formulation (Murashige and Skoog, 1962) are most frequently used (see Chapter 3). Cytokinins or auxins are most frequently added to Stage I media to enhance explant survival and shoot development (Hu and Wang, 1983). The types and levels of growth regulators used in Stage I media are dependent on the species, genotype, and explant size.

Knowledge of the specific sites of hormone biosynthesis in intact plants provides insight into the relationship between explant size and dependence on exogenous growth regulators in the medium. Endogenous cytokinins and auxins are synthesized primarily in root tips and leaf primordia, respectively. Consequently, smaller explants, especially cultured apical meristem domes, exhibit greater dependence on medium supplementation with exogenous cytokinin and auxin for maximum shoot survival and development (Shabde and Murashige, 1977). Larger shoot tip explants usually do not require the addition of auxin in Stage I medium for establishment. Rapid adventitious rooting of shoot tip explants often provides a primary endogenous cytokinin source. Most Stage I media are agar-solidified and supplemented with at least a cytokinin (Wang and Charles, 1991). The most frequently used cytokinins are N^6-benzyladenine (BA), kinetin (Kin), and N^6-(2-isopentenyl)-adenine (2-iP). Due to its low cost and high effectiveness, the cytokinin BA is most widely used (Gaspar et al., 1996). Substituted urea compounds, such as thidiazuron, exhibit strong cytokinin-like activity and have been used to facilitate the shoot culture of recalcitrant woody species (Huetteman and Preece, 1993).

Many types of auxins are used. The naturally occurring auxin indole-3-acetic acid (IAA) is the least active, whereas the stronger and more stable compounds α-naphthalene acetic acid (NAA), a synthetic auxin and indole-3-butyric acid (IBA), a naturally occurring auxin, are more often used. Stage I medium PGR levels and combinations that promote explant establishment and shoot growth, but limit formation of callus and adventitious shoot formation, are selected.

A commonly held misconception is that primary explants exhibit immediate and predictable growth responses following inoculation. For many species, particularly herbaceous and woody perennials, consistency in growth rate and shoot multiplication is achieved only after multiple subculture on Stage I medium. Physiological stabilization may require from 3 to 24 months and 4 to 6 subcultures. Failure to allow culture stabilization, before transfer to a Stage II medium containing higher cytokinin levels, may result in diminished shoot multiplication rates or production of undesirable basal callus and adventitious shoots. With some species, the time required for stabilization can be reduced by initial culture in liquid medium.

In many commercial laboratories, stabilized cultures, verified as being specific pathogen tested and free of cultivable contaminants, are often maintained on media that limit shoot production to maintain genetic stability. These cultures, called mother blocks, serve as sources of shoot tips or nodal segments for initiation of new Stage II cultures (Figure 11.2).

STAGE II: PROLIFERATION OF AXILLARY SHOOTS

Stage II propagation is characterized by repeated enhanced formation of axillary shoots from shoot tips or lateral buds cultured on a medium supplemented with a relatively higher cytokinin level to disrupt apical dominance of the shoot tip. A subculture interval of 4 weeks with a three- to eightfold increase in shoot number is common for many crops propagated by shoot culture. Given these multiplication rates, conservatively, more than 4.3×10^7 shoots could be produced yearly from a single starting explant.

Stage II cultures are routinely subdivided into smaller clusters, individual shoot tips or nodal segments that serve as propagules for further proliferation (Figure 11.4a). Additionally, axillary shoot clusters may be harvested as individual unrooted Stage II microcuttings or multiple shoot clusters for *ex vitro* rooting and acclimatization (Figure 11.2). Clearly, Stage II represents one of the most costly stages in the production process.

Both source and orientation of explants can affect Stage II axillary shoot proliferation. Subcultures inoculated with explants that had been shoot apices in the previous subculture often exhibit higher multiplication rates than lateral bud explants. Inverting shoot explants in the medium can double or triple the number of axillary shoots produced on vertically oriented explants per culture period in some species.

The number of subcultures possible before initiation of new Stage II cultures from the mother block is required depending on the species or cultivar and its inherent ability to maintain acceptable multiplication rates while exhibiting minimal genetic variation and off-types (Kurtz et al., 1991). Some species can be maintained with monthly subculture from 8 to 48 months in Stage II. In contrast, for some ferns (*Nephrolepsis*), only three subcultures may be possible before the frequency of off-types increases to unacceptable levels.

Increased production of off-types is often attributed to production of adventitious shoots particularly through an intermediary callus stage (Jain and De Klerk, 1998). Stage II cultures, originally regenerating from axillary shoots, often begin producing adventitious shoots at the base of axillary shoot clusters after a number of subcultures on the same medium. These so-called mixed cultures

FIGURE 11.4 (a) Stage II shoot multiplication is achieved by repeated formation of axillary shoot clusters from explants containing lateral buds (*Aronia arbutifolia* L.). Depending on the species, individual microcuttings or shoot clusters may be rooted and acclimatized *ex vitro*. Scale bar = 1 cm. (b) For maximum survival, Stage III rooting may be required prior to acclimatization to *ex vitro* conditions. (c) Rooted and acclimatized Stage IV plantlets.

can develop without any morphological differences being apparent. Selecting only terminal shoots of axillary origin for subculture, instead of shoot bases, decreases the frequency of off-types, including the segregation of periclinal chimeras.

Selection of Stage II cytokinin type and concentration is made on the basis of shoot multiplication rate, shoot length, and frequency of genetic variation. Although shoot proliferation is enhanced at higher cytokinin concentrations, the shoots produced are usually smaller and may exhibit symptoms of hyperhydricity. Depending on the species, exogenous auxins may or may not enhance cytokinin-induced axillary shoot proliferation (Figure 11.5). Addition of auxin in the medium often mitigates the inhibitory effect of cytokinin on shoot elongation, thus increasing the number of usable shoots of sufficient length for rooting (Figure 11.6). This benefit must be weighed against the increased chance of callus formation. Similarly, shoot elongation in Stage II cultures may be achieved by adding gibberellic acid to the medium (Martin et al., 2002).

The possibility of adverse carryover effects on rooting of plantlets and survivability in Stage IV should be evaluated when selecting a Stage II cytokinin. For example, with some tropical foliage plant species, the use of BA in Stage II can significantly reduce Stage IV plantlet survival and rooting to as low as 10% (Griffis et al., 1981). Use of Kin or 2-iP instead of BA yields survival rates in excess of 90%. In some species, the adverse effect of BA on Stage IV survival and rooting has been attributed to production of an inhibitory BA metabolite that can be reduced by substituting with *meta*-topolin, a BA analogue (Werbrouck et al., 1995; 1996).

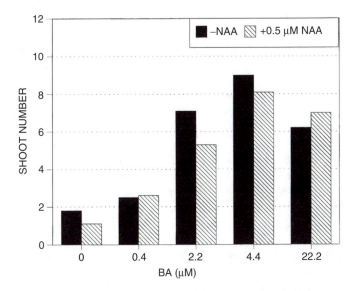

FIGURE 11.5 Effect of BA concentration on Stage II axillary shoot proliferation, produced from two-node explants of *Aronia arbutifolia* after 28-day culture, in presence and absence of 0.5 μ*M* NAA.

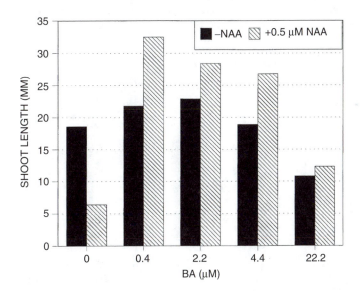

FIGURE 11.6 Inclusion of auxin may reduce the inhibitory effect of BA on axillary shoot elongation. Data shown for shoots generated from two-node explants of *Aronia arbutifolia* after 28-day culture, in presence and absence of 0.5 μ*M* NAA.

STAGE III: PRETRANSPLANT (ROOTING)

This step is characterized by preparation of Stage II shoots or shoot clusters for successful transfer to soil. The process may involve the following:

1. Elongation of shoots prior to rooting
2. Rooting of individual shoots or shoot clumps
3. Fulfilling dormancy requirements of storage organs by cold treatment
4. Prehardening cultures to increase survival

Where possible, commercial laboratories have developed procedures to transfer Stage II microcuttings to soil, thus bypassing Stage III rooting (Figure 11.2).

There are several reasons for eliminating Stage III rooting. Estimated costs for Stage III range from 35% to 75% of the total production costs. This reflects the significant input of labor and supplies required to complete Stage III rooting. Considerable cost savings can be realized if Stage III is eliminated. Furthermore, it has been observed that *in vitro*-formed root systems are largely nonfunctional, and die following transplanting. This results in a delay in transplant growth prior to production of new adventitious roots.

For various reasons, however, it may not always be feasible to transplant Stage II microcuttings directly to soil. Given the aforementioned limitations of Stage III rooting, Debergh and Maene (1981) proposed using Stage III solely to elongate Stage II shoots clusters prior to separation and rooting *ex vitro*. Elongated shoots may be further pretreated in an aqueous auxin solution prior to transplanting. Usually, Stage III rooting of herbaceous plants can be achieved on medium in the absence of auxins. However, with many woody species, the addition of an auxin (IBA or NAA) in Stage III medium is required to enhance adventitious rooting (Figure 11.4b) and plant performance *ex vitro*. Optimal auxin concentration is determined based upon percent rooting, root number and length (Figure 11.7). It is critical that the roots not be allowed to elongate to prevent root damage during transplanting. Care must be taken when selecting an auxin. For example, compared to IBA, use of NAA for Stage III rooting has been shown to decrease survival rates or suppress post-transplant growth (Conner and Thomas, 1981; Figure 11.8). A better understanding of the developmental and molecular basis of rooting and the mode of action of auxin will lead to more efficient rooting procedures, especially for problematic species (De Klerk, 2002).

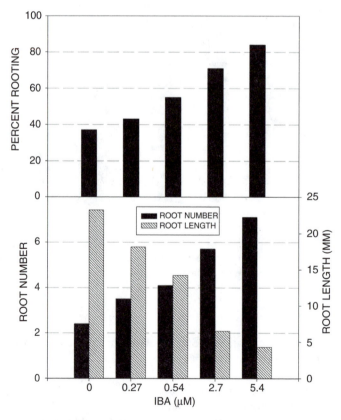

FIGURE 11.7 Effects of IBA on Stage III rooting of 10 mm *Aronia arbutifolia* microcuttings after 28-day culture. Increased IBA concentrations enhance rooting percentage but inhibit root elongation.

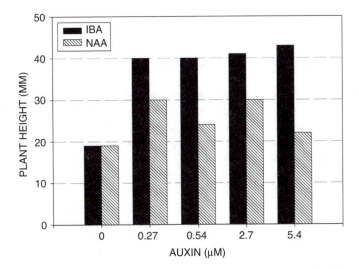

FIGURE 11.8 Comparative effects of Stage III rooting of *Aronia arbutifolia* microcuttings with IBA or NAA on post-transplant *ex vitro* growth, after 28 days.

STAGE IV: TRANSFER TO NATURAL ENVIRONMENT

The ultimate success of shoot or node culture depends on the ability to transfer and reestablish vigorously growing plants from *in vitro* to greenhouse conditions (Figure 11.4c). This involves acclimatizing or hardening off plantlets to conditions of significantly lower relative humidity and higher light intensity. Even when acclimatization procedures are carefully followed, poor survival rates are frequently encountered. Micropropagated plants are difficult to transplant for two primary reasons: a heterotrophic mode of nutrition, and poor control of water loss.

Plants cultured *in vitro* in the presence of sucrose and under conditions of limited light and gas exchange exhibit no or extremely reduced capacities for photosynthesis. Reduced photosynthetic activity is associated with low RubPcase activity. During acclimatization, there is a need for plants to make a rapid transition from the heterotrophic to the photoautotrophic state for survival (Preece and Sutter, 1991). Unfortunately, this transition is not immediate. For example, in cauliflower, no net increase in CO_2 uptake is achieved until 14 days after transplantation. This occurs only following development of new leaves, since the leaves produced *in vitro* in the presence of sucrose never develop photosynthetic competency. Interestingly, before senescencing, these older leaves function as "lifeboats" by supplying stored carbohydrate to the developing and photosynthetically competent new leaves. This is not the rule with all micropropagated plants because the leaves of some species become photosynthetic and persist after acclimatization.

A composite of anatomical and physiological features, characteristic of plants produced *in vitro* under 100% relative humidity, contribute to the limited capacity of micropropagated plants to regulate water loss immediately following transplanting. These features include reductions in leaf epicuticular wax, poorly differentiated mesophyll, abnormal stomate function, and poor vascular connection between shoots and roots.

To overcome these limitations, plantlets are transplanted into a well-drained "sterile" growing medium and maintained initially at high relative humidity and reduced light (40–160 $\mu M \times m^{-2} \times s^{-1}$) at 20–27°C. Relative humidity may be maintained with humidity tents, single tray propagation domes, intermittent misting, or fog systems. However, use of intermittent mist often results in slow plantlet growth following waterlogging of the medium and excessive leaching of nutrients. Transplants are acclimatized by gradually lowering the relative humidity over a 1- to 4-week period. Plants are incrementally moved to higher light intensities to promote vigorous growth. During this stage, container size and growing medium type can have a profound effect on the quality of the plants produced (Kane et al., 1990).

CONCLUSION

Propagation from preexisting meristems through shoot and node culture is the most reliable and widely used procedure. However, the need for multiple subcultures on different media makes shoot and node culture extremely labor-intensive. Total labor costs, typically ranging from 50 to 70% of production costs, limit expansion of the micropropagation industry. Current application of the technology is restricted to high-value horticultural crops, such as ornamental plants. Expansion of the industry to include production of vegetable, plantation, and forest crops depends on development of more efficient micropropagation systems. Cost-reduction strategies, including the elimination of production steps, and the development of reliable automated micropropagation systems, including bioreactor technology, will facilitate this expansion (Aitken-Christie et al., 1995).

LITERATURE CITED AND SUGGESTED READINGS

Aitken-Christie, J., T. Kozai, and S. Takayama. 1995. Automation in tissue culture — general introduction and overview. In: *Automation and Environmental Control in Plant Tissue Culture.* (Ed.) J. Aitken-Christie, T. Kozai, and M.A.L. Smith. Kluwer Academic Publishers, Dordrecht. 1–18.

Ball, E.A. 1946. Development in sterile culture of shoot tips and subjacent regions of *Tropaeolum majus* L. and of *Lupinus albus* L. *Amer. J. Bot.* 33:301–318.

Brand, M. 1993. Initiating cultures of Halesia and Malus: influence of flushing stage and benzyladenine. *Plant Cell Tissue Organ Cult.* 33:129–132.

Conner, A.J. and M.D. Thomas. 1981. Re-establishing plantlets from tissue culture: a review. *Comb. Proc. Intl. Plant Prop. Soc.* 31:342–357.

Debergh, P.C. and L.J. Maene. 1981. A scheme for commercial propagation of ornamental plants by tissue culture. *Sci. Hort.* 14:335–345.

Debergh, P.C. and P.E. Read. 1991. Micropropagation. In: *Micropropagation Technology and Application.* (Ed.) P.C. Debergh and R.H. Zimmerman. Kluwer Academic Publishers, Boston. 1–13.

De Klerk, G.J. 2002. Rooting of microcuttings: theory and practice. In Vitro *Cell. Dev. Biol.-Plant* 38:415–422.

Fahn, A. 1974. *Plant Anatomy.* Pergamon Press, New York.

Gaspar, T. et al. 1996. Plant hormones and plant growth regulators in plant tissue culture. In Vitro *Cell. Dev. Biol.-Plant* 32:272–289.

George, E.F. 1993. *Plant Propagation by Tissue Culture, Part 1: The Technology.* Exegetics, Ltd., London.

Griffis, J.L., Jr., G. Hennen, and R.P. Oglesby. 1981. Establishing tissue-cultured plant in soil. *Comb. Proc. Intl. Plant Prop. Soc.* 33:618–622.

Hu, C.Y. and P.J. Wang. 1983. Meristem, shoot tip and bud cultures. In: *Handbook of Plant Cell Culture, Volume 1: Techniques for Propagation and Breeding.* (Ed.) D.F. Evans et al. Macmillan Publishing, New York. 177–227.

Huetteman, C.A. and J.E. Preece. 1993. Thidiazuron: a potent cytokinin for woody plant tissue culture. *Plant Cell Tissue Organ Cult.* 33:105–119.

Jain, S.M. and G.J. De Klerk. 1998. Somaclonal variation in breeding and propagation of ornamental crops. *Plant Tissue Culture Biotech.* 4:63–75.

Kane, M.E. et al. 1990. Micropropagation of the aquatic plant *Cryptocoryne lucens. HortScience* 25:687–689.

Kurtz, S., R.D. Hartmann, and I.Y.E. Chu. 1991. Current methods of commercial micropropagation. In: *Scale-Up and Automation in Plant Propagation, Volume 8: Cell Culture and Somatic Cell Genetics of Plants.* (Ed.) I.K. Vasil. Academic Press, San Diego. 7–34.

Lloyd, G. and B. McCown. 1980. Commercially-feasible micropropagation of Mountain laurel, *Kalmia latifolia*, by use of shoot-tip culture. *Intl. Plant Prop. Soc. Proc.* 30:421–427.

Martin, G., M.E. Kane, and T. Yeager. 2002. Micropropagation of Sweet Viburnum (*Viburnum odoratissimum*). *Intl. Plant Prop. Soc. Proc.* 51: 614–619.

Miller, L.R. and T. Murashige. 1976. Tissue culture propagation of tropical foliage plants. *In Vitro* 12:797–813.

Morel, G. 1960. Producing virus-free Cymbidium. *Amer. Orchid Soc. Bull.* 29:495–497.

Morel, G. 1965. Clonal propagation of orchids by meristem culture. *Cymbidium Soc. News* 20:3–11.

Morel, G. and Martin, C. 1952. Guerison de Dahlias atteints d'une maladie à virus. *C.R. Acad. Sci. Ser. D.* 235:1324–1325.

Murashige, T. 1974. Plant propagation through tissue culture. *Annu. Rev. Plant Physiol.* 25:135–166.

Murashige, T. and F. Skoog. 1962. A revised medium for rapid growth and bioassays with tobacco tissue cultures. *Physiol. Plant.* 15:473–497.

Preece, J.E. and E.G. Sutter. 1991. Acclimatization of micropropagated plants to greenhouse and field. In: *Micropropagation Technology and Application.* (Ed.) P.C. Debergh and R.H. Zimmerman. Kluwer Academic Publishers, Boston. 71–93.

Read, P.E. 1988. Stock plants influence micropropagation success. *Acta. Hort.* 226:41–52.

Shabde, M. and T. Murashige. 1977. Hormonal requirements of excised *Dianthus caryophyllus* L. shoot apical meristem *in vitro*. *Amer. J. Bot.* 64:443–448.

Styer, D.J. and C.K. Chin. 1984. Meristem and shoot-tip culture for propagation, pathogen elimination, and germplasm preservation. *Hort. Rev.* 5:221–277.

Wang, P.J. and A. Charles. 1991. Micropropagation through meristem culture. In: *Biotechnology in Agriculture and Forestry, Volume 17: High-Tech and Micropropagation I.* (Ed.) Y.P.S. Bajaj. Springer-Verlag, Berlin. 32–52.

Werbrouck, S.P.O. et al. 1995. The metabolism of benzyladenine in *Spathiphyllum floribundum* 'Schott Petite' in relation to acclimatisation problems. *Plant Cell Rep.* 14:662–665.

Werbrouck, S.P.O. et al. 1996. *Meta*-topolin, an alternative to benzyladenine in tissue culture? *Physiol. Plant.* 98:291–297.

Wickson, M. and K.V. Thimann. 1958. The antagonism of auxin and kinetin in apical dominance. *Physiol. Plant.* 11:62–74.

12 Propagation from Nonmeristematic Tissues: Organogenesis

Otto J. Schwarz, Anjuna R. Sharma, and Robert M. Beaty

CHAPTER 12 CONCEPTS

- Organogenesis is the *de novo* formation of plant organs *in vivo* or *in vitro*, using either meristematic or nonmeristematic tissues.

- Some plant cells retain the ability to dedifferentiate from their current structural and functional state and begin a new developmental path toward other morphogenetic endpoints.

- Organogenesis can occur via two developmental sequences — indirect or direct.

- The process of organogenesis involves three main phases: dedifferentiation, induction, and differentiation.

- Adventitious root initiation can be divided into four stages: formation of meristematic locus, multiplication of cells to form a spherical cluster, cell multiplication resulting in a bilateral root meristem, and cell elongation resulting in emergence of the root.

- Root initiation and subsequent development result from a controlled and differential expression of genes.

INTRODUCTION

Organogenesis, the ability of plant tissues to form various organs *de novo*, has long been an object of interest and practical utility. The ancient Chinese successfully cloned selected genotypes of forest trees through the organogenic process of adventitious rooting of woody cuttings. In a recent historical review of adventitious rooting research, the practice of "cuttage," or the rooting of cuttings, was said to predate the time of Aristotle (384–322 BC) and Theophrastus (371–287 BC) (Haissig and Davis, 1994). The process of organogenesis provides the basis for asexual plant propagation largely from nonmeristematic somatic tissues. One needs only to visit a local garden center in the early spring or fall to witness the many varieties of woody and herbaceous plants offered for sale to the gardening public. Many of these plants are propagated exclusively via asexual means. The cloning of these selected genotypes probably included the *de novo* organ formation of roots on cuttage; the *in vitro* multiplication of preexisting shoot meristems followed by *de novo* root formation on the resultant microshoots; or possibly *de novo* shoot regeneration on *in vitro* cultured explant tissues followed by adventive rooting of the microshoots. Whether accomplished *in vivo* or *in vitro*, using either meristematic or nonmeristematic tissues, the *de novo* genesis of plant organs is broadly defined here as organogenesis. As stated above, our focus is specifically on "organogenesis *in vitro*" on nonmeristematic plant tissues. The reviews by Dore (1965) and Haissig and Davis (1994) can provide a more thorough account of the problems to be encountered in obtaining an all-encompassing definition of the terms associated with plant organ regeneration *in vitro* or *in vivo*.

The material presented in this chapter represents only a small fraction of the information available in the scientific literature concerning the developmental process of organogenesis. It is hoped that the concepts selected for discussion provide a model framework upon which to build an understanding of the morphogenic events that accompany organogenesis and, in addition, to begin to gain insight into the complex process that directs the morphogenetic process. The production of adventitious apical meristems on conifer cotyledonary tissue is presented in some detail as an example of the organogenic process, followed by a brief discussion of the process of adventitious root initiation.

ORGANOGENESIS AS A DEVELOPMENTAL PROCESS

In its broadest sense, development is the process that results in a functional, mature organism. According to Fosket (1994), it includes "all of those events during the life of a plant or animal that produce the body of the organism and give it the capacity to obtain food, to reproduce and to exploit the opportunities and deal with the hazards of its environment." Organogenesis is a developmental process that is in some ways unique to plants. Animal cells follow developmental paths that normally involve irreversible differentiation toward a specific cell or tissue type. In other words, animal cells and tissues remain structurally and functionally committed to an initial developmental endpoint. Plant cells, however, may retain the ability to dedifferentiate from their current structural and functional state and to begin a new developmental path toward a number of other morphogenetic endpoints.

Plant tissues *in vitro* may produce many types of primordia, including those that will eventually differentiate into embryos, flowers, leaves, shoots, and roots. For the purposes of this chapter, nonzygotic embryogenesis will not be discussed (see Chapters 14 and 15). Although the botanical definition of an organ is quite restrictive, in this discussion, we regard the development of any of the aforementioned primordia to be encompassed by the definition of organogenesis. These primordia originate *de novo* from a cellular dedifferentiation process followed by initiation of a series of events that results in their formation. The cell or cells thought to be the direct progenitors are somehow stimulated to undergo a number of rapid cell divisions leading to the formation of a meristemoid. Meristemoids are characterized as being an aggregation of meristem-like cells (i.e., smaller, isodiametric, thin-walled, micro-vacuolated cells with densely staining nuclei and cytoplasm; Thorpe, 1978; 1982). Early in their development, meristemoids are thought to be morphogenetically plastic and capable of developing into a number of different primordia (root, shoot, etc.). This unique developmental flexibility has been widely used by plant propagators. Two organogenic events comprise a common approach to *in vitro* plant propagation. The first event is to regenerate multiple shoot meristems; this is followed by their growth and development into microshoots of a size suitable for the second event, the induction of *de novo* root meristem production. It is believed that under *in vitro* culture conditions, these organogenic events can be the result of two differing ontogenic pathways.

According to Hicks (1980), two developmental sequences leading to organogenesis emerge from the literature. They differ in the presence or absence of a callus stage in the organogenic sequence of events. A developmental sequence involving an intervening callus stage is termed indirect organogenesis.

Primary Explant → Callus → Meristemoid → Organ Primordium

De novo organ formation via indirect organogenesis may increase the possibility of introducing variation in the chromosomal constitution (e.g., ploidy change) of the cells in the callus stage and, hence, the possibility for both physiological and morphogenic variation in the resulting organs. This type of variation has been termed somaclonal variation (Chapter 26). Although some plants are more prone to this problem than others, a general rule of thumb for plant propagation is to minimize any stage in the organogenic sequence containing "unorganized" callus growth. For the research scientist concerned with determining the physicochemical mechanisms driving the organogenic process, there is a second important reason to avoid or at least minimize this type of developmental event. The presence of a proliferating callus stage has the potential to complicate greatly the analysis of ongoing molecular events that accompany and perhaps drive *de novo* organ production (Chapter 13). Those particular cells or groups of cells that are destined to become organogenic are in relatively small number among the much larger mass of dividing callus cells, reducing the ability to analyze their particular chemistry and physiology.

Direct organogenesis is accomplished without an intervening proliferative callus stage. For *de novo* organogenesis (i.e., shoot, root, and flower), the sequence of events is described as follows:

Primary Explant → Meristemoid → Organ Primordium

The major distinction between these two *de novo* pathways is, of course, the presence or absence of a discernible callus stage early in the morphogenic sequence of events. The callus tissue, containing cells that have dedifferentiated into a less determined, morphologically more flexible form, serves as the starting point for *de novo* organogenesis. In the absence of a callus intermediate stage, cells present within the explant have the capacity to act as the direct precursors of the new primordium. Hicks (1980) presents a substantial listing of *in vitro* culture systems that provide examples of direct and indirect organogenesis from both meristematic and nonmeristematic explant tissues.

Researchers have utilized the organogenic process as a model system for asking questions about the causal events that direct shoot or root production and, by generalization, other morphogenic processes. In one such model, the process of organogenesis has been broken into several phases, illustrated in the following diagram (adapted from Christianson and Warnick, 1985).

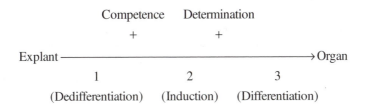

Key points of interest to developmental biologists are the two initial phases that precede the differentiation phase. These phases encompass events that begin with dedifferentiation, which results in the attainment of "competence," followed by induction, which culminates in the fully "determined" state. The morphological differentiation and development of the nascent shoot or root then proceed, eventually resulting in a functional organ. The first two phases of the model encompass those events that occur prior to morphogenesis. As shown by Hicks (1980), embryos, seedling parts, apical meristems, primordial organs, cell layers from mature organs, complex fragments from mature organs, intact organs, and protoplasts have been used for primary explants.

DEDIFFERENTIATION

The process of dedifferentiation involves reversion to a less committed, more flexible or plastic developmental state that may or may not give rise to callus tissue. In the case of direct organogenesis, competent cells that have not produced callus tissue are thought to be the sole organ progenitors, implying that they undergo the dedifferentiation process and proceed individually or perhaps in small groups to produce a new primordia. Leaf explants of *Convolvulus arvensis* L. produce roots or shoots via an indirect organogenic pathway that involves a dedifferentiation process that produces limited callus growth from which the new organs are generated (Christianson and Warnick, 1983). The result of completion of this first phase is that the primary explant acquires a state of competence, which is defined by its ability to respond to organogenic stimuli.

Attainment of tissue competence is not always a single-step process involving, for example, the application of plant growth regulators (PGRs). When cultured for 4 days on relatively high kinetin and low 2,4-dichlorophenoxyacetic acid (2,4-D) levels and then moved to a PGR-free medium, alfalfa (*Medicago sativa* L.) callus produces roots. Shoots are produced when callus is treated to the same culture regime, except that the medium contains high 2,4-D and low kinetin levels. However, the PGR pretreatment followed by transfer to basal, PGR-free medium is not sufficient to bring all of the tissue through the second (induction) phase that is required for eventual root or shoot production (Walker et al., 1979). An additional size requirement imposed a lower limit of 105 µm as the minimal diametric size for callus cellular aggregates to be competent for morphogenetic induction. Cellular aggregates allowed to grow to the minimum size became competent for organogenic induction and root or shoot formation.

INDUCTION

The induction phase occurs between the time the tissue becomes competent and the time it becomes fully determined for primordia production. This phase has been shown to encompass a number of "phenocritical" times, postulated to result from the function of an integrated gene pathway that guides the developmental process and precedes morphological differentiation. Landauer (1958) suggested that certain chemical and physical stimuli could interrupt a genetically determined developmental pathway, modifying the morphogenic result, which produces a mutant-like phenotype or "phenocopy." Christianson and Warnick (1984) have identified several chemical agents that intervene during the induction phase of *C. arvensis*. Agents that induce the production of a phenocopy are called phenocopying agents. These phenocopying agents were shown to be only effective during certain experimentally determined times during the induction phase. These phenocritical times were postulated to define key steps in the genetic pathway for orchestrating the developmental process. The phenocopying agents, or stage-specific inhibitors, were thought to block, in some direct way, the action of a gene or gene product of a currently active "orchestrating gene." In the *C. arvensis* system, this meant blocking shoot organogenesis and producing the modified phenotype (i.e., phenocopy); in this case, callus.

The end of the induction process is defined as the point when a cell or group of cells becomes fully committed to the production of shoots or roots. Operationally, this endpoint is attained when explant tissue can be removed from the root- or shoot-inducing medium; can be placed on basal medium without PGR, containing mineral salts, vitamins, and a source of carbon; and can proceed to produce the desired organs. The tissue has now completed the induction process and can be considered fully determined.

There is another term of interest that is used to describe or perhaps measure the extent of developmental commitment toward a particular endpoint (e.g., bud formation) that a tissue has reached during the induction phase. This term is "canalization," and it is defined by Waddington (1940) as "the property of developmental pathways of achieving a standard phenotype in spite of genetic or environmental disturbances." *C. arvensis* explant tissue that has become fully canalized for shoot production can be placed on media designed to induce roots and still form shoots (Christianson and Warnick, 1983). If the explant is removed from the shoot-inducing medium before it becomes fully canalized, shoot production is severely inhibited, with root production becoming the prominent endpoint. This developmental flexibility illustrates the degree of morphogenic plasticity that can be exhibited by plant tissue under *in vitro* culture conditions.

The extent that explant tissues become competent and eventually determined is, in part, a function of the physical and chemical environment to which they have been exposed. Under *in vitro* culture conditions, the properties of the medium and associated culture environment play a central role in delivering these organogenic signals. Skoog and Miller (1957) first demonstrated that a key system for the chemical regulation of *in vitro* organogenesis was the ratio of auxin to cytokinin present in the culture medium. They summarized their findings by concluding, "these tend to show that quantitative interactions between growth factors, especially between IAA and kinetin (auxins and kinins) and between these and other factors, provide a common mechanism for the regulation of all types of growth investigated from cell enlargement to organ formation." A great many other variables, such as day length, light quality and quantity, explant tissue age and size, genotype, and mineral nutrition, to mention a few, have been shown to affect regeneration ability. It is certainly possible that, as Christianson and Warnick (1988) suggest, "at least some cases of non-responsiveness to culture *in vitro* are simply due to the failure of one 'happy medium' to elicit competence for induction and to induce organogenesis." In its broadest sense, this idea suggests that a lack of tissue organogenic responsiveness under any particular set of experimental conditions is the result of failure of the explant tissues to achieve the state of competence for induction. This is not a particularly comforting situation for an investigator's state of mind, if one calculates all of the possible combinations and permutations involved in testing even a few of these parameters. Until appropriate biochemical or genetic markers are discovered that clearly indicate the current developmental disposition of the primary explant tissue, and until methods are discovered that can override the recalcitrant tissues' current state of genetic programming, our approach will have to remain an educated guess.

DIFFERENTIATION

Of the three phases proposed by the model, the third or differentiation phase is the best documented. It is in this phase of the process that morphological differentiation and development of the nascent organ begins. In describing the initial differentiation events of organ initiation, McDaniel (1984) states, "the general picture that emerges is that organ initiation involves an abrupt shift in polarity followed by a smoothing of this shift into a radially symmetrical organization and the concurrent growth along the new axis to form the bulge characteristic of organogenesis." These initial events, and those events leading to bud formation, are shown in pictorial series in Figures 12.1, 12.2, and 12.3. There has been a great deal of discussion in the literature concerning these initial differentiation events with respect to the tissue types involved and, perhaps more pointedly, to the precise number of cells that take part in meristem initiation. Again, it seems that there is not an absolute answer. Christianson (1987), basing his comments on his own experiments and those of Marcotrigiano and Gouin (1984a; 1984b) using plant chimeras, states, "shoots formed *in vitro* can arise from more than one cell in an explant, but they usually do not." Based upon extensive cytological investigation of the cellular events surrounding organogenic shoot initiation on immature zygotic embryos of sunflower (*Helianthus annuus* L.), Bronner et al. (1994) concluded that shoots developed from the "simultaneous divisions of several cells" and, therefore, were multicellular in origin. The epidermis and outer layers of the cortex of the immature embryos were found to contribute these germinal cells. The tissue origin and cell number involved in these very early differentiative events seem to depend upon a number of, as yet, poorly understood variables. Recognizable centers of cell division can occur deep within the explant tissues or, as just described for sunflower, can be more superficial in origin.

One can follow, through the careful use of microscopy, the internal histological and surface architectural changes that occur during the differentiation phase of the organogenic process (Figures 12.1, 12.2, and 12.3). The morphological sequence of events that is depicted tracks the *de novo* production of shoot buds by the process of direct organogenesis on cotyledons of the Central American pine (*Pinus oocarpa* Schiede). This series is presented as a more or less typical example of the organogenic process. Keep in mind that this is a singular example, and that all manner of variations of the developmental theme have been described for *in vitro* cultured plant species. However, the sequence of events that leads to organogenic shoot bud formation in the cotyledons of this conifer is probably generally applicable to most plant species producing adventitious buds (Thorpe, 1980; Yeung et al., 1981).

The final differentiation phase of organogenesis provides the first opportunity to observe the structural genesis of the new organ made possible by the developmental program that was put in place during the preceding induction phase. The sequence of morphogenic events observed in *de novo* bud formation on cotyledons of *P. oocarpa* included changes in surface morphology, the appearance of meristemoids, vertical and/or horizontal expansion of the meristematic region, protrusion of the meristematic region above the surrounding epidermis, the appearance of an organized meristem with leaf primordia, and full development of an adventitious bud. The explant was placed on induction medium containing the PGR benzyladenine (BA) for 10–14 days. It was then transferred to basal medium without PGR for the remainder of the experiment. In this particular experiment, the organogenic process was observed for approximately 40 days.

Histological sections of newly cultured cotyledons from day 0 illustrate the anatomy of untreated control tissue. In transverse section (Figure 12.1A), the cotyledons are triangular in outline, and the cells of the epidermis are more or less regular in conformation on all three sides of the cotyledon. The round mesophyll cells surrounding the central vascular tissue are compact and exhibit very little intercellular space. Scanning electron microscopy (SEM) micrographs of day 0 control cotyledonary tissue (Figure 12.2A) echo the regularity of the surface and near surface tissues shown in the histological sections. Even ranks of long, smooth-surfaced rectangular cells lie between irregular double and triple rows of stomata. The subsidiary cells surrounding the stomata are noticeably smaller in surface area than the long ranks of cells between the double and triple rows of epidermal cells. After 3 days on induction medium, the epidermal cells of the epistomatic cotyledonary surfaces have become irregular (Figure 12.1B). Small groups of meristemoids, mainly clustered at or just under the epidermis, are evident by 15–18 days (Figures 12.1A, C, D, E, and G). These meristematic zones enlarge both perpendicularly and parallel to the cotyledon surface, taking on a nodular appearance. During this time, as a result of mitotic activity, the epidermal cells lose much of their regularity and become uneven in length, forming small cell clusters (Figures 12.2C and 12.2D). These localized areas of meristematic activity begin to form meristematic domes (Figures 12.1F and 12.1H) that eventually protrude from the surface of the explant (Figure 12.3A). In this organogenic scheme, the ontogeny of the development of the dome structures involves incorporation of cells of the epidermis. In some conifers, the meristematic domes have been observed to push through the surface epidermal layer, causing the epidermis to rupture. The dome soon achieves a higher degree of organization and begins to produce leaf primordia (Figure 12.3B). Both periclinal and anticlinal divisions occur to the right and left of the central apical portion of the newly developed meristem, allowing for the successive production of leaf primordia (Figure 12.1H). The apex continues to grow, giving rise to numerous leaf primordia and culminating in the formation of an adventitious bud (Figures 12.1I and 12.3C).

The cytohistological zones typically associated with conifer seedling apical domes (Sacher, 1954) and apical meristem domes in developing conifer embryos (Fosket and Miksche, 1966) are clearly mirrored in this fully developed shoot bud (see legend, Figure 12.1I) produced *in vitro* by *de novo* organogenesis. The simultaneous development of numerous meristemoids on a single explant (Figure 12.3D) is not uncommon in *in vitro* organogenic culture systems. The degree of

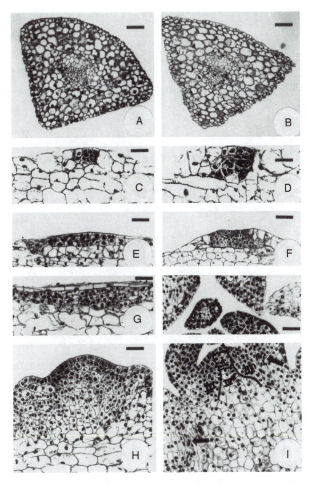

FIGURE 12.1 Histological micrographs of the differentiation phase of adventitious bud formation on seedling cotyledonary explants of a Central American pine (*Pinus oocarpa*). (A) Transverse section of the cotyledonary explant before exposure to a shoot bud induction medium. The epidermis has a regular, well-ordered appearance on all three sides. The more central mesophyll tissue is compact and has little intercellular space. Bar = 46 μm. (B) After 3 days on hormone-containing induction medium, the cells of the epistomatic surfaces have become irregular in size and shape, resulting in an uneven cotyledonary surface. Bar = 46 μm. (C) The presence of meristemoids is easily observed after 12–18 days and is denoted by the darkly stained nuclei in the group of small cells clustered at the top center of the micrograph. Bar = 28 μm. (D) The meristematic region rapidly increases in size. In this case, initial growth extends the meristemoid downward into the mesophyll tissue. Bar = 28 μm. (E) Meristematic growth may also occur laterally along the plane of the explant surface. Bar = 28 μm. (F) The rapidly developing bud meristem begins to protrude above the surrounding epidermis. Bar = 46 μm. (G) The dividing cells of the meristematic regions sometimes do not involve the epidermis. These regions may or may not develop into functional bud meristems. (H) A meristematic dome with several leaf primordia protrudes high above the surrounding epidermis. Note the numerous anticlinal divisions in the epidermis just to the left of the center of the dome and the periclinal divisions just to the right of the center of the dome. Bar = 25 μm. (I) Fully developed adventitious buds, complete with enclosing outer primordial leaves. Note the procambial strands in the area of the leaf primordia (arrows). The cytohistological zones associated with pine apical meristems are represented as follows: I: apical zone initials; II: central mother cell zone; III: peripheral tissue zone; IV: rib meristem. Bar = 28 μm. (From Bajaj, Y.P.S., [Ed.] *Biotechnology in Agriculture and Forestry and Trees*, Vol. III. Springer-Verlag, Berlin. 312. Used with permission.)

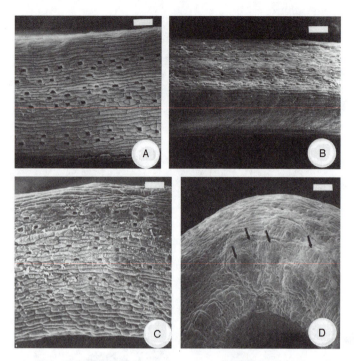

FIGURE 12.2 Scanning electron micrographs (SEMs) of the differentiation phase of adventitious bud formation on seedling cotyledonary explants of a Central American pine (*Pinus oocarpa*), showing developmental events prior to meristematic dome protrusion. (A) SEM of the explant prior to exposure to shoot bud induction medium. Stomata occur in rows of two and three surrounded by small subsidiary cells. Rows of elongated rectangular epidermal cells are interspersed between the rows of stomata. Bar = 83 μm. (B) The explant after 3 days of exposure to the induction medium. The abaxial side with almost no stomata and one of the epistomatic sides with its numerous stomata are just beginning to mirror the changes that have already begun in the subsurface tissue by a slight increase in roughness of the epidermal surface. Bar = 114 μm. (C) The cells of the epidermis have begun to divide causing an obvious disruption of the regular rows of cells between rows of stomata. Bar = 96 μm. (D) The explant surface has now become nodular and uneven. Disruption of the regular rows of cells has increased as the meristematic areas increase in size, producing nodules. Bar = 192 μm.

differentiation of these nascent bud meristems is highly variable. Buds in early to fully mature stages can be present on a single explant (Figure 12.3D).

The series of morphogenic events chronicled for *P. oocarpa* seems to fit the model presented for organogenesis via a direct developmental pathway. In practice, well-developed buds are removed from the cotyledons and are placed on a medium formulated to foster their continued growth and development. Elongated shoots, larger than 5 mm for *P. oocarpa* or 1 cm for other *Pinus* species, are transferred to either *in vitro* or *ex vitro* conditions designed to promote adventitious rooting, hence completing plantlet regeneration. The initiation of adventitious roots on *in vitro*-produced microcuttings is often much more than a trivial process. Because of its practical importance, this process is discussed in the following section.

ROOT SPECIFIC ORGANOGENESIS

As mentioned in the opening paragraph of this chapter, the production of root meristems from somatic tissues has a long history, which was driven strongly by the need to produce large quantities of high-quality plant materials. Horticulturists select specific plant genotypes for their outstanding

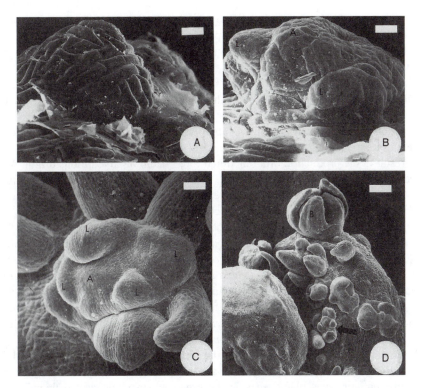

FIGURE 12.3 Scanning electron micrographs of the differentiation phase of adventitious bud formation on seedling cotyledonary explants of a Central American pine (*Pinus oocarpa*), showing meristematic dome protrusion to a well-developed adventitious bud. (A) Meristematic domes protrude well above the surrounding epidermis of the explant 2 to 3 weeks after beginning exposure to the shoot bud induction medium. Note the continuity of the epidermal surface over the entire dome. Bar = 83 μm. (B) A slightly more differentiated dome with newly formed leaf primordia. L indicates leaf primordia. A indicates the shoot apex. Bar = 71 μm. (C) Shoot bud with numerous leaf primordia and elongating leaves about 5 weeks after culture initiation. L indicates leaf primordia. A indicates the shoot apex. Bar = 156 μm. (D) A fully developed shoot bud (B), enclosed by outer primordial leaves. Elongation along the bud's central axis has begun. Arrow indicates another bud at an earlier stage of development. Bar = 714 μm.

floral or vegetative beauty, and foresters mark "plus trees" for their outstanding growth characteristics. Both of these types of selections have the common goal of capturing individual genotypes in order to produce large numbers of economically valuable clones. The ability to produce clones by taking conventional cuttings or by means of *in vitro*-based propagation systems, such as the one presented in this chapter, is often controlled by the efficiency with which one is able to induce adventitious rooting of these materials. Certain plant species, as well as many developmentally mature plant materials, are extremely recalcitrant with respect to the production of adventitious roots. Some of the challenges faced by those who wish to asexually propagate woody plants, whether by the ancient art of cuttage or by rooting microcuttings produced via *in vitro* propagation systems, are discussed in detail in several recent review articles (Blakesley et al., 1991; Blakesley and Chaldecott, 1993; Haissig et al., 1992; Wilson, 1994; Wilson and Van Staden, 1990). For those readers interested in this topic, these reviews can provide a window into the extensive literature covering this subject area.

The process of *de novo* root initiation that is adventitious in origin (arising from sites in the plant other than their normal sites in the embryo or primary root) has been generally described as consisting of the following four stages:

1. The formation of a meristematic locus by the dedifferentiation of a stem cell or cells
2. Multiplication of the stem cells into a spherical cluster
3. Further cell multiplication, with initiation of planar cell divisions, to form a recognizable bilateral root meristem
4. Cell elongation of those cells located in the basal part of the developing meristem, resulting in the eventual emergence of the newly formed root (Blakesley et al., 1991)

The selection of a microcutting of the appropriate developmental stage, coupled with its sequential exposure to an *in vitro* environment designed to initiate the sequence of events described above, may result in the production of a functional root system of adventitious origin. An example of the successful induction of adventitious roots on hypocotylary explants of pine cultured under *in vitro* conditions is shown in Figure 12.4. The explants were induced to root as the result of being co-cultured under *in vitro* conditions with a bacterium (root-stimulating bacterium [RSB]) found to stimulate rooting in a variety of woody and herbaceous plants (Schwarz and Burns, 1997). The control explants, cultured without the rooting bacterium, remained rootless 7 weeks after culture initiation. RSB shows similarity to the *Sphingomonas* group of bacteria, which have also been observed to induce rooting in mung bean (*Vigna radiata* L.) seedlings (A. Sharma and O. Schwarz, unpublished data). These bacteria synthesize indole compounds, including indole-3-acetic acid (IAA) and secrete them into the culture media from where they may play a role in the induction of adventitious roots in cuttings. Auxin is known to enter a plant through the cut surfaces by pH trapping (Rubery and Sheldrake, 1973) and influx carriers (Delbarre et al., 1996).

Not all cells are found to be competent for the formation of adventitious roots, and the site of root meristem initiation varies widely; however, according to Esau (1960), it usually occurs "in the vicinity of differentiating vascular tissues in the organ which gives rise to them." This site of adventitious root initiation places the newly formed root close to both the xylem and phloem of

FIGURE 12.4 Adventitious roots produced on conifer (*P. strobus*) explants after 7 weeks of *in vitro* co-culture with an adventitious root-stimulating bacterium (RSB), compared to untreated controls.

the parental axis, which, again according to Esau, "facilitates the establishment of vascular connection between the two organs." During the dedifferentiation phase, starch grains accumulate in cells from which the root primordia originate (De Klerk et al., 1999). It is during this time that the cells become competent to respond to external stimuli such as auxin. Following the dedifferentiation phase is the induction phase, during which the cells are most responsive to external rooting stimuli and undergo cell division giving rise to a meristemoid of about 30 cells. De Klerk et al. (1999) have observed a rapid degradation of starch grains during this phase in apple microcuttings. External stimulus, such as the presence of auxin, required for root formation, is not required beyond this stage and could be inhibitory for further growth. The meristemoid forms a dome-shaped root primordium during the differentiation phase and grows to form the root that emerges out of the stem. Recent articles by De Klerk et al. (1999; 2002) summarize the current progress in adventitious rooting.

Root primordia may also occur in relatively undifferentiated cells, such as those found in callus tissue sometimes produced at the basal/cut end of a microcutting. Usually these roots do not develop the vascular connections needed to sustain their above-ground microshoot partner. Production of adventitious roots in callus tissue almost always results in a regenerated microplant with little or no prospect for survival when transferred to *ex vitro* conditions.

If the four stages that chronicle the organogenesis of functional adventitious roots are considered in the light of the developmental model proposed by Christianson and Warnick (1985), both the dedifferentiation and induction phases of the model can be fit into the first and perhaps the second of the four stages. The extent to which the newly induced root meristem division centers (stage 2) are canalized has not often been shown experimentally. However, in a careful study of adventitious root production in *P. radiata*, Smith and Thorpe (1975a; 1975b) showed that the presence of the root stimulating agent (indolebutyric acid [IBA]) was required until meristemoid appearance (this is similar to stage 2, as described above). After this point, IBA was no longer required to complete adventitious root production.

It is apparent that the *Convolvulus* leaf disk system, described earlier in the chapter, is not completely analagous to adventive root initiation in *P. radiata*. The adventitious process in the *P. radiata* is thought to represent an example of direct organogenesis, its root meristem arising without an intervening callus stage. Remember that the *Convolvulus* example of Christianson and Warnick (1983) involved an indirect organogenic pathway in which the dedifferentiation process produced a limited callus growth from which the new organs were generated. Another interesting difference is that the *Convolvulus* tissues became fully determined prior to the initiation of any observable changes in the morphology of the dedifferentiated cells of the callus.

This is not the case in *P. radiata*, as the first event of root initiation was the formation of a discernable meristematic locus defined by the expansion of a single cell and the swelling of its nucleus. One or more of the cells surrounding the meristematic locus (peripheral cell) repartitions its cytoplasm and initiates unequal cell division, cutting off the smallest of the two daughter cells toward the meristematic locus. These anatomical/morphological events take place prior to the point of full canalization, which may be associated with the appearance of fully developed meristemoids. These meristemoids or spheres of meristematic tissues are the result of the continued division of the peripheral cells. Continued development of the meristemoid (stages 3 and 4) gives rise to fully developed adventitious roots. As with the genesis of shoot apical meristems, much work remains to be done in uncovering the mechanisms that control the sequence of morphological events required to produce a functional adventitious root.

A recent approach to solving the riddle of whole-plant root development involves the use of the step-by-step process of genetic analysis of single gene developmental mutants of *Arabidopsis* (Schiefelbein et al., 1997). These studies have lead to the hypothesis that the patterns observed for root development are "largely guided by positional cues that are established during embryogenesis." Simply stated, the cells within the developing root are reacting to cues resulting from the cells' position in relation to the other cells of the newly forming root. The molecular nature of these signals and their regulation is not clearly understood at this time. To better understand rooting activity and associated metabolic activities, several studies have focused on endogenous levels of IAA, polyamine, and peroxidase activity during different stages of adventitious root formation (Gasper et al., 1994; Nag et al., 2001). High levels of IAA and putrescein have been observed in cells during the inductive phase of organogenesis, followed by a decline during the later stages of root initiation. Low peroxidase activity was found in mung bean seedlings (Nag et al., 2001) during the inductive phase, followed by a reduced activity thereafter. Compounds released as a result of wounding are also thought to be involved in the dedifferentiation phase of rooting, enhancing the competence of the tissue to respond to plant hormones and possibly facilitating the uptake of auxin (Van der Krieken et al., 1997).

Development of roots results from a controlled and differential expression of several genes. A few genes specifically involved in lateral and adventitious root formation have been characterized. A hydroxyproline-rich glycoprotein gene (*HRGPnt3*) is expressed transiently in tobacco cells that are involved in root initiation (Vera et al., 1994). HRGPnt3 is a cell wall protein and is thought to modify the cell wall of those nascent root cells that penetrate the outer cell layers before root emergence. Expression of the gene lateral root primordium 1 (*LRP1*) has been observed in *Arabidopsis* during the early stages of root development and is turned off before the lateral roots emerge from the plant (Smith and Federoff, 1995). Thus the *LRP1* and *HRGPnt3* genes could serve as useful molecular markers for studying the early stages of lateral and adventitious root development.

Similarly, the root meristemless mutant (*rml*) of *Arabidopsis* is characterized by the growth arrest of lateral and adventitious roots at a size of less than 2 mm, and thus the *RML* gene could serve as an important marker for studying the later stages of root development (Cheng et al., 1995). Hemerly et al. (1993) observed a high expression of the gene *cdc2* in areas of active cell division where lateral or adventitious roots are formed. This expression is also seen during normal root development. To better understand the rooting process, Ermel et al. (2000) have initiated investigations that combine histological studies in walnut (*Juglans regia* L.) cotyledonary explants along with differential expression pattern of genes such as lateral root primordium (*LRP*), tonoplast induced protein (γ-*TIP*) and chalcone synthase gene (*CHS*), which serve as molecular markers for root primordium formation and differentiation. Histologically, the development of roots from cotyledonary explants without any hormone treatment by Ermel et al. (2000) were found to be similar to auxin stimulated roots of cotyledonary explants by Gutmann et al. (1996).

Molecular-based genetic analysis, applied to the genesis of adventitious roots, may be most fruitful with respect to increasing our understanding of the mechanisms controlling this process.

SUMMARY

The vast majority of published information concerning *in vitro* plant organogenesis has been based upon the intuitive application of technical procedures established through the venerable technique called trial and error. Although this process is by necessity of a descriptive nature, many careful investigations have had as their underlying purpose the elucidation of the "basic causes of cellular differentiation and organized development" (Earle and Torrey, 1965). For the developmental biologist, the investigative goal is, therefore, no less than the establishment of a mechanism for this inducible organogenic process. A great many of the almost infinite number of combinations and permutations of experimental variables have been tried, and results have been largely positive. That is, a great many plant species have responded with the *de novo* production of shoot and/or root meristems. There remain a good number of instances, however, in which experimental trial and error have been to no avail. Discovery of the key needed to unlock the developmental potential of these recalcitrant explant tissues must await the uncovering of the link between genetics and the production of three-dimensional form — i.e., the phenomenon of morphogenesis.

LITERATURE CITED AND SUGGESTED READINGS

Blakesley, D. and M.A. Chaldecott. 1993. The role of auxin in root initiation. Part II: Sensitivity, and evidence from studies on transgenic plant tissues. *Plant Growth Regul.* 13:77–84.

Blakesley, D., G.D. Weston, and J.F. Hall. 1991. The role of endogenous auxin in root initiation. Part I: Evidence from studies on auxin application, and analysis of endogenous levels. *Plant Growth Regul.* 10:341–353.

Bronner, R., G. Jeannin, and G. Hahne. 1994. Early cellular events during organogenesis and somatic embryogenesis induced on immature zygotic embryos of sunflower (*Helianthus annuus*). *Can. J. Bot.* 72:239–248.

Cheng, J.C., K.A. Seeley, and Z.R. Sung. 1995. *RML1* and *RML2, Arabidopsis* genes required for cell proliferation at the root tip. *Plant Physiol.* 107:365–376.

Christianson, M.L. 1987. Causal events in morphogenesis. In: *Plant Tissue and Cell Culture.* (Ed.) Alan R. Liss, New York, 45–55.

Christianson, M.L. and D.A. Warnick. 1983. Competence and determination in the process of *in vitro* shoot organogenesis. *Dev. Biol.* 95:288–293.

Christianson, M.L. and D.A. Warnick. 1984. Phenocritical times in the process of *in vitro* shoot organogenesis. *Dev. Biol.* 101:382–390.

Christianson, M.L. and D.A. Warnick. 1985. Temporal requirement for phytohormone balance in the control of organogenesis *in vitro*. *Dev. Biol.* 112:494.

Christianson, M.L. and D.A. Warnick. 1988. Organogenesis *in vitro* as a developmental process. *Hortic. Sci.* 23:515.

Delbarre, A., P. Muller, V. Imhoff and J. Guern. 1996. Comparison of mechanisms controlling uptake and accumulation of 2,4-dichlorophenoxy acetic acid, naphthalene-1-acetic acid, and indole-3-acetic acid in suspension-cultured tobacco cells. *Planta* 198:532–541.

De Klerk, G. J. 2002. Rooting of microcuttings: theory and practice. In Vitro *Cell Dev. Biol. Plant.* 38:415–422.

De Klerk, G.J., W.V.D. Krieken, and J.C. Jong. 1999. The formation of adventitious roots: New concepts, new possibilities (Review). In Vitro *Cell Dev. Biol. Plant.* 35:189–199.

Dore, J. 1965. Physiology of regeneration in carmophytes. In: *Encyclopedia of Plant Physiology*, Vol. XV/2. (Ed.) W. Ruhland. Springer-Verlag, Berlin. 1–91.

Earle, E.D. and J.G. Torrey. 1965. Morphogenesis in cell colonies grown from *Convolvulus* cell suspensions plated on synthetic media. *Amer. J. Bot.* 52:891–899.

Ermel, F. F. et al. 2000. Mechanisms of primordium formation during adventitious root development from walnut cotyledon explants. *Planta* 211:563–574.

Esau, K. 1960. *Plant Anatomy.* John Wiley & Sons, New York. 735.

Fosket, D.E. 1994. *Plant Growth and Development: A Molecular Approach.* Academic Press, New York. 580.

Fosket, D.E. and J.P. Miksche. 1966. A histochemical study of the seedling shoot meristem of *Pinus lambertiana. Amer. J. Bot.* 53:694–702.

Gasper, T. et al. 1994. Peroxidase activity and endogenous free auxin during adventitious root formation. In: *Physiology, growth and development of plants in culture.* (Ed.) P.J. Lumsden, J.R. Nicholas, and W.J. Davis. Kluwer Academic Publishers, Dordrecht, the Netherlands. 289–298.

Gutmann, M. et al. 1996. Histological studies of adventitious root formation from in-vitro walnut cotyledon fragments. *Plant Cell Rep.* 15:345–349.

Haissig, B.E. and T.D. Davis. 1994. A historical evaluation of adventitious rooting research to 1993. In: *Biology of Adventitious Root Formation.* (Ed.) T.D. David and B.E. Haissig. Plenum, New York. 275–332.

Haissig, B.E., T.D. Davis, and D.E. Riemenschneider. 1992. Researching the controls of adventitious rooting. *Physiol. Plant.* 84:310–317.

Hemerly, A. S. et al. 1993. *cdc2a* expression in *Arabidopsis thaliana* is linked with competence for cell division. *Plant Cell,* 5:1711–1723.

Hicks, G.S. 1980. Patterns of organ development in tissue culture and the problem of organ determination. In: *The Botanical Review*, Vol. 46. (Ed.) A. Cronquist. New York Botanical Garden, New York. 1–23.

Landauer, W. 1958. On phenocopies, their developmental physiology and genetic meaning. *Amer. Nat.* 92:201–213.

Marcotrigiano, M. and F.R. Gouin. 1984a. Experimentally synthesized plant chimeras. 1. In vitro recovery of *Nicotiana tabacum* L. chimeras from mixed callus cultures. *Ann. Bot.* 54:503–511.

Marcotrigiano, M. and F.R. Gouin. 1984b. Experimentally synthesized plant chimeras. 2. A comparison of *in vitro* and *in vivo* techniques for the production of interspecific *Nicotiana* chimeras. *Ann. Bot.* 54:513–521.

McDaniel, C.N. 1984. Competence, determination, and induction in plant development. In: *Pattern Formation, A Primer in Developmental Biology*. (Ed.) G.M. Malacinski and S.V. Bryant. Macmillan, New York. 393–411.

Nag, S., K. Saha, and M.A. Choudhuri. 2001. Role of auxin and polyamines in adventitious root formation in relation to changes in compounds involved in rooting. *J. Plant Growth Regul.* 20:182–194.

Rubery, P. H. and A. R. Sheldrake. 1973. Effect of pH and surface charge on cell uptake of auxin. *Nat. New Biol.* 244: 285–288.

Sacher, J.A. 1954. Structure and seasonal activity of the shoot apicies of *Pinus lambertiana* and *Pinus ponderosa*. *Amer. J. Bot.* 41:749–759.

Schiefelbein, J.W., J.D. Masucci, and H. Wang. 1997. Building a root: The control of patterning and morphogenesis during root development. *Plant Cell* 9:1089–1098.

Schwarz, O.J. and J.A. Burns. 1997. Bacterial stimulation of adventitious rooting in conifers. U.S.A. Patent #5,629,468.

Skoog, F. and C.O. Miller. 1957. Chemical regulation of growth and organ formation in plant tissues cultured *in vitro*. *Symp. Soc. Exp. Biol.* 11:118–140.

Smith, D. L. and N. V. Federoff. 1995. *LRP1*, a gene expressed in lateral and adventitious root primordial of Arabidopsis. *Plant Cell* 7:735–745.

Smith, D.R. and T.A. Thorpe. 1975a. Root initiation in cuttings of *Pinus radiata* seedlings I. Developmental sequence. *J. Exp. Bot.* 26:184–192.

Smith, D.R. and T.A. Thorpe. 1975b. Root initiation in cuttings of *Pinus radiata*. II. Growth regulator interactions. *J. Exp. Bot.* 26:193–202.

Thorpe, T.A. 1978. Physiological and biochemical aspects of organogenesis *in vitro*. In: *International Congress of Plant Tissue and Cell Culture, University of Calgary*. International Association for Plant Tissue Culture, Calgary. 49–58.

Thorpe, T.A. 1980. Organogenesis *in vitro*: Structural, physiological, and biochemical aspects. In: *Intl. Rev. Cyt. Suppl. 11a*. (Ed.) I.K. Vasil. Academic Press, New York. 71–111.

Thorpe, T.A. 1982. Physiological and biochemical aspects of organogenesis *in vitro*. In: *Proc. 5th Intl. Cong. Plant Tissue and Cell Culture*. Kluwer, Dordrecht. 121–124.

Van der Krieken, W. M. et al. 1997. Increased induction of adventitious rooting by slow release auxins and elicitors. In: *Biology of Root Formation and Development*. (Ed.) A. Altman and Y. Waisel. Plenum, New York and London. 95–105.

Vera, P., C. Lamb, and P.W. Doerner. 1994. Cell-cycle regulation of hydroxyproline-rich glycoprotein *HRGPnt3* gene expression during the initiation of lateral root meristems. *Plant J.* 6:717–727.

Waddington, C.H. 1940. The genetic control of wing development of *Drosophila*. *J. Genet.* 41:75–139.

Walker, K.A., M.L. Wendeln, and E.G. Jaworski. 1979. Organogenesis in callus tissue of *Medicago sativa*, the temporal separation of induction processes from differentiation processes. *Plant Sci. Lett.* 16:23.

Wilson, P.J. 1994. The concept of a limiting rooting morphogen in woody stem cuttings. *J. Hort. Sci.* 69:591–600.

Wilson, P.J. and J. Van Staden. 1990. Rhizocaline, rooting co-factors, and the concept of promoters and inhibitors of adventitious rooting — A review. *Ann. Bot.* 66:479–490.

Yeung, E.C. et al. 1981. Shoot histogenesis in cotyledon explants of *radiata* pine. *Bot. Gaz.* 142:494–501.

13 Molecular Aspects of *In Vitro* Shoot Organogenesis

Shibo Zhang and Peggy G. Lemaux

CHAPTER 13 CONCEPTS

- *In vitro* shoot organogenesis initiates from differentiated somatic cells.

- The shoot apical meristem (SAM) is located at the growing tip of the plant.

- Adventitious shoot meristems develop directly from plant tissues or *in vitro* cultures.

- Genes have been identified that function during *in vitro* organogenesis.

- Critical genes have been identified in hormone response, cell division, and meristem development.

INTRODUCTION

Shoot organogenesis is one of the pathways for the *in vitro* regeneration of a plant. An adventitious shoot is formed first, then adventitious roots are induced from the shoot, and finally, an entire plant develops. The induction and *in vitro* development of shoots via organogenesis have been reviewed from a developmental biology perspective (Hick, 1994) and is also described in Chapter 12. This chapter will focus mainly on the molecular aspects of *in vitro* shoot organogenesis.

In vitro shoot organogenesis is unique in that shoot meristems are initiated from differentiated somatic cells, not from embryonic cells as occurs during embryogenesis. *In vitro* shoot organogenesis includes the following main processes: response of somatic cells to plant hormones, cell division of the responding cells, and initiation and development of new shoots from the responding cells. With the appropriate response to exogenous plant hormones — i.e., plant growth regulators (PGRs) — somatic cells are able to turn on or speed up the timing of the cell cycle and to program progeny cells with a new developmental fate. Within the past 10 years, genetic and molecular analyses have identified a few of the critical genes involved in these various aspects of *in vitro* shoot organogenesis, namely the cytokinin signal transduction pathway, the regulation of cell cycle and cell division, and shoot meristem development and maintenance.

Cytokinin is the most efficient PGR (Chapter 8), in some cases combined with auxin, for induction of *in vitro* shoot organogenesis. Recently, a few critical molecular components involved in cytokinin signal transduction pathways have been identified, including *CRE1*, the cytokinin receptor, and its downstream genes, and *AHPs* and *ARRs* (Haberer and Kieber, 2002). In addition, genetic and molecular approaches have led to the identification of *ESR1* (Banno et al., 2001), the expression of which appears to be sufficient to induce adventitious shoot formation, even in the absence of PGR. The regulatory control of the cell cycle in plants involves, as it does in animals, the $p34^{cdc2}$ and cyclin gene products (Shaul, 1996). In addition, several genes, critical to regulation of development and maintenance of the shoot meristem, have been found. These include maize

KNOTTED1 (Vollbrecht et al., 1991) and the *Arabidopsis STM* (Long et al., 1996), *WUS* (Laux et al., 1996; Mayer et al., 1998), and *CLV1–3* (Fletcher et al., 1999) genes. There is also accumulating evidence that molecular interactions exist among the cytokinin, cell cycle, and shoot meristem developmental pathways.

MOLECULAR ASPECTS OF SHOOT MERISTEM DEVELOPMENT *IN VIVO*

The shoot meristem is initiated and develops within an embryo during the process of embryogenesis that occurs from a zygote. The initial shoot meristem becomes the shoot apical meristem (SAM) during later developmental stages because of its position at the growing tip of the plant. Except for the root, all lateral organs in a plant derive from the flanks of the SAM. As a result of histological analyses, the SAM is subdivided into three distinct zones, the peripheral, central, and rib zones (Figure 13.1). Lateral organs are derived from the peripheral zone, stem tissue from the rib zone. The central zone acts as a type of stem cell resource, which replenishes both the peripheral and rib zones while maintaining itself as the source of stem cells.

Genetic and molecular analyses of shoot meristem development in maize, *Arabidopsis* and other plants have led to identification of several genes that are critical to initiation and maintenance of the SAM. Maize *KNOTTED1* (*KN1*) was the first gene identified as being specifically expressed in the SAM during embryo development (Smith et al., 1995) and in the growing plant tip, but not in leaf primordia (Jackson et al., 1994). The expression pattern of the *KN1*-homolog in other cereals is the same, for example in barley (Figure 13.2). The gene *KN1* encodes a homeodomain protein (Vollbrecht et al., 1991), and ectopic expression of this protein or *KN1*-like homologs in tobacco and *Arabidopsis* causes the induction of adventitious shoot formation from leaf tissues (Lincoln et al., 1994; Sinha et al., 1993). In maize, loss-of-function mutations in *KN1* are defective in shoot meristem maintenance (Kerstetter et al., 1997).

The *SHOOT MERISTEMLESS* (*STM*) gene in *Arabidopsis* encodes a KN1-type homeodomain protein, and its expression is restricted in the SAM (Long et al. 1996). Loss-of-function mutations in *STM* in *Arabidopsis* result in the failure of shoot meristem development during embryogenesis. Long's study showed that *STM*, like *KN1* in maize, is critical in the development and maintenance of the shoot meristem. In addition to *STM,* a few more genes have been identified in *Arabidopsis* as being involved in shoot meristem development and maintenance. The gene *WUSCHEL* (*WUS*) encodes a novel subtype of homeodomain transcription factor (Mayer et al., 1998). In *WUS* mutants, shoot meristem development is arrested after formation of a few leaves during embryogenesis (Laux et al., 1996). Expression analysis showed that *WUS* is first expressed in *Arabidopsis* in subepidermal cells of the apex at the 16-cell stage of embryogenesis, and that gradually its expression is confined to subepidermal cells in the central zone of the post-embryonic SAM (Mayer et al., 1998). These

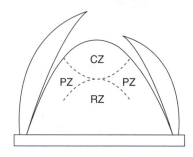

FIGURE 13.1 Diagram of a shoot apical meristem (SAM) showing the central zone (CZ), peripheral zone (PZ), and rib zone (RZ).

(a)   (b)

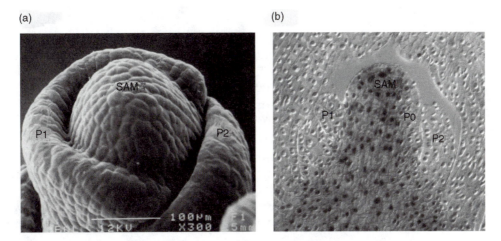

FIGURE 13.2 Expression of KNOTTED1-homolog in a barley shoot apical meristem (SAM). (a) Scanning electron micrograph of a barley shoot apex including the SAM and the two leaf primordia (P1, P2). (b) Immunolocalization of KNOTTED1-homolog in barley shoot apex using KNOTTED1 antibody shows expression of KNOTTED1-homolog only in the SAM, not in leaf primordia (P0, P1, and P2).

studies thus suggest it is possible that *WUS* is responsible for the initiation and maintenance of stem cells in the central zone of the SAM in *Arabidopsis*. *WUS* and *STM* in *Arabidopsis* are activated independently; however, *STM* expression is lost in *WUS* mutants and vice versa (Lenhard et al., 2002).

In addition to *STM* and *WUS*, which promote shoot meristem activity, a different group of genes in *Arabidopsis* has been identified as being responsible for restricting cell proliferation activity in the SAM; they are the *CLAVATA1*, *2*, and *3* (*CLV1–3*). Loss-of-function mutations at these loci cause an enlargement in the SAM and floral meristem (Leyser and Furner, 1992; Clark et al., 1993; Clark et al., 1995; Kayes and Clark, 1998). Genetic analyses indicate that products of *CLV1* and *CLV3* act in the same pathway to regulate cell proliferation in the SAM (Clark et al., 1995). The product of *CLV2* may act in the same meristem control pathway as those of *CLV1* and *CLV3*; however, it appears to function more broadly in plant development (Kayes and Clark, 1998). The *CLV1* gene encodes a leucine-rich region (LRR) transmembrane receptor serine/threonine kinase, which is expressed in the central region of the SAM and floral meristem (Clark et al., 1997). The nature of its protein-binding domain suggests that the *CLV1* receptor binds to an extracelluar protein or peptide ligand. Recently, *CLV3* has been found to encode a small protein, predicted to be extracellular, which is expressed at apex of the SAM and floral meristem, predominantly in L1 and L2 cell layers (Fletcher et al., 1999). Genetic, molecular and biochemical studies have provided strong evidence that *CLV3* acts as, or in the production of, the ligand for the *CLV1* receptor kinase (Rojo et al., 2002).

Therefore, the *CLV1–3* suite of genes and *WUS* in *Arabidopsis* act in a feeback loop to control stem cells in the SAM, in order to maintain its integrity as a continuous source of stem cells during plant development (Figure 13.3). Besides *STM*, *WUS*, and *CLV1–3*, many more genes have been found in *Arabidopsis* that are involved in shoot meristem development and maintenance, such as *UFO* and *CUC2*. The identity and details of these genes and their functions are already discussed in reviews (e.g., Fletcher and Meyerowitz, 2000; Lenhard and Laux, 1999).

In summary, shoot meristem development and maintenance is a complicated but well-regulated process with a large number of genes involved. Although certain aspects of the molecular mechanisms involved in shoot meristem initiation during embryo development are already known, much remains to be learned about the pathways in order for them to be fully elucidated.

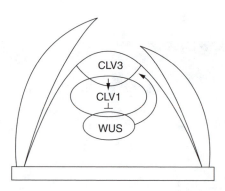

FIGURE 13.3 Diagram showing feedback loop controlling stem cell maintenance in shoot meristem. CLV3 is signalling through the CLV1 receptor complex to restrict WUS-expressing stem cell population. WUS provides a feedback signal to the overlying cells to maintain CLV3-expressing cell population in the superficial layers.

MOLECULAR ASPECTS OF PLANT CELL DIVISION

The basic molecular mechanisms controlling cell division in plants and animals appear to share functional similarities. The key regulatory gene products functioning during the plant and animal cell cycle are the cyclin-dependent kinases (CDKs) and the cyclins (CYCs) (Shaul, 1996).

CYCLIN-DEPENDENT KINASES

The first CDK gene to be identified, *cdc2,* was found in *Schizosaccharomyces pombe*, a yeast (Hindley and Phear, 1984). Recently, plant homologs were cloned from pea (Feiler and Jacobs, 1990), alfalfa (Hirt et al., 1991), maize (Colasanti et al., 1991), rice (Hashimoto et al., 1992), petunia (Bergounioux et al., 1992), soybean (Miao et al., 1993), *Arabidopsis* (Ferreira et al., 1991; Hirayama et al., 1991; Imajuku et al., 1992), and *Antirrhinum* (Fobert, 1994). Certain *cdc2* homologs from alfalfa, maize, rice, and soybean were able to complement, at least in part, yeast cell-cycle mutants, thereby proving functional equivalence; however, the plant genes could only complement mutations affecting certain phases of the yeast cell division cycle. This occurred because, unlike yeast, where the product of a single *cdc2* functions in all cell-cycle phases, plants have distinct *cdc2* genes that are active in different cell-cycle phases. The developmental regulation of the expression of one *cdc2* gene in *Arabidopsis* (*cdc2aAt*) was extensively analyzed and its expression correlated with actual cell division, and with competence for cell division (Ferreira et al., 1991; Martinez et al., 1992; Hemerly et al., 1993). Using *in situ* hybridization, expression of *cdc2aAt* in wild-type plants was found in root and shoot apical meristems, in developing first leaves, and at the root–shoot junction from which adventitious roots initiate; nearly undetectable levels of expression were found in fully expanded leaves. In the vegetative shoot apex, *cdc2aAt* expression corresponded precisely to the pattern of mitotic activity; in the root beyond the apical meristem, *cdc2aAt* expression was restricted to division-competent parenchymal and pericycle cells. A similar expression pattern of *cdc2Zm* was observed in maize (Colasanti et al., 1991).

Cell-cycle genes might play a role in mediating the effects of various PGRs. In leaf mesophyll protoplasts from transgenic tobacco plants containing a *cdc2aAt* promoter-*uidA* (β-glucuronidase [GUS]) fusion, expression of GUS was studied during dedifferentiation (Hemerly et al., 1993). Cultivation in the presence of either auxin or cytokinin alone caused *cdc2aAt*-mediated induction of GUS expression and resulted in division competence (despite the lack of observable cell division), presumably mediated through the hormonal action on the *cdc2aAt* gene. In wounded transgenic *Arabidopsis* leaves, rapid induction of *cdc2aAt* promoter-driven GUS expression occurred around damaged surfaces. Although no significant cell proliferation occurred, this induction might represent increased division competence, as was observed in tobacco. Correlation between *cdc2* transcript abundance and the increased proliferative state of tissue was also demonstrated in maize (Colasanti et al., 1991) and alfalfa (Hirt et al., 1991). Although *cdc2* expression is involved in proliferative competence, additional factors are likely required for cell division, e.g., an activating cyclin subunit.

CYCLINS

Cyclins were first identified in embryos of marine invertebrates (Evans et al., 1983); presently in mammals, eight different groups of cyclins (A–H) have been isolated. The A and B types, required for the cell to enter into mitosis, are designated mitotic cyclins. The cyclins C, D, and E are identified as G_1 cyclins (Lew et al., 1991). Among them, D-type cyclins are implicated in the exit from the quiescent state (G_0) and re-entry into the cell cycle (Quelle et al., 1993); E-type cyclins are required for progression from G_1 to S phases (Koff et al., 1992).

Mitotic-like cyclins (A and B types) have been isolated from several plant species, including soybean (Hata et al., 1991), carrot (Hata et al., 1991), alfalfa (Hirt et al., 1992), *Antirrhinum* (Fobert et al., 1994), maize (Renaudin et al., 1994), and tobacco (Qin et al., 1995). In *Arabidopsis* (Hemerly et al., 1992), soybean (Hata et al., 1991), and maize (Renaudin et al., 1994), the A- and B-type cyclins are believed to be functional homologs of animal mitotic cyclins because of their ability to induce maturation of *Xenopus* oocytes. *In situ* hybridization studies of seedling roots that looked at the spatial and temporal expression patterns of the mitotic-like *Arabidopsis* cyclins, particularly *cyc1At*, revealed that the *cyc1At* transcript is restricted primarily to the root apical meristem (Hemerly et al., 1993). Expression in pericycle cells was found only in zones of lateral root initiation even before morphological change occurred. This is similar to results in *Arabidopsis* plants containing the *cyc1At* promoter fused to GUS (Ferreira et al., 1994), in which a close correlation was found between GUS expression and mitotic activity in the shoot apical meristem, developing flowers and embryos. When GUS expression driven by the *cyc1At* promoter was studied during dedifferentiation of mesophyll protoplasts, expression of GUS was induced only when treatment with appropriate combinations of auxin and cytokinin resulted in cell division. However, it should be noted that various cyclins are expressed during the cell cycle and each cyclin might fulfill a different function (Ferreira et al., 1994; Fobert et al., 1994; Hirt et al., 1992).

The D-type cyclins in *Arabidopsis* (Soni et al., 1995) exhibit the closest similarity to mammalian D-type cyclins, which mediate the exit from the quiescent state (G_0) and re-entry into the cell cycle. To determine the timing of expression of the D3 cyclin during the cell cycle, *Arabidopsis* suspension cultures were treated with hydroxyurea or colchicine to block cells at the early S phase or metaphase, respectively (Soni et al., 1995). In both cases, a dramatic reduction in D3 cyclin expression was observed. When the hydroxyurea block was removed, induction of D3 cyclin occurred slightly before histone H4 induction. It remained high until the end of the S phase and during this time no significant induction of the mitotic *cyc1At* gene occurred. The transcriptional response of the D3 cyclin genes to PGRs was studied in *Arabidopsis* suspension cultures by depleting the medium of hormones and carbon source; transcript levels of D cyclin genes after readdition of these compounds varied depending on the plant mitogenic signal. In contrast to mitotic-like cyclins (A or B type), expression of the D type is not restricted to dividing cells and might be expressed earlier.

Based on the work on *cdc2aAt* and *cyc1At* in *Arabidopsis*, a simplified model for the role of cyclins during plant development and dedifferentiation was proposed (Shaul et al., 1996). This model proposes that the cell cycle genes are highly expressed in young, dividing tissues; however, the process of differentiation involves a reduction (but not elimination) in *cdc2aAt* expression and a cessation in *cyc1At* expression. In differentiated tissues, the level of *cdc2aAt* expression reflects the degree of division competence. Dedifferentiation and reacquisition of the potential to divide involves activation of *cdc2aAt*; however, other signals are probably required for division to occur.

MOLECULAR ASPECTS OF CYTOKININ: ITS RELATION TO CELL DIVISION AND SHOOT MERISTEM DEVELOPMENT

Cytokinin is the most effective PGR for induction of *in vitro* shoot organogenesis. It should be noted, however, that, as revealed by Skoog and Miller 45 years ago (Skoog and Miller, 1957), it is the ratio of cytokinin to auxin, rather than the absolute amount of either hormone, that is probably the critical factor in triggering developmental events. Within the past few years, significant progress has been made in understanding cytokinin metabolism, transport, perception, and signal transduction (for review, see Haberer and Kieber, 2002) and in unraveling the relationship of cytokinin with cell division and shoot meristem development.

CYTOKININ SIGNAL TRANSDUCTION

The first step in the response of cells to cytokinin involves the cytokinin receptor. Using an activation-tagging mutagenesis scheme, an *Arabidopsis* gene, *cytokinin-independent1* (*CKI1*), was cloned based on its ability to promote *in vitro* a typical cytokinin response, including rapid cell division and shoot formation without the application of exogenous cytokinin when *CKI1* was overexpressed (Kakimoto, 1996). *CKI1* encodes a protein similar to the two-component regulators (Chang and Stewart, 1998) with a histidine kinase domain and a receiver domain of histidine kinase, and was therefore suggested as a possible cytokinin receptor. However, the fact that it is a gain-of-function mutation does not lead to a definitive conclusion about *CKI1* function. Recently, a more confirmed cytokinin receptor, *CRE1*, has been identified in *Arabidopsis* from the screening of 19,000 ethyl methane sulfonate-mutagenized *Arabidopsis* plants. Cells from the *cre1* mutant displayed reduced sensitivity to cytokinin *in vitro* (Inoue et al., 2001). CRE1 is identical to the genes of *AHK4* (Suzuki et al., 2001; Ueguchi et al., 2001) and *WOL* (Mahonen et al., 2000) in *Arabidopsis*. The amino acid sequence of CRE1, which shares some weak homology to CKI1, is an unusual type of hybrid kinase with two C-terminal response regulator domains. It also contains a histidine kinase domain and two transmembrane regions flanking a novel, what is presumed to be an extracellular, domain. A series of experiments in yeast and *E. coli* provided evidence that cytokinin does act as a ligand for CRE1/AHK4 (Inoue et al., 2001; Suzuki et al., 2001; Yamada et al., 2001). In addition to CRE1, additional components involved in the cytokinin signal transduction pathway have been identified. The *Arabidopsis* histidine phosphotransfer protein (AHP) intermediates act as signaling shuttles between the cytoplasm and nucleus and this process is dependent on cytokinin (Miyata et al., 1998; Suzuki et al., 1998; Suzuki et al., 2000). The *Arabidopsis* response regulators (ARRs), based on their structures and cytokinin-inducible expression patterns, are classified into two distinct subtypes: type A, with 10 members, and type B, with 11 members (Imamura et al., 1999; D'Agostino and Kieber, 1999). A type-B ARR was found to function as a transcriptional activator for a type A (Hwang and Sheen, 2001; Sakai et al., 2001). These studies show that a multiple-step phosphorelay pathway is involved in cytokinin signal transduction:

$$\text{CRE1/AHK4/WOL} \rightarrow \text{AHPs} \rightarrow \text{ARRs}$$

Cytokinin transport might also play a role during the *in vitro* response to cytokinins, because in some cases the tissues or cells that respond by induction of shoot organogenesis were not those that were in contact with the induction media. For example, during *in vitro* culture of maize and barley vegetative shoots on a medium with a high ratio of cytokinin to auxin, the responding cells were those from the axillary shoot meristematic domes (Zhang et al., 1998) or in some maize inbreds in nodal regions (Zhang et al., 2002). The cells in contact with the induction medium at the cut edge of the stem did not respond. The existence of a cytokinin transporter is supported by studies of its role during plant growth. Based on the fact that cytokinin exists in the xylem sap and yet the root tip is the major site of cytokinin biosynthesis leads to the assumption that cytokinins are transported through the xylem to exert their effects on aerial parts of a plant. A putative cytokinin transporter gene, *AtPUP1*, has been isolated in *Arabidopsis* based on functional complementation of a yeast mutant deficient in adenine uptake (Gillissen et al., 2000).

CYTOKININ AND CELL DIVISION

Cytokinin was first identified as a factor that promoted cell division (Miller et al., 1955; Miller et al., 1956); however, the molecular mechanisms by which this occurs have not yet been fully elucidated. Genetic and molecular studies so far have identified a few probable links between cytokinin and cell division. A few studies have suggested that cytokinins play a role in the G2→M transition, although a decisive link has not yet been demonstrated. For example, cytokinin, by stimulating tyrosine dephosphorylation and activation of $p34^{cdc2}$-like H1 histone kinase, controlled the cell cycle at mitosis in tobacco protoplasts (Zhang et al., 1996). Also, cytokinins were shown to induce expression of *cdc2* in *Arabidopsis* roots and other tissues (Hemerly et al., 1993).

Recently, cytokinin was found to regulate the G1→S transition through a D type of cyclin (*CycD3*) (Riou-Khamlichi et al., 1999). The steady-state level of *CycD3* mRNA was induced both in seedlings and *in vitro* cultured cells by application of cytokinin. Also, expression of *CycD3* in the intact plant is localized in meristematic tissues, including the meristems of the shoot, leaf primordial, and axillary shoots. In the *Arabidopsis* mutant *amp1* (Chaudhury et al., 1993), there is an increased level of endogenous cytokinins, coupled with a higher steady-state level of *CycD3* mRNA. More intriguing to the study of *in vitro* plant development, leaf explants overexpressing *CycD3* were able to form healthy green calli without cytokinin application in contrast to explants from wild-type plants, which formed calli only in the presence of cytokinin. This study demonstrated that overexpression of *CycD3* can bypass the requirement for cytokinin to turn on the cell cycle, which implies that cytokinin regulates the cell cycle through *CycD3*.

CYTOKININ AND SHOOT MERISTEM DEVELOPMENT

Cytokinin or, more precisely, a higher ratio of cytokinin to auxin, is usually required to induce shoot organogenesis *in vitro*. Recently, genetic and molecular studies have suggested that cytokinin induces shoot meristem development through regulation of the expression of certain key genes, such as *KN1*-type homeobox genes. Transgenic plants overexpressing a cytokinin biosynthetic gene from bacteria, *ipt*, had increased levels of *KNAT1*, the *Arabidopsis* homolog of *KN1* and *STM* (Rupp et al., 1999). Also, elevated levels of *KNAT1* and *STM* were detected in the *Arabidopsis* mutant, *amp1*, with a higher level of endogenous cytokinin.

A seemingly converse relationship exists between cytokinin and *KN1*-type gene expression. Expression of KN1 under control of a senescence-specific promoter, SAG12, resulted in a delay of senescence, similar to that which occurs from expression of *ipt* under the control of the same promoter (Ori et al., 1999). Endogenous cytokinin levels were 15 times higher in older *SAG:KN1* leaves than in wild-type leaves, suggesting that KN1 may inhibit senescence by increasing cytokinin levels. In transgenic rice plants overexpressing *OSH1*, the rice *KN1* homolog, there is also an increased level of the active form of cytokinin (Kusaba et al., 1998). All these studies suggest that the levels of cytokinin and the expression of *KN1*-type genes may positively affect each other in an interdependent fashion. The relationship between the cytokinin and *KN1*-like genes was further investigated using a fusion protein between the homeodomain of KNAT2, a KN1-like protein in *Arabidopsis*, and the hormone-binding domain of the glucocorticoid (GR) receptor (Hamant et al., 2002). Upon activation of the KNAT2-GR fusion, delayed leaf senescence and a higher rate of shoot initiation were observed following cytokinin induction. This suggests that cytokinin acts synergistically with KNAT2 to regulate meristem activity in the shoot meristem.

MOLECULAR ASPECTS OF *IN VITRO* SHOOT MERISTEM DEVELOPMENT

The identification of critical genes involved in shoot meristem development and maintenance *in vivo*, e.g., maize *KN1*, *Arabidopsis STM*, *WUS*, and *CLV1–3*, raised the question of whether these genes also play a role during the process of *in vitro* shoot meristem development. Expression of KN1 or its homolog was studied in maize and barley during *in vitro* axillary shoot meristem proliferation and adventitious shoot meristem formation (Zhang et al., 1998).

Vegetative shoot segments from germinated seedlings of maize cv. H99 and barley cv. Golden Promise were cultured *in vitro* on a medium with higher levels of cytokinin (2.0 mg/l 6-benzyl-aminopurine [6-BA]) and lower levels of auxin (0.5 mg/l 2,4-dichlorophenoxyacetic acid [2,4-D]). After 2 to 4 weeks in culture, the normally well-regulated control of cell division within the axillary shoot meristems of the cultured shoots was broken and the cells within the meristematic domes divided quickly to produce an enlarged shoot meristematic dome (Figures 13.1b and 13.2b in Zhang et al., 1998). Adventitious shoot meristems (ADMs) were initiated directly from the enlarged shoot meristematic domes (Figures 13.1c and 13.2c in Zhang et al., 1998).

Expression analysis of KN1 in maize or the KN1-homolog in barley showed that the expression of the two genes was maintained within the shoot meristematic cells during *in vitro* cell proliferation of the axillary shoot meristems (Figures 13.1h and 13.2h in Zhang et al., 1998). The ADMs appeared to derive directly from KN1-expressing shoot meristematic cells (see Figure 13.4; also Figures 13.1i and 13.2i in Zhang et al., 1998). Also, the expression pattern of KN1 within the ADMs is the same as that of KN1 in the normal shoot apical meristem, in which KN1 is expressed throughout the entire meristematic dome, except the cells involved in leaf initiation. These experiments were the first to demonstrate that *in vitro* shoot meristem development likely follows the same pathways as *in vivo* shoot meristem development. Therefore, *KN1*-type genes can be useful as molecular markers to study *in vitro* shoot meristem induction and development.

The usefulness of the *KN1*-type genes as molecular markers during *in vitro* shoot organogenesis was confirmed further by studying the expression pattern of *Brostm*, a KN1-homolog in *Brassica*, during *in vitro* shoot induction in *Brassica oleracea* (Teo et al., 2001). Using an *in vitro* culture system in the presence of exogenous cytokinin, adventitious shoots could be induced directly from internodal stem segments (3–5 mm in length) of *B. oleracea* without intermediate callus formation. This induction system permitted an investigation into how cells in the stem segment could directly switch their developmental fate to become shoot meristematic cells. *Brostm* and *Brocyc* (a cyclin-type gene in *Brassica*) were used during various stages of shoot induction as molecular tags to

(a)   (b)

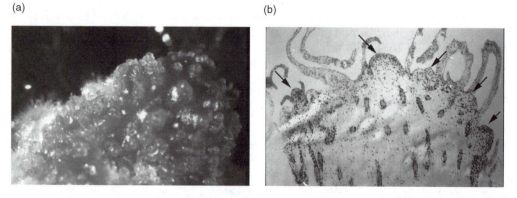

FIGURE 13.4 Expression of KNOTTED1 in the *in vitro* adventitious shoot meristems (ADMs) of maize. (a) Multiple ASMs induced *in vitro* from an enlarged shoot meristematic dome. (b) Expression of KNOTTED1 (arrows) in the *in vitro* ADMs.

mark, respectively, shoot meristematic cells and dividing cells. Analysis of the results showed that the process of developmental fate switching started in groups of 5–7 phloem parenchyma cells subtending the vascular bundles in the stem segment and before any cell division occurred. Expression of *Brostm* was induced specifically in response to the application of the cytokinin, 6-BA, within 4 hours of treatment, and its expression persisted during the cell proliferation process that leads to shoot formation. The cells expressing *Brostm* were found to be distinct from those expressing *Brocyc* during adventitious shoot formation.

The study of *Brassica* also demonstrated at the molecular level that there is a time gap in establishing cell determination. The developmental switching of cells in the stem segment was initiated within 4 hours of cytokinin treatment, as demonstrated by the initiation of expression of *Brostm*. However, 8 hours of cytokinin treatment was required for an actual shoot to develop. This result supports the hypothesis that *de novo* organogenesis involves a three-step process: competence acquisition, induction, and organogenesis determination (Christianson and Warnick, 1983; Christianson and Warnick, 1984; Christianson and Warnick, 1985). It also indicates that expression of KN1 or a KN1-homolog is not always sufficient to induce shoot meristem formation, as shown in earlier studies of overexpression of KN1 in maize leaf tissue (Zhang et al., 1998) and in barley awn tissue (Williams-Carrier et al., 1997).

Genetic and molecular approaches were recently used to identify genes that either regulate or are involved in adventitious shoot organogenesis. Using leaf explants undergoing adventitious shoot formation, a novel MADS box cDNA, *PkMADS1*, was isolated from a cDNA library of a woody tree species, *Paulownia kawakamii* (Prakash and Kumar, 2002). Analysis of the deduced amino acid sequence, corresponding to its MADS domain, revealed high similarity (90%) to the AGL24 (AGAMOUS-like) protein from *Arabidopsis* (Hartmann et al., 2000) and to the STMADS16 protein from potato (Carmona et al., 1998). Expression from *PkMADS1* was only detected in shoot-forming cultures of *Paulownia*, but not in callus-forming cultures. Transcripts from *PkMADS1* in plants were detected only in shoot apices, not in root apices, flowers, or the initial leaf explants. Transgenic *P. kawakamii* plants overexpressing the *PkMADS1* protein had phenotypic abnormalities, such as axillary shoot development. In the knockout, anti-sense transformants, shoots were stunted and had altered phyllotaxy, and in some cases, the shoot apical meristem did not develop. Leaf explants from the antisense transgenic plants had a tenfold decrease in shoot regeneration compared with explants from the overexpressing transformants or the wild type. These results indicate that expression from *PkMADS1* is necessary for shoot formation both *in vitro* and *in vivo*.

Because cytokinin is the most efficient PGR that can induce axillary or adventitious shoot organogenesis, genes involved in cytokinin metabolism or in the signal transduction pathway leading from cytokinin would be likely to have either a positive or negative effect on shoot organogenesis. Recently, an *Arabidopsis* gene, *enhancer of shoot regeneration 1* (*ESR1*), was identified by screening an *Arabidopsis* cDNA library for a gene that, when overexpressed in *Arabidopsis* root explants, confers the ability for cytokinin-independent shoot induction (Banno et al., 2001). Overexpression of the *ESR1* gene product also greatly enhanced the efficiency of shoot regeneration, following application of cytokinin (2-isopentyl adenine [2-iP]) on root explants, coupled with a shift in the optimal concentration of cytokinin to a lower level. *ESR1* encodes a putative transcription factor with an AP2/EREBP domain (Ohme-Takagi and Shinshi, 1995; Buttner et al., 1997; Hao et al., 1998; Fujimoto et al., 2000). When cultured on cytokinin-containing shoot induction medium, wild-type root explants had transiently elevated *ESR1* transcript levels during shoot regeneration. This transient increase occurred before *STM* expression was detected but after acquisition of competence for shoot regeneration on a 2,4-D-containing callus-induction medium. These results led to the suggestion that *ESR1* may be a downstream effector for cytokinin, enhancing the initiation of shoot regeneration after acquisition of competence for shoot organogenesis. Whether endogenous cytokinin levels were altered in the transgenic plants overexpressing the product of *ESR1* was not discussed. Over production of cytokinin in plant tissues, as occurs with the overexpression of the *ipt* gene product in tobacco (Hagen and Guilfoyl, 1992), usually enhances shoot regeneration capacity. High shoot regeneration capability was observed in the *Arabidopsis* mutant, *HOC*, which was shown to overproduce cytokinin (Catterou et al., 2002). Root explants of the *HOC* mutant, cultured *in vitro*, were able to develop many shoots in the absence of exogenous growth regulators. The *HOC* locus is likely involved in cytokinin metabolism, leading to cytokinin overproduction in this mutant.

SUMMARY

Molecular analysis of *in vitro* shoot organogenesis has so far supported the developmental model proposed by Christianson and Warnick (1985) in which shoot organogenesis is conceptually divided into three phases: competence, induction, and determination. There is a competence stage before cells can be induced to form a shoot. However, reliable molecular markers are not available yet to mark this stage. To date, the best candidates available are probably CDC2 and CycD-3, which indicate the potential for cell division. For the stages of induction and determination, the maize *KN1*-type homologs are useful molecular markers to identify the early stages in the induction of shoot organogenesis *in vitro* and its expression is maintained during later shoot development. At present, we understand much more about how cytokinin promotes cell division. Plant cells are responding to cytokinin through a multiple-step signal transduction pathway, which involves CRE1/AHK4/WOL as cytokinin receptors, AHPs as phosphotransmitters, and ARRs as the response regulators. To turn on or speed up the cell cycle, cytokinin likely acts through activation of CycD-3; however, the connection between the signal transduction pathway and activation of CycD-3 has still not been elucidated. Molecular evidence exists as to why cytokinin, or a higher ratio of cytokinin to auxin, favors induction of shoot organogenesis, such as the positive relationship between the level of cytokinin and the expression of KN1-type homeobox genes, which are involved in shoot meristem development. Future advances in the study of the molecular mechanisms involved in the dedifferentiation of somatic cells and the acquisition of competence for shoot organogenesis *in vitro* will truly contribute to our understanding of the flexibility of plant cell development.

LITERATURE CITED AND SUGGESTED READINGS

Banno, H. et al. 2001. Overexpression of *Arabidopsis* ESR1 induces initiation of shoot regeneration. *Plant Cell* 13:2609–2618.

Bergounioux, C. et al. 1992. A Cdc2 gene of petunia-hybrida is differentially expressed in leaves protoplasts and during various cell cycle phases. *Plant Molec. Biol.* 20:1121–1130.

Buttner, M. and K.B. Singh. 1997. *Arabidopsis thaliana* ethylene-responsive element binding protein (AtEBP), an ethylene-inducible, GCC box DNA-binding protein interacts with an ocs element binding protein. *Proc. Natl. Acad. Sci. U.S.A.* 94:5961–5966.

Carmona, M.J., N. Ortega, and F. Garcia-Maroto. 1998. Isolation and molecular characterization of a new vegetative MADS-boxgene from *Solanum tuberosum* L. *Planta* 207:181–188.

Catterou, M. et al. 2002. hoc: An *Arabidopsis* mutant overproducing cytokinins and expressing high *in vitro* organogenic capacity. *Plant J.* 30:273–287.

Chang, C. and R.C. Stewart. 1998. The two-component system. *Plant Physiol.* 117:723–731.

Chaudhury, A.M. et al. 1993. amp1: A mutant with high cytokinin levels and altered embryonic pattern, faster vegetative growth, constitutive photomorphogenesis and precocious flowering. *Plant J.* 4:907–916.

Christianson, M.L. and D.A. Warnick. 1983. Competence and determination in the process of *in vitro* shoot organogenesis. *Dev. Biol.* 95:288–293.

Christianson, M.L. and D.A. Warnick. 1984. Phenocritical times in the process of *in vitro* shoot organogenesis. *Dev. Biol.* 101:382–390.

Christianson, M.L. and D.A. Warnick. 1985. Temporal requirement for phytohormone balance in the control of organogenesis *in vitro*. *Dev. Biol.* 12:494–497.

Clark, S.E., M.P. Running, and E.M. Meyerowitz. 1993. CLAVATA1, a regulator of meristem and flower development in *Arabidopsis*. *Development* 119:397–418.

Clark, S.E., M.P. Running, and E.M. Meyerowitz. 1995. CLAVATA3 is a specific regulator of shoot and floral meristem development affecting the same processes as CLAVATA1. *Development* 121:2057–2067.

Clark, S.E., R.W. Williams, and E.M. Meyerowitz. 1997. Control of shoot and floral meristem size in *Arabidopsis* by a putative receptor-kinase encoded by the *CLAVATA1* gene. *Cell* 89:575–585.

Colasanti, J., M. Tyers, and V. Sundaresan. 1991. Isolation and characterization of complementary DNA clones encoding a functional p34[cdc2] homologue from *Zea mays*. *Proc. Natl. Acad. Sci. U.S.A.* 88:3377–3381.

Colasanti, J. et al. 1993. Localization of the functional p34[cdc2] homolog of maize in root tip and stomatal complex cells: Association with predicted division sites. *Plant Cell* 5:1101–1111.

D'Agostino, I.B. and J.J. Kieber. 1999. The emerging family of plant response regulators. *Trends Biochem. Sci.* 24:452–456.

Evans, T. et al. 1983. Cyclin: a protein specified by maternal mRNA in sea urchin eggs that is destroyed at each cleavage division. *Cell* 33:389–396.

Feiler, H.S. and T.W. Jacobs. 1990. Cell division in higher plants: A cdc2 gene, its 34-kDa product, and histone H1 kinase activity in pea. *Proc. Natl. Acad. Sci. U.S.A.* 87:5397–5401.

Ferreira, P.C.G. et al. 1991. The *Arabidopsis* functional homolog of the p34[cdc2] protein kinase. *Plant Cell* 3:531–540.

Ferreira, P.C.G. et al. 1994. Developmental expression of the *Arabidopsis* cyclin gene cyc1At. *Plant Cell* 6:1763–1774.

Fletcher, J.C. et al. 1999. Communication of cell fate decisions by CLAVATA3 in *Arabidopsis* shoot meristems. *Science* 283:1911–1914.

Fletcher, J.C., and E.M. Meyerowitz. 2000. Cell signaling within the shoot meristem. *Curr. Opin. Plant Biol.* 3:23–30.

Fobert, P.R. et al. 1994. Patterns of cell division revealed by transcriptional regulation of genes during the cell cycle in plants. *EMBO* 13:616–624.

Fujimoto, S. Y. et al. 2000. *Arabidopsis* ethylene-responsive element binding factors act as transcriptional activators or repressors of GCC box-mediated gene expression. *Plant Cell* 12:393–404.

Gillissen, B. et al. 2000. A new family of high-affinity transporters for adenine, cytosine, and purine derivatives in *Arabidopsis*. *Plant Cell* 12:291–300.

Haberer, G. and J.J. Kieber. 2002. Cytokinins: New insights into a classic phytohormone. *Plant Physiol.* 128:354–362.

Hagen, L.Y. and T.J. Guilfoyl. 1992. Altered morphology in transgenic tobacco plants that overproduced cytokinins in specific tissues and organs. *Dev. Biol.* 153:386–395.

Hamant, O. et al. 2002. The KNAT2 homeodomain protein interacts with ethylene and cytokinin signaling. *Plant Physiol.* 130:657–665.

Hao, D., M. Ohme-Takagi, and A. Sarai. 1998. Unique mode of GCC box recognition by the DNA-binding domain of ethylene-responsive element-binding factor (ERF domain) in plant. *J. Biol. Chem.* 273:26857–26861.

Hartmann, U. et al. 2000. Molecular cloning of SVP: A negative regulator of the floral transition in *Arabidopsis*. *Plant J.* 21:351–360.

Hashimoto, J. et al. 1992. Isolation and characterization of cDNA clones encoding cdc2 homologues from *Oryza sativa*: A functional homologue and cognate variants. *Molec. Gen. Genet.* 233:10–16.

Hata, S. et al. 1991. Isolation and characterization of complementary DNA clones for plant cyclins. *EMBO* 10:2681–2688.

Hemerly, A. et al. 1992. Genes regulating the plant cell cycle: Isolation of a mitotic-like cyclin from *Arabidopsis thaliana*. *Proc. Natl. Acad. Sci. U.S.A.* 89:3295–3299.

Hemerly, A.S. et al. 1993. Cdc2a Expression in *Arabidopsis* is linked with competence for cell division. *Plant Cell* 5:1711–1723.

Hick, G. S. 1994. Shoot induction and organogenesis *in vitro*: A developmental perspective. *In vitro Cell. Devel. Biol. Plant* 30:10–15.

Hindley, J. and G.A. Phear. 1984. Sequence of the cell division gene *CDC2* from *Schizosaccharomyces pombe*: patterns of splicing and homology to protein kinases. *Gene* 31:129–134.

Hirayama, T. et al. 1991. Identification of two cell-cycle-controlling cdc2 gene homologs in *Arabidopsis thaliana*. *Gene* (Amsterdam) 105:159–166.

Hirt, H. et al. 1991. Complementation of a yeast cell cycle mutant by an alfalfa complementary DNA encoding a protein kinase homologous to p34[cdc2]. *Proc. Natl. Acad. Sci. U.S.A.* 88:1636–1640.

Hirt, H. et al. 1992. Alfalfa cyclins differential expression during the cell cycle and in plant organs. *Plant Cell* 4:1531–1538.

Hwang, I. and J. Sheen. 2001. Two-component circuitry in *Arabidopsis* cytokinin signal transduction. *Nature* 413:383–289.

Imajuku, Y. et al. 1992. Exon-intron organization of the *Arabidopsis thaliana* protein kinase genes CDC2a and CDC2b. *FEBS Lett.* 304:73–77.

Imamura, A. et al. 1999. Compilation and characterization of *Arabidopsis thaliana* response regulators implicated in his-asp phosphorelay signal transduction. *Plant Cell Physiol.* 40:733–742.

Inoue, T. et al. 2001. Identification of CRE1 as a cytokinin receptor from *Arabidopsis*. *Nature* (London) 409:1060–1063.

Jackson, D., B. Veit, and S. Hake. 1994. Expression of maize *KNOTTED*1 related homeobox genes in the shoot apical meristem predicts patterns of morphogenesis in the vegetative shoot. *Development* (Cambridge) 120:405–413.

Kakimoto, T. 1996. CKI1, a histidine kinase homolog implicated in cytokinin signal transduction. *Science* 274:982–985.

Kayes, J. M. and S. E. Clark. 1998. *CLAVATA2*, a regulator of meristem and organ development in *Arabidopsis*. *Development* 125:3843–3851.

Kerstetter, R.A. et al. 1997. Loss of function mutations in the maize homeobox gene, *knotted1*, are defective in shoot meristem maintenance. *Development* 124:3045–3054.

Koff, A. et al. 1992. Formation and activation of a cyclin E-cdk2 complex during the G1 phase of the human cell cycle. *Science* 257:1689–1694.

Kusaba, S. et al. 1998. Alteration of hormone levels in transgenic tobacco plants overexpressing the rice homeobox gene *OSH*1. *Plant Physiol.* 116:471–476.

Laux, T. et al. 1996. The *WUSCHEL* gene is required for shoot and floral meristem integrity in *Arabidopsis*. *Development* 122:87–96.

Lenhard, M., G. Juergens, and T. Laux. 2002. The *WUSCHEL* and *SHOOTMERISTEMLESS* genes fulfill complementary roles in *Arabidopsis* shoot meristem regulation. *Development* 129:3195–3206.

Lenhard, M. and T. Laux. 1999. Shoot meristem formation and maintenance. *Curr. Opin. Plant Biol.* 2:44–50.

Lew, D. J., V. Dulic, and S. I. Reed. 1991. Isolation of three novel human cyclins by rescue of G1 cyclin (Cln) function in yeast. *Cell* 66:1197–1206.

Leyser, H.M.O. and I.J. Furner. 1992. Characterization of three shoot apical meristem mutants of *Arabidopsis thaliana*. *Development* 116:397–403.

Lincoln, C. et al. 1994. A knotted1-like homeobox gene in *Arabidopsis* is expressed in the vegetative meristem and dramatically alters leaf morphology when overexpressed in transgenic plants. *Plant Cell* 6:1859–1876.

Long, J. A. et al. 1996. A member of the *KNOTTED* class of homeodomain proteins encoded by the STM gene of *Arabidopsis*. *Nature* 379:66–69.

Mahonen, A.P. et al. 2000. A novel two-component hybrid molecule regulates vascular morphogenesis of the *Arabidopsis* root. *Genes Dev.* 14:2938–2943.

Martinez, M.C. et al. 1992. Spatial pattern of cdc2 expression in relation to meristem activity and cell proliferation during plant development. *Proc. Natl. Acad. Sci. U.S.A.* 89:7360–7364.

Mayer, K.F.X. et al. 1998. Role of *WUSCHEL* in regulating stem cell fate in the *Arabidopsis* shoot meristem. *Cell* 95:805.

Miao, G.-H., Z. Hong, and D.P.S. Verma. 1993. Two functional soybean genes encoding p34[cdc2] protein kinases are regulated by different plant developmental pathways. *Proc. Natl. Acad. Sci. U.S.A.* 90:943–947.

Miller, C.O. et al. 1955. Kinetin, a cell division factor from deoxyribonucleic acid. *J. Amer. Chem. Soc.* 77:1392.

Miller, C.O. et al. 1956. Isolation, structure and synthesis of kinetin, a substance promoting cell division. *J. Amer. Chem. Soc.* 78:1345–1350.

Miyata, S. et al. 1998. Characterization of genes for two-component phosphorelay mediators with a single HPt domain in *Arabidopsis thaliana*. *FEBS Lett.* 437:11–14.

Ohme-Takagi, M. and H. Shinshi. 1995. Ethylene-inducible DNA binding proteins that interact with an ethylene-responsive element. *Plant Cell* 7:173–182.

Ori, N. et al. 1999. Leaf senescence is delayed in tobacco plants expressing the maize homeobox gene knotted 1 under the control of a senescence-activated promoter. *Plant Cell* 11:1073–1080.

Prakash, A.P. and P.P. Kumar. 2002. PkMADS1 is a novel MADS box gene regulating adventitious shoot induction and vegetative shoot development in *Paulownia kawakamii*. *Plant J.* 29:141–151.

Qin, L.-X. et al. 1995. Identification of a cell cycle-related gene, cyclin, in *Nicotiana tabacum* (L.). *Plant Physiol.* 108:425–426.

Quelle, D. E. et al. 1993. Overexpression of mouse D-type cyclins accelerates G-1 phase in rodent fibroblasts. *Genes Dev.* 7:1559–1571.

Renaudin, J.-P. et al. 1994. Cloning of four cyclins from maize indicates that higher plants have three structurally distinct groups of mitotic cyclins. *Proc. Natl. Acad. Sci. U.S.A.* 91:7375–7380.

Riou-Khamlichi, C. et al. 1999. Cytokinin activation of *Arabidopsis* cell division through a D-type cyclin. *Science* 283:1541–1544.

Rojo, E. et al. 2002. CLV3 is localized to the extracellular space, where it activates the *Arabidopsis* CLAVATA stem cell signaling pathway. *Plant Cell* 14:969–977.

Rupp, H.-M. et al. 1999. Increased steady state mRNA levels of the STM and KNAT1 homeobox genes in cytokinin overproducing *Arabidopsis thaliana* indicate a role for cytokinins in the shoot apical meristem. *Plant J.* 18:557–563.

Sakai, H. et al. 2001. ARR1, a transcription factor for genes immediately responsive to cytokinins. *Science* 294:1519–1521.

Shaul, O. et al. 1996. Regulation of cell division in *Arabidopsis*. *Crit. Rev. Plant Sci.* 15:97–112.

Sinha, N.R., R.E. Williams, and S. Hake. 1993. Overexpression of the maize homeobox gene, *KNOTTED*-1, causes a switch from determinate to indeterminate cell fates. *Genes Dev.* 7:787–795.

Skoog, F. and C.O. Miller. 1957. Chemical regulation of growth and organ formation in plant tissues cultured *in vitro*. *Symp. Soc. Exp. Biol.* 11:118–140.

Smith, L.G., D. Jackson, and S. Hake. 1995. Expression of knotted1 marks shoot meristem formation during maize embryogenesis. *Dev. Genet.* 16:344–348.

Soni, R. et al. 1995. A family of cyclin D homologs from plants differentially controlled by growth regulators and containing the conserved retinoblastoma protein interaction motif. *Plant Cell* 7:85–103.

Suzuki, T. et al. 1998. Histidine-containing phosphotransfer (HPt) signal transducers implicated in His-to-Asp phosphorelay in *Arabidopsis*. *Plant Cell Physiol.* 39:1256–1268.

Suzuki, T. et al. 2001. The *Arabidopsis* sensor His-kinase, AHK4, can respond to cytokinins. *Plant Cell Physiol.* 42:107–113.

Suzuki, T. et al. 2000. Compilation and characterization of histidine-containing phosphotransmitters implicated in His-to-Asp phosphorelay in plants: AHP signal transducers of *Arabidopsis* thaliana. *Biosci. Biotechnol. Biochem.* 64:2486–2489.

Teo, W. et al. 2001. The expression of Brostm, a *KNOTTED*1-like gene, marks the cell type and timing of *in vitro* shoot induction in *Brassica oleracea*. *Plant Mol. Biol.* 46:567–580.

Ueguchi, C. et al. 2001. Novel family of sensor histidine kinase genes in *Arabidopsis thaliana*. *Plant Cell Physiol.* 42:231–235.

Vollbrecht, E. et al. 1991. The developmental gene Knotted-1 is a member of a maize homeobox gene family. *Nature* 350:241–243.

Williams-Carrier, R. E. et al. 1997. Ectopic expression of the maize *kn1* gene phenocopies the hooded mutant of barley. *Development* 124:3737–3745.

Yamada, H. et al. 2001. The *Arabidopsis* AHK4 histidine kinase is a cytokinin-binding receptor that transduces cytokinin signals across the membrane. *Plant Cell Physiol.* 42:1017–1023.

Zhang, K., D.S. Letham, and P.C. John. 1996. Cytokinin controls the cell cycle at mitosis by stimulating the tyrosine dephosphorylation and activation of p34[cdc2]-like H1 histone kinase. *Planta* 200:2–12.

Zhang, S. et al. 1998. CDC2Zm and KN1 expression during adventitious shoot meristem formation from *in vitro* — proliferating axillary shoot meristems in maize and barley. *Planta* 204:542–549.

Zhang, S. et al. 2002. Similarity of expression patterns of *knotted1* and *ZmLEC1* during somatic and zygotic embryogenesis in maize (*Zea mays* L.). *Planta* 215:191–194.

14 Propagation from Nonmeristematic Tissues: Nonzygotic Embryogenesis*

Dennis J. Gray

CHAPTER 14 CONCEPTS

- Nonzygotic embryogenesis occurs among widely disparate cell and tissue types and may be regarded as a universal capability of higher plants.

- Nonzygotic embryos appear to arise from single cells, not as a result of sexual reproduction, but otherwise are identical to their zygotic counterparts.

- Nonzygotic embryogenesis conclusively demonstrates cellular totipotency and represents a highly efficient method of plant propagation.

- Embryogenic cell cultures are widely used for genetic manipulation such as *in vitro* selection and transgenic technologies.

INTRODUCTION

An embryo can be defined as the earliest recognizable multicellular stage of an individual that occurs before it has developed the structures or organs characteristic of a given species. In most organisms, embryos are morphologically distinct entities that function as an intermediate stage in the transition between the gametophytic to sporophytic life cycle (Figure 14.1). For example, in higher plants, we are most familiar with embryos that develop within seeds; such embryos usually arise from gametic fusion products (zygotes) following sexual reproduction and are termed zygotic embryos, although seed-borne embryos also can develop apomictically (i.e., without sexual reproduction). However, plants are unique in that morphologically and functionally correct nonzygotic embryos also can arise from widely disparate cell and tissue types at a number of different points from both the gametophytic and sporophytic phases of the life cycle (Figure 14.2).

The first demonstration that plants could produce nonzygotic embryos *in vitro* was published in 1958 by Steward et al. Subsequently, Reinert (1959) observed bipolar embryos to differentiate in a culture of carrot roots after transfer from one medium to another. While carrot was the first species in which *in vitro* nonzygotic embryogenesis was reported, in subsequent years many species of angiosperms and gymnosperms have been added to the list of successes. In fact, demonstrations of nonzygotic embryogenesis are so widespread that it may be regarded to be a universal capability of higher plants.

A plethora of terminology has arisen to designate nonzygotic embryos. Such embryos originally were termed "embryoids" to denote perceived significant differences from zygotic embryos and, unfortunately, this term persists in some literature. However, differences in embryogenic cell origins

---

* Florida Agricultural Journal Series No. R-10166

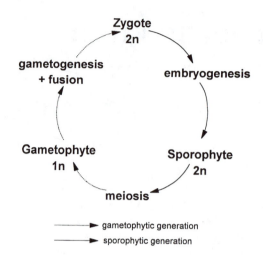

FIGURE 14.1 Typical angiosperm life cycle showing the natural role of embryogenesis in sporophyte development.

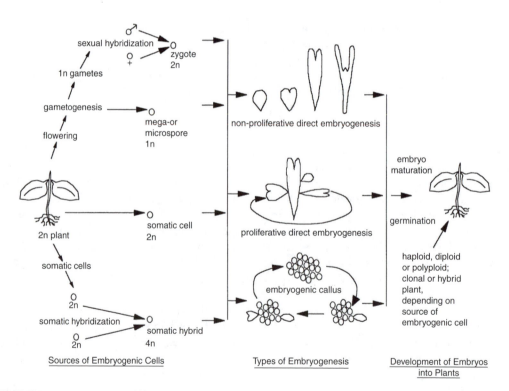

FIGURE 14.2 Sources of embryogenic cells, types of embryogenesis, and plants obtained via *in vitro* culture of higher plants.

notwithstanding, distinctions between zygotic and nonzygotic embryos become blurred as our understanding of embryo development increases. Nonzygotic embryos are now shown to be functionally equivalent to zygotic embryos and the suffix "oid" should be dropped. Other terms for nonzygotic embryos are based primarily upon differences in their specific sites of origin and often are interchangeable, leading to some inconsistencies in the literature. For example, nonzygotic embryos can arise from plant vegetative cells, reproductive tissues, zygotic embryos, or callus cells

A.

B.
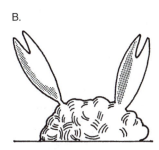

FIGURE 14.3 Comparison of typical zygotic embryo development in seed (A) with nonzygotic embryo development from callus (B). Zygotic embryos typically develop within nutritive seed tissues (shaded area) and are connected to the mother plant by a suspensor (arrow). In contrast, nonzygotic embryos often develop perched above subtending tissue, the only source of nutrition being the narrowed suspensor. (From Gray, D.J. and A. Purohit. 1991. Somatic embryogenesis and the development of synthetic seed technology. *Crit. Rev. Plant Sci.* 10:33–61. With permission.)

derived from any of these. Thus, "somatic embryos" grow from somatic cells; "haploid or pollen embryos" are derived from pollen grains or microspore mother cells; "nucellar embryos" are formed from nonzygotic nucellar seed tissue; "direct secondary embryos" develop from previously formed embryos; and so on.

Most of these embryo types share the common trait of being able to be manipulated via *in vitro* culture. Such embryogenic culture systems form the basis for many biotechnological approaches to plant improvement because they allow not only clonal plant propagation, but also specific and directed changes that can be introduced into desirable, elite individuals by genetic engineering of somatic cells. Modified individual cells and embryos then can be efficiently multiplied *in vitro* to very high numbers prior to plant development. This approach to genetic improvement bypasses the unwanted consequences of sexual reproduction (mass genetic recombination and required cycles of selection) inherent in conventional breeding technology.

Because of the many commonalities exhibited by these embryo types as well as the various methods available to study and manipulate them, for convenience, this chapter will only utilize the term "nonzygotic embryo" to designate the embryos that develop through *in vitro* culture, regardless of origin. Thus, the developmental processes and experimental procedures described will tend to be applicable to many of the embryo types mentioned above.

IN VIVO VERSUS *IN VITRO* GROWTH CONDITIONS

In seeds, nutritive tissues directly encase the developing embryo. Early in embryogenesis, nutritive substances enter the zygotic embryo either through the suspensor or the embryo body (Figure 14.3A). The relative reliance of the developing embryo on obtaining nutrition via the suspensor versus the endosperm differs greatly depending on the species. In contrast, nonzygotic embryos develop naked, not encased in seed endosperm, and as such, are not subjected to specialized and highly regulated nutritional regimes (Gray and Purohit, 1991). Often, a suspensor is the only link between the embryo and growth medium (Figure 14.3B). This demonstrates that the suspensor can serve as the pathway for all needed nutrition and that endosperm is not absolutely necessary for embryogenesis and germination to occur.

COMMON ATTRIBUTES OF EMBRYOGENIC CULTURE PROTOCOLS

Given the range of different explants, culture conditions and media used to initiate and maintain embryogenic cultures, certain generalities can be made regarding basic methodology. The use of specific plant growth regulators (PGRs), which are required in most instances, will be discussed later with regard to embryogenic cell initiation.

The choice of genotype and explant often is crucial in obtaining an embryogenic response (see Gray, 1990, and Gray and Meredith, 1992, for reviews). For example, with corn, only a limited range of genotypes are capable of producing embryogenic cultures and, with few exceptions, only immature embryo explants can be used. With corn, the age of immature embryos also is important because embryos that are too young do not survive culture, and those that are too old do not produce embryogenic callus. Interestingly, most species tested to date can be induced to produce embryogenic cultures from at least some genotypes and tissues. Additionally, the embryogenic response is heritable, such that it can be bred from embryogenic lines into nonembryogenic lines via sexual hybridization, albeit with a certain degree of difficulty and transfer of other, often undesirable, traits via sexual hybridization.

Environmental requirements for optimum culture growth often are quite specific, but not unusual. Cultures may require growth in either dark or light or a combination of both over time, the optimization of which can be experimentally quantified. Culture in darkness may be necessary in order to prevent the triggering effect of light on many plant biological processes that may adversely affect the growth of embryogenic cell populations. Culture in darkness suppresses unwanted tissue differentiation in explant tissue, for example, by limiting the development of plastids into chloroplasts. Similarly, dark conditions may inhibit precocious germination of young embryos. Temperature requirements also are specific, but tend to be in the range of "room temperature" (i.e., 23–27°C). Despite the similarity in culture requirements, improvements in nonzygotic embryo development and maturation are increasingly being obtained by optimization of medium composition.

ORIGIN OF NONZYGOTIC EMBRYOS

It is generally accepted that, like zygotic embryos, nonzygotic embryos arise from a single cell, in contrast to budding from a cell mass. This distinction is important in considering efficient genetic engineering because modification of a single embryogenic cell might eventually result in a modified plant, compared to genetic modification of a cell within a bud, which would result in a chimeric plant.

Nonzygotic embryos growing from isolated cells, such as microspores or protoplasts, clearly develop from single cells. However, the origin of nonzygotic embryos that develop from complex intact primary explant tissue or callus is more difficult to resolve because the action of microscopic single cells cannot be readily followed. Nonzygotic embryos often develop with a well-defined suspensor apparatus identical to that of zygotic embryos, somewhat suggesting a single cell origin; whereas others can develop with a broad basal attachment, suggesting a multicellular budding phenomenon (Figure 14.4).

INITIATION OF EMBRYOGENIC CELLS

In complex explants, nonzygotic embryos typically can be initiated only from the more juvenile or meristematic tissues. For example, immature zygotic embryos or zygotic embryo cotyledons and hypocotyls dissected from ungerminated seeds are commonly used explants. The young leaves, shoot tips, or even the roots of established plants sometimes are used to initiate embryogenic cultures. However, the explant response is highly genotype dependent, so that, for any given species, only a certain type or range of explant can be used to initiate embryogenic cultures.

There are several pathways by which nonzygotic plant cells become embryo initials (Christensen, 1985). In instances where the explant consists of undifferentiated embryonic tissue, such as an immature zygotic embryo, initiation and maintenance of an embryogenic callus is akin to

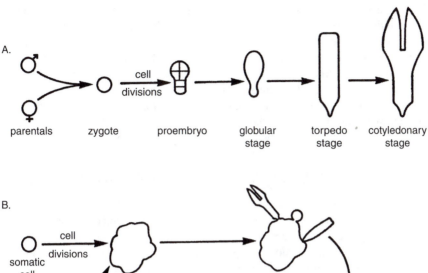

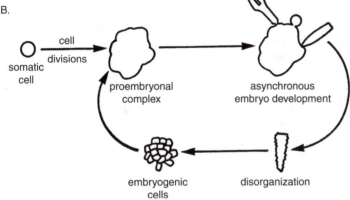

FIGURE 14.4 Comparison of typical zygotic embryogenesis occurring within seeds (A) with proliferative nonzygotic embryogenesis occurring *in vitro* (B). Zygotic embryogenesis is characterized as being very regulated, such that embryos synchronously pass through distinct developmental stages; whereas nonzygotic embryogenesis is often nonuniform, with many stages present at a given time. Nonzygotic embryos may bypass maturation and become disorganized, adding to the proembryonic tissue mass. (From Gray, D.J. and A. Purohit. 1991. Somatic embryogenesis and the development of synthetic seed technology. *Crit. Rev. Plant Sci.* 10:33–61. With permission.)

culturing and propagating a preexisting proembryonal complex. Thus, the embryogenic cells present in explant tissue prior to culture initiation are simply propagated and otherwise manipulated *in vitro*. However, in many instances, embryogenic cells are induced from nonembryogenic cells; this represents a dramatic change in their presumptive fate. The shift in developmental pattern involves a dedifferentiation away from the cell's "normal" fate followed by redetermination toward an embryogenic cell type. For example, cells in leaf explants, which normally would develop into constituents of relatively short-lived parenchymatous tissue, instead become embryogenic under certain conditions. This is a pivotal change in development because cells that normally would be capable of only a few divisions, at most, before senescence, instead become redirected to become totipotent and capable of possibly unlimited divisions. Such embryogenic cells become immortal in the sense that they reinstate the germ line by being capable of developing into mature, reproductive individuals.

The fact that isolated somatic cells can develop normally into embryos irrevocably demonstrates that the developmental program for embryogenesis is contained within and controlled by the cell itself and not by external factors. However, the exact nature of the triggering mechanisms of embryogenesis, whether it be a physical, biochemical, and/or genetic event or events, is unknown. Early attempts to identify genes that became activated as a direct consequence of embryogenic cell induction failed due to inadequate understanding of the basic induction phenomenon, which resulted in faulty experimental designs. Because the cell cultures used were already induced, any genes critical to the induction step already had been activated. Hence, the genes that were observed in such studies actually were those activated later in embryogenesis such that they were controlling downstream developmental processes and were not related to the triggering event. Despite the technical difficulty of identifying critical induction-controlling genes, a number of other genes and their products have been identified during nonzygotic embryo development (e.g., Zimmerman, 1993). In some instances, the genetic mechanisms resolved for nonzygotic embryos can be related directly to that of the corresponding zygotic embryo. Current information concerning molecular aspects of nonzygotic embryo development is presented in Chapter 15.

INDUCTIVE PLANT GROWTH REGULATORS

In practice, the initiation of embryogenic cells requires *in vitro* culture of the appropriate explant on or in a medium that contains specific PGRs. In fact, a preponderance of reports of embryogenic culture initiation employed a very narrow range of PGRs that typically are added to culture medium. Synthetic auxins, notably 2,4-dichlorophenoxyacetic acid (2,4-D), are added to medium in the predominance of reported protocols. Similar auxins used include dicamba, indolebutyric acid (IBA), naphthoxyacetic acid (NOA), picloram, and others. In addition, several weaker auxins, such as indoleacetic acid (IAA), a natural plant hormone, and napthyleneacetic acid (NAA) have been utilized in a few culture systems. Auxins serve to induce the formation of embryogenic cells, possibly by initiating differential gene activation, as noted earlier, and also appear to promote increase of embryogenic cell populations through repetitive cell division, while simultaneously suppressing cell differentiation and growth into embryos. However, an auxin often is not required in instances where the explant consists of preexisting embryogenic cells, as described above, possibly because a discrete induction step is not required.

In addition to auxin-like PGRs, cytokinins also are required to induce embryogenesis in many dicotyledonous (dicot) species. In a few instances, only a cytokinin is required to cause embryogenic cultures to develop. The most commonly used cytokinin is benzyladenine (BA), but others such as thidiazuron (TDZ) and kinetin, and the natural cytokinin, zeatin, also are utilized.

Actual PGR concentration is important for an optimum response because concentrations that are too low may not trigger the inductive event, and concentrations that are too high, particularly when considering phenoxy-auxins, may become toxic. Typically, following the induction period, resulting culture material is transferred to medium lacking PGRs, which removes the auxin-induced suppression of embryo development and allows embryogenesis to occur; however, not all culture systems require this two-step procedure. For example, as mentioned above, certain species do not require any induction step at all and others, particularly poaceous monocotyledonous (molocots), become induced and undergo complete embryogenesis in the continuous presence of auxin. Examples of different PGR regimes used in culture systems are illustrated via the laboratory exercises provided after this chapter.

EMBRYO DEVELOPMENT

The physical, observable transition from a nonembryogenic to an embryogenic cell may occur when the progenitor cell undergoes an unequal division, resulting in a larger vacuolate cell and a smaller, densely cytoplasmic (embryogenic) cell. Embryogenic cells are readily distinguished by their small size, isodiametric shape, and densely cytoplasmic appearance (Figure 14.5). This type of unequal cell division is identical to that observed in zygotes and may be an early indication of developmental polarity. The embryogenic cell then either continues to divide irregularly to form a proembryonal complex, or divides in a highly organized manner to form a somatic embryo (Figures 14.2 and 14.4). However, an often undesirable difference exhibited by nonzygotic embryos is that

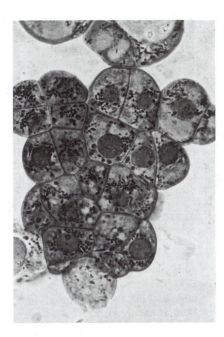

FIGURE 14.5 Typical embryogenic cells of grape. Note the three-cell stage embryo in the upper center part of the cell mass. (From Gray, D.J. 1995. *Somatic Embryogenesis in Woody Plants*, Kluwer Academic Publishers, Dordrecht, the Netherlands. Reprinted by permission of Kluwer Academic Publishers.)

they frequently deviate from the normal pattern of development either by producing callus undergoing direct secondary embryogenesis or germinating precociously. This tendency toward erratic development clearly is due to environmental factors as discussed next.

Zygotic and nonzygotic embryos share the same gross pattern of development, with both typically passing through globular, scutellar, and coleoptilar stages for monocots, or globular, heart, torpedo, and cotyledonary stages for dicots and conifers. Generally, the anatomy and morphology of well-developed nonzygotic embryos is faithful to the corresponding zygotic embryo type, such that they easily can be identified by eye or with the aid of a stereomicroscope (Figure 14.6). For example, nonzygotic embryos of grass and cereal species typically possess a scutellum, coleoptile, and embryo axes, which are distinctive embryonic organs of monocots (Figure 14.7A). Embryos of dicot species have a distinct hypocotyl and cotyledons (usually two) (Figure 14.7B); those of conifers also exhibit a hypocotyl and numerous cotyledons (Figure 14.7C), the number of which is similar to that of the given species.

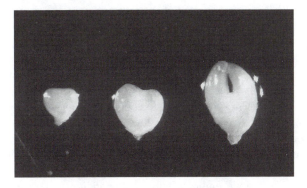

FIGURE 14.6 Somatic embryos of grape at heart and early cotyledonary stages, illustrating morphologies highly faithful to those of zygotic embryos.

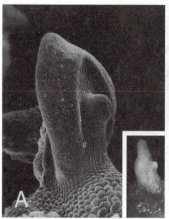
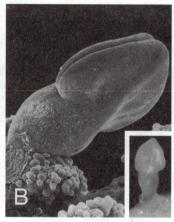
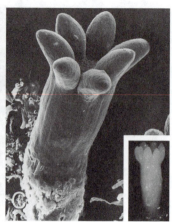

FIGURE 14.7 Comparison of monocot (A), dicot (B), and gymnosperm (C) nonzygotic embryos. Scanning electron micrographs, with corresponding stereomicrographs (inset). (A) Somatic embryo of orchardgrass growing from an embryogenic callus. Note the protruding coleoptile and notch, through which the first leaves will emerge after germination. The prominent large flattened body of the embryo is the scutellum. (From Gray, D.J et al. 1984. Somatic embryogenesis in suspension and suspension-derived callus cultures of *Dactylis glomerata. Protoplasma* 122:196–202.) (B) Somatic embryo of grape growing from an embryogenic callus. Note two flattened cotyledons and subtending hypocotyls. (From Gray, D.J. 1995. *Somatic Embryogenesis in Woody Plants.* Kluwer Academic Publishers, Dordrecht, the Netherlands. Reprinted by premission of Kluwer Academic Publishers.) (C) Somatic embryo of Norway Spruce. Note multiple cotyledons and elongated hypocotyls. (From Fowke, L.C. et al. 1994. Scanning electron microscopy of hydrated and desiccated mature somatic embryos and zygotic embryos of white spruce (*Picea glauca* (Moench) Voss.). *Plant Cell Rep.* 13:612–618. Used with permission.)

Embryo development occurs through an exceptionally organized sequence of cell division, enlargement, and differentiation. During early development, the embryo assumes a clavate, globular shape and remains essentially an undifferentiated but organized mass of dividing cells with a well-defined epidermis. Subsequent heart through early torpedo stages are characterized by cell differentiation and polarized growth, notably elongation and initiation of rudimentary cotyledons in dicots (Figure 14.6) and development of the scutellum with initiation of the coleoptilar notch in poaceous monocots. At the same time, obvious tissue differentiation begins with the development of embryonic vasculature (Figure 14.8) and accumulation of intracellular storage substances. Final stages of development toward maturation are distinguished by overall enlargement, increase in cotyledon size in dicots (Figure 14.9) and coleoptilar enlargement in monocots. At the same time, the embryonic axis becomes increasingly developed. In dicots, the root apical meristem becomes well established, embedded in tissue located above the suspensor apparatus and at the base of the hypocotyl, whereas the shoot apical meristem develops externally between the cotyledons. In monocots, the embryo axis develops laterally and parallel to the scutellum. The root apical meristem is embedded and the shoot apical meristem develops externally, but is protected by the coleoptile. All of the events described above occur in concert with each other in a manner that is essentially identical to that of zygotic embryos. But, for a number of reasons, as described below, nonzygotic embryos often differ somewhat from their zygotic counterparts in morphology and performance.

An obvious difference in gross morphology between nonzygotic embryos growing *in vitro* and zygotic embryos in seeds is simply caused by the physical constraint on zygotic embryos imposed by the developing seed coat, often causing them to become compressed and/or flattened into a shape and size distinct for a given species or variety. This is made apparent by comparing the morphology of seed-borne zygotic embryos with corresponding nonzygotic embryos of a given species (Figure 14.10). Zygotic embryos excised from seed typically exhibit a highly compressed

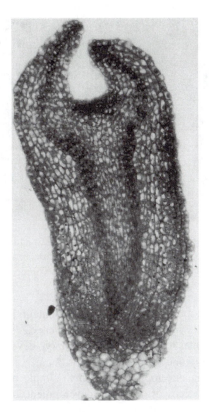

FIGURE 14.8 Longitudinal section through a grape somatic embryo showing typical vascular system. (From Gray, D.J. 1995. *Somatic Embryogenesis in Woody Plants*. Kluwer Academic Publishers, Dordrecht, the Netherlands. Reprinted with permission of Kluwer Academic Publishers.)

FIGURE 14.9 Cotyledonary-stage somatic embryos of cantaloupe growing from a cultured cotyledon. Note that the fine suspensor is the only connection to explant. (From Gray, D.J. et al. 1993. High-frequency somatic embryogenesis from quiescent seed cotyledons of *Cucumis melo* cultivars. *J. Amer. Soc. Hort. Sci.* 118:425–432. With permission.)

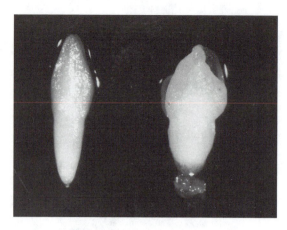

FIGURE 14.10 Comparison of grape zygotic embryo, compressed and flattened by development within a seed (left), with grape somatic embryo, which is not flattened (right). (From Gray, D.J. and A. Purohit. 1991. Somatic embryogenesis and the development of synthetic seed technology. *Crit. Rev. Plant Sci.* 10:33–61. With permission.)

shape because the embryos become flattened during development. In contrast, nonzygotic embryos tend to be larger and have wider hypocotyls and fleshier cotyledons. It is possible that pressure exerted by the seed coat contributes to other aspects of embryo development that are lacking during nonzygotic embryogenesis (Gray and Purohit, 1991).

In addition to differences in development related to simple physical constraints, for several reasons, significantly more instances of abnormal development are known to occur during nonzygotic embryogenesis when compared to zygotic embryogenesis. For example, as mentioned above, species that normally produce zygotic embryos with suspensors often produce clusters of nonzygotic embryos from a proembryonal cell complex (Figure 14.4). This basic change in developmental pattern is likely due to differences between the seed and *in vitro* environments, since immature zygotic embryos dissected from seeds and cultured also often develop abnormally (Gray and Purohit, 1991).

Nonzygotic embryos growing in mass from proembryonal complexes tend to develop asynchronously so that several stages are present in cultures at any given time (Figures 14.2 and 14.4). Nonzygotic embryos initiated over time are subjected to different nutrient regimes as medium becomes depleted, then replenished, between and during subcultures. This leads to differences in development even among embryos from a single culture. With such variable and unregulated environmental conditions, nonzygotic embryos often bypass maturation altogether, becoming disorganized, forming new embryogenic cells, and contributing to asynchrony (Figure 14.4). Nonzygotic embryos also often exhibit structural anomalies such as extra cotyledons (Figure 14.11) and poorly developed apical meristems.

EMBRYO MATURATION

Maturation is the terminal event of embryogenesis and is characterized by the attainment of mature embryo morphology; accumulation of storage carbohydrates, lipids, and proteins (see the section above on *in vivo* versus *in vitro* growth conditions); reduction in water content; and, often, a gradual decline or cessation of metabolism. Nonzygotic embryos typically do not mature properly when compared to zygotic embryos. In fact, in the preponderance of instances, rapid growth continues to occur, leading to precocious germination.

Although complete maturation is not absolutely necessary in order to obtain plants from nonzygotic embryos, it is required to achieve high rates of plant recovery. As such, factors that

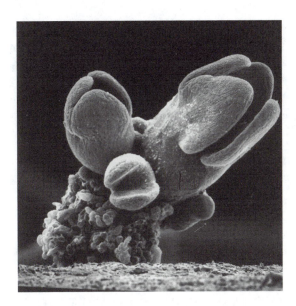

FIGURE 14.11 Asynchronous and abnormal somatic embryo development from embryogenic callus of grape. Increasing levels of development (clockwise from lower right globular embryo) are evident in a single group of embryos. In addition, the normal cotyledon number of two (lower center embryo) contrasts with supernumery cotyledon development of three and four on the two upper embryos. (From Gray, D.J. and J.A. Mortensen. 1987. Initiation and maintenance of long term somatic embryogenesis from anthers and ovaries of *Vitis longii* 'Microsperma'. *Plant Cell Tissue Organ Cult.* 9:73–80. Reprinted with permission from Kluwer Academic Publishers.)

influence or enhance nonzygotic embryo maturation have been explored. A number of culture medium components have been shown to promote maturation. In particular high sucrose levels (9–12%), timed pulses of abscisic acid (ABA), a naturally occurring plant hormone, and polyethylene glycol (PEG), an osmotically active compound, in conjunction with certain amino acids, notably glutamine, have been remarkably successful. For example, an 8-week treatment with ABA and PEG caused white spruce nonzygotic embryos to accumulate more lipids and to better resemble zygotic embryos morphologically (Attree and Fowke, 1993). Similarly, a short pulse of ABA in conjunction with glutamine resulted in alfalfa nonzygotic embryos that could withstand dehydration as mentioned next (Senaratna et al., 1989).

QUIESCENCE AND DORMANCY

Perhaps the most obvious developmental difference between zygotic and nonzygotic embryos is that the latter lack a quiescent resting phase. By comparison, zygotic embryos of many species begin a resting period during seed maturation. Zygotic embryo quiescence in conjunction with the protective and nutritive tissues that comprise a seed are the major factors allowing seeds to be stored conveniently. Thus, the ability to withstand dehydration appears to be a normal step in embryo development.

However, nonzygotic embryos tend to grow and germinate without normal maturation, to become disorganized into embryogenic tissue, or to die. They rarely enter a resting stage. Although quiescence and dormancy have only recently been documented in somatic embryos, it is possible that dormancy occurs in instances where plants cannot be obtained from well-developed embryos (see Gray, 1986, and Gray and Purohit, 1991, for reviews).

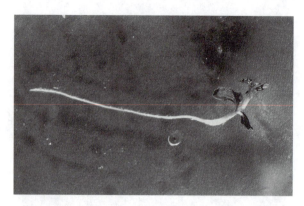

FIGURE 14.12 Plant development from a germinated melon somatic embryo. Note shoot and root development. (From Gray, D.J. et al. 1993. High-frequency somatic embryogenesis from quiescent seed cotyledons of *Cucumis melo* cultivars. *J. Amer. Soc. Hort. Soc.* 118:425–432. With permission.)

EMBRYO GERMINATION AND PLANT DEVELOPMENT

Obtaining plants from nonzygotic embryos often is more difficult than would be expected. The early literature concerning first reports of nonzygotic embryogenesis for a given species or cultivar often did not include information on plant recovery. When plants were obtained, the recovery rate was very low or not reported, suggesting that a majority of nonzygotic embryos were too abnormal to germinate. Typically, plant recovery from nonzygotic embryos ranges from 0 to 50%. This is very low compared to zygotic embryos in commercial seed, for which germination and plant development typically exceed 90% in soil. In all but a few instances, nonzygotic embryos develop very poorly if at all when planted in soil-like seed. Only recently has research been conducted to raise plant recovery rates.

Progress in culture methodology has resulted in better development and germination of nonzygotic embryos. As mentioned earlier regarding maturation, pulse treatments with various amino acids, osmotica, and PGRs, particularly ABA, have resulted in nonzygotic embryos with better maturation, including the ability to be dehydrated and stored like seeds, as well as improved germination characteristics (Attree and Fowke, 1993). This demonstrates that careful attention to culture conditions and nutrition, especially with regard to timed pulses of certain factors, results in plant recovery rates from nonzygotic embryos equivalent to those of zygotic embryos. In general, it can be concluded that conditions favoring embryo maturation also favor the recovery of plants.

After the event of germination, the embryo begins development into a plant. Typically, the storage reserves present (lipids, proteins, and/or starch, depending on the species) become depleted, concomitant with the start of increased mitotic activity in shoot and root meristems. Eventually, a young, photosynthetically competent plant develops (Figure 14.12), which can be gradually acclimated to ambient conditions.

USES FOR EMBRYOGENIC CULTURES

Embryogenic culture systems are used for a number of purposes. First, they constitute an important tool for the study of plant development, both due to the unique convenience of *in vitro* culture over *en planta* growth and the contrasts that can be drawn from differences between nonzygotic embryogenesis *in vitro* and zygotic embryogenesis in seeds.

Second, embryogenic cultures are an efficient vehicle for genetic engineering because they often produce many embryos per volume of cell mass, and isolated genes integrated into single embryogenic cells can become incorporated into the genome of the plant that ultimately develops.

Third, many nonzygotic embryos can be produced from a single desirable plant for efficient clonal propagation. Certain potential uses of embryogenic systems as research tools have been described or suggested throughout this chapter and an in-depth discussion of genetic engineering applications is presented in Chapter 19. The use of nonzygotic embryogenesis as a cloning vehicle is discussed in more detail next.

SYNTHETIC SEED TECHNOLOGY

An active research front has emerged over the past decade with the goal of developing nonzygotic embryogenesis into a commercially useful method of plant propagation. The technology that has emerged is termed synthetic (or artificial) seed technology. A synthetic seed is defined as a somatic embryo that is engineered to be of practical use in commercial plant production (Gray and Purohit, 1991). Applications for synthetic seed vary depending on the relative sophistication of existing production systems for a given crop and the opportunities for improvement. Whether or not a cost advantage results from synthetic seed will ultimately determine its commercial use. For seed-propagated agronomic crops, quiescent, nonzygotic embryos produced in bioreactors and encapsulated in synthetic seed coats will be necessary. Certain vegetable crops with seeds that are expensive to produce, such as seedless watermelon, are attractive candidates for synthetic seed technology, since the per-plant cost might be reduced.

Similarly, conifers, which are difficult to improve by breeding due to a long life cycle, would benefit from application of synthetic seed technology if elite individuals could be cloned and planted with seed efficiency (Farnum et al., 1983). For vegetatively propagated crops, particularly those with a high per-plant value, naked, hand-manipulated, nonquiescent embryos may be cost effective. For example, many ornamental crops are commercially micropropagated painstakingly in tissue culture via adventitious bud proliferation with the major expense being the labor needed for multiple culture and rooting steps. Substitution of embryogenic culture systems for such crops would greatly reduce labor costs, since mass-produced somatic embryos easily could be hand selected, placed directly into planting flats, and rooted (Gray and Purohit, 1991).

Ongoing efforts to develop synthetic seed technology constitute one of the most active areas of nonzygotic embryogenesis research. The potential benefits of using "clonal seeds" in agricultural production is great enough to stimulate continued investigation of factors regulating nonzygotic embryo maturation.

CONCLUSION

In vitro embryogenesis from nonzygotic cells must be regarded as a universal property of higher plants, albeit one that generally does not occur in nature. When induced from isolated cells, nonzygotic embryogenesis conclusively demonstrates not only cellular totipotency, but also the presence of a highly conserved developmental program that is innate to a wide range of cell types. For the student, nonzygotic embryogenesis represents an opportunity to study in a culture vessel one of the more developmentally complex aspects of the plant life cycle.

Embryogenic cell culture allows the entire genome of a plant to be manipulated with microbiological techniques. Once the cells are proliferated, genetically engineered, or otherwise managed, intact plants can easily be reconstituted. From such simple manipulation comes numerous possibilities for better understanding and ultimately improving our use of plants.

LITERATURE CITED AND SUGGESTED READINGS

Attree, S.M. and L.C. Fowke. 1993. Embryogeny of gymnosperms: advances in synthetic seed technology of conifers. *Plant Cell Tissue Organ Cult.* 35:1–35.

Christensen, M.L. 1985. An embryogenic culture of soybean: towards a general theory of embryogenesis. In: *Tissue Culture in Forestry and Agriculture.* (Ed.) R.R. Henke et al. Plenum, New York. 83–103.

Farnum, P., R. Timmis, and J.L. Kulp. 1983. Biotechnology of forest yield. *Science* 219:694–702.

Fowke, L.C., S.M. Attree, and P.J. Rennie. 1994. Scanning electron microscopy of hydrated and desiccated mature somatic embryos and zygotic embryos of white spruce (*Picea glauca* (Moench) Voss.). *Plant Cell Rep.* 13:612–618.

Gray, D.J. 1986. Quiescence in monocotyledonous and dicotyledonous somatic embryos induced by dehydration. *HortScience* 22:810–814.

Gray, D.J. 1990. Somatic embryogenesis and cell culture in the Poaceae. In: *Biotechnology in Tall Fescue Improvement.* (Ed.) M.J. Kasperbauer. CRC Press, Boca Raton, FL. 25–57.

Gray, D.J. 1995. Somatic embryogenesis in grape. In: *Somatic Embryogenesis in Woody Plants.* (Ed.) S.M. Jain, P.K. Gupta, and R.J. Newton. Kluwer Academic Publishers, Dordrecht, the Netherlands. 191–217.

Gray, D.J., B.V. Conger, and G.E. Hanning, 1984. Somatic embryogenesis in suspension and suspension-derived callus cultures of *Dactylis glomerata. Protoplasma* 122:196–202.

Gray, D.J., D.W. McColley, and M.E. Compton. 1993. High-frequency somatic embryogenesis from quiescent seed cotyledons of *Cucumis melo* cultivars. *J. Amer. Soc. Hort. Sci.* 118:425–432.

Gray, D.J. and C.P. Meredith. 1992. Grape. In: *Biotechnology of Perennial Fruit Crops.* (Ed.) F.A. Hammerschlag and R.E. Litz. CAB International, Wallingford, UK.

Gray, D.J. and J.A. Mortensen. 1987. Initiation and maintenance of long term somatic embryogenesis from anthers and ovaries of *Vitis longii* 'Microsperma'. *Plant Cell Tissue Organ. Cult.* 9:73–80.

Gray, D.J. and A. Purohit. 1991. Somatic embryogenesis and the development of synthetic seed technology. *Crit. Rev. Plant Sci.* 10:33–61.

Reinert, J. 1959. Ueber die Kontrolle der Morphogenese und die Induktion von Adventiveembryonen an Gewebekulturen aus Karotten. *Planta* 53:318–333.

Senaratna, T., B.D. McKersie, and S.R. Bowley. 1989. Desiccation tolerance of alfalfa (*Medicago sativa* L.) somatic embryos — influence of abscisic acid, stress pretreatments and drying rates. *Plant Sci.* 65:253–259.

Steward, F.C., M.O. Mapes, and K. Mears. 1958. Growth and organized development of cultured cells. II. Organization in cultures grown from freely suspended cells. *Amer. J. Bot.* 45:705–708.

Zimmerman, L.J. 1993. Somatic embryogenesis: A model for early development in higher plants. *Plant Cell* 5:1411–1423.

15 Some Developmental and Molecular Aspects of Somatic Embryogenesis (Nonzygotic Embryogenesis)

Andreas Mordhorst, Erika Charbit, and Sacco C. de Vries

CHAPTER 15 CONCEPTS

- In general, two different systems or modes of somatic embryogenesis are recognized. The first is referred to as indirect somatic embryogenesis, where embryos originate from totipotent callus cells. The second is direct somatic embryogenesis; this involves formation of somatic embryos directly within a differentiated tissue without an intermediate culturing or callus phase.

- Somatic embryos have been shown to share several developmental parameters with their zygotic counterparts. Much if not all of the morphology, physiology, biochemistry, and molecular biology are conserved between the two processes.

- Markers of embryogenic competence can be obtained by comparing the expression profiles of embryogenic cultures before and after culturing onto a medium containing auxin. They can also be identified comparing embryogenic and nonembryogenic cultures, both before induction of embryo development, or by comparing cultures with and without embryo competence.

- The change of the cell fate from somatic to embryo competence can be achieved either by overexpression of various genes in the absence of any hormonal stimulus (e.g., *LEC1*, *LEC2*, *BBM*, *WUS*, *AGL15*) or in combination with an auxin treatment (*AtSERK*). Conversely, recessive mutations in several genes (*pkl*, *pt/amp1*, *clv1*, *clv3*) make embryogenic culture initiation possible using germinating seeds under conditions where it is not possible with wild-type seeds.

INTRODUCTION

The ability of plant cells to recapitulate embryo development via somatic embryogenesis (nonzygotic embryogenesis; see Chapter 14) has been utilized extensively to identify genes involved in plant embryogenesis. Proteins or genes expressed by embryos, embryogenic cells, and nonembryogenic cells have been compared. More recently, genetic approaches, mainly using *Arabidopsis*, have shown that mutations in single genes can lead to enhanced somatic embryogenesis directly *in planta* or in tissue culture. In *Arabidopsis*, other genes have been identified that upon ectopic expression result in spontaneous formation of somatic embryos directly from fully differentiated tissues. The purpose of this chapter is to discuss several of these recent studies. Most of the earlier work has been summarized in recent reviews (e.g., Mordhorst et al., 1997), and will only be briefly mentioned here.

In general, two different systems or modes of somatic embryogenesis are recognized. The first is referred to as indirect somatic embryogenesis, and involves *in vitro* culturing of explants in the presence of growth regulators. The resulting cell or callus cultures contain totipotent cells capable of forming somatic embryos. Alternatively, direct somatic embryogenesis involves the formation of somatic embryos directly within a differentiated tissue, without an intermediate culturing or callus phase. Direct somatic embryogenesis can occur in response to exogenous growth regulators or spontaneously.

The ability of a cell to express its totipotency is called "competence" (Halperin, 1966). Other definitions have refined this concept to acknowledge the requirement for exogenous growth regulators (Choi and Sung, 1989; Carman, 1990) in order to achieve the status of embryogenic competence (De Jong et al., 1993). Induction of indirect somatic embryogenic competence takes a certain period of time in the presence of auxins (Finstad et al., 1993; Dudits et al., 1995). It is assumed that during this period extensive dedifferentiation of cells and tissues needs to take place. In carrot, the embryogenic cultures contain so-called proembryogenic masses (PEMs), described by Halperin in 1966. PEMs are small masses of 10–20 small and tightly adhering cells formed in culture from competent single cells (Toonen et al., 1994). At this point, embryogenic cells are considered to be visually distinct from nonembryogenic cells.

In direct somatic embryogenesis systems, such a period of dedifferentiation is apparently not required or is not visible due to the absence of callus development. This implies that certain cells in differentiated tissues can switch fate within a much shorter time or employ different mechanisms compared to those of indirect embryogenesis.

DEVELOPMENTAL PROGRAM OF SOMATIC EMBRYOGENESIS

In earlier work (see, for example, Mordhorst et al., 1997; Zimmerman, 1993) somatic embryos were shown to share several developmental parameters with their zygotic counterparts. Much if not all of the morphology, physiology, biochemistry and molecular biology are conserved between the two processes. Somatic embryos pass through the same stereotypical morphological stages, both embryo types are desiccation tolerant, they accumulate the same embryo specific storage products and they apparently share a common gene expression program. In an attempt to also provide genetic evidence for these correlations, Mordhorst et al. (2002) produced somatic embryos of mutants such as *shoot meristemless* (*stm*; Barton and Poethig, 1993; Endrizzi et al., 1996; Long et al., 1996), *wuschel* (*wus*; Laux et al., 1996; Mayer et al., 1998) and *zwille/pinhead* (*zll/pnh*; McConnell and Barton, 1995; Moussian et al., 1998; Lynn et al., 1999). Formation of the embryonic shoot apical meristem (SAM) is affected in all of these mutants. As predicted, somatic embryos of all mutants recapitulated the mutant phenotype of zygotic embryos with respect to the absence of the SAM (Mordhorst et al., 2002). Because the *stm* mutation prevents embryonic as well as post-embryonic (organogenic) shoot meristem formation (Barton and Poethig, 1993), the appearance of a SAM in somatic embryos would have been very unlikely in that mutant. Also in *wus* mutant embryos and regenerated mutant plants, the mutant phenotype remained unchanged. In accordance with seed grown plants, defective meristems were repetitively reinitiated.

The defect due to the *zll* mutation only affects embryonic SAM formation, whereas adventitious functional SAMs can be formed on seedlings and *in vitro* from root explants (McConnell and Barton, 1995; Moussian et al., 1998). The role of *zll* was proposed to be involved in creating positional information required to acquire embryonic shoot meristem fate (Moussian et al., 1998). The finding that *zll* somatic embryos exhibited the same defects as observed in *zll* zygotic embryos suggests that the same embryo specific positional information is required for the development of somatic embryos. Apparently the intrinsic capability of SAM formation in *zll* mutants is not employed under tissue culture conditions. Thus, embryos from somatic cells and from the zygote use the same developmental program.

INDUCTION OF THE EMBRYOGENIC PATHWAY *IN VITRO*

Because auxins are generally used to induce indirect somatic embryogenesis, one of the important questions is whether auxins used in tissue culture experiments are specific inducers of developmental embryo programs. Alternatively, their effect could simply be to erase the existing differentiation and organization of the explant tissue. Reinitiating an embryogenic pathway can then be viewed as just one of the effects of disorganization.

In the carrot model system and for most of the other studied species, induction of the embryogenic state takes place after exposure to a synthetic auxin, in general the herbicide 2,4-dichlorophenoxyacetic acid (2,4-D). Removal of auxins is then required to prevent inhibition of postglobular stages of embryo development (Fujimura et al., 1980). In carrot (Kitamiya et al., 2000), alfalfa (Gyorgyey et al., 1991; Dudits et al., 1993), *Solanum melongena* (Magioli et al., 2001), and *Picea glauca* (Dong and Dunstan, 1996), a high concentration of 2,4-D supplied during a short period to the competent cultures can induce embryogenesis. These high concentrations of auxin trigger asymmetric divisions leading to embryo formation (Dudits et al., 1993).

Although only in few cases is direct evidence available to demonstrate that the effect of alternative treatments indeed acts by increasing embryogenic competence, several publications highlight the importance of extracellular glycosylated proteins. The formation of competent cells allows for the acceleration of acquisition of embryogenic competence as well as the PEM formation in young cultures (De Vries et al., 1988). Reciprocally, adding arabino-galactan proteins (AGPs) from nonembryogenic cultures to embryogenic ones makes them nonembryogenic (Kreuger and van Holst, 1993). It is necessary to point out that the cells that produce these AGPs are not the embryogenic cells (McCabe et al., 1997; Toonen et al., 1997). More recently, a class of small peptides, the phytosulfokines, was shown to improve somatic embryogenesis in carrot cultures, presumably through enhancing proliferation of embryogenic cells (Hanai et al., 2000).

Other methods to induce embryogenic competence make use of a change in the carbon source (Kunitake et al., 1997; Blanc et al., 1999; 2002) or the application of a heat shock, such as is applied to induce embryogenesis in *Brassica napus* microspores (Cordewener et al., 1995). Apparently, attainment of embryogenic competence in culture is highly dependent on both the specific treatment use for induction and the plant species.

GENES CONTROLLING EMBRYOGENIC COMPETENCE

Markers of embryogenic competence can be obtained by comparing the expression profiles of embryogenic cultures before and after culturing onto a medium containing auxin (De Vries et al., 1988). They can also be identified comparing embryogenic and nonembryogenic cultures, both before induction of embryo development, or by comparing cultures with and without embryo competence. A leucine-rich region (LRR) containing receptor kinase designated somatic embryogenesis receptor kinase (SERK) was identified that is expressed in single carrot cells that are competent to form embryos (Schmidt et al., 1997).

The protein AtGRP-5 is highly expressed not only in *A. thaliana* embryogenic calli just before induction of somatic embryogenesis, but also in cells actively dividing in the explants that give rise to embryogenic cells (Magioli et al., 2001). However, in searches for genes marking embryogenic competence, it has been quite difficult to distinguish between those that were induced more generally and those that were induced specifically. An alternative approach has been to use genes expressed very early during zygotic embryogenesis and to ask whether these were capable of improving embryogenic competence. One example is the *Arabidopsis* version of the *SERK* gene, which is designated as *AtSERK1* (Hecht et al., 2001). This gene was expressed in the ovule primordium, the entire female gametophyte, and the early embryo up until the heart stage of development. Upon ectopic expression, resulting seedlings looked morphologically similar to the wild type, but produced more embryogenic cells after exprosure to auxins (Hecht et al., 2001).

A second example is the *AGL15* gene. It is a member of the homeotic MADS-box gene family. *AGL15* is expressed in the *Arabidopsis* embryo from the first division up to the eight-cell stage (Rounsley et al., 1995). The expression level after this stage is very low, and is nil outside the embryo.

The expression of genes homologous to *AGL15* has been studied in the following three species: *Taraxacum officinalis*, *Medicago sativa*, and *B. napus.* In these three cases, the gene is primarily expressed in the tissues descended from the double fertilization event (Perry et al., 1996). It is not only expressed in embryos, but also the expression level outside the embryos is ten times lower (Heck et al., 1995; Rounsley et al., 1995). More recently, the *AGL15* gene was shown to be able to confer embryogenic potential under *in vitro* conditions after ectopic expression. Interestingly, this also resulted in upregulation of the *AtSERK1* gene, indicating that the *AGL15* gene could be a controlling factor for embryogenic competence in *Arabidopsis* (Harding et al., 2003).

Using a different approach in *Arabidopsis*, the initiation of somatic embryogenesis has now been described in several mutants. In some mutants, embryos developed spontaneously on tissue, reminiscent of direct somatic embryogenesis; whereas, in others, a prolonged induction with exogenous auxins was required (indirect somatic embryogenesis). *Arabidopsis pickle* (*pkl*) mutant root segments spontaneously formed somatic embryo-like structures upon dissection. This remark-able finding has been explained by assuming that *pkl* roots retain the embryonic characteristic of totipotency after germination, while it is normally lost in the wild type. Support of this hypothesis is seen in the prolonged synthesis of particular seed-specific fatty acids (Ogas et al., 1997). The *PKL* gene encodes a CHD3 protein, possibly involved in a general gene repression mechanism (Ogas et al., 1999). Consistent with the view that in *pkl* embryonic traits was not switched off upon germination was the observation that *LEC1* gene expression was not repressed in *pkl* seedlings (Ogas et al., 1999). *LEC1* is a putative transcriptional regulator controlling the embryogenic to vegetative transition (see below).

Another *Arabidopsis* mutant was identified that facilitated the efficient initiation of liquid embryogenic cell cultures. By using germinating seeds of the *primordia timing* mutant (*pt*, member of the *häuptling* (*hpt*)/*constitutive photomorphogenic2* (*cop2*)/*altered meristem program1* (*amp1*)/*pt* complementation group), embryogenic cell lines could be established reproducibly (Mordhorst et al., 1998). The *AMP1/PT* gene encodes a carboxypeptidase believed to be involved in signalling (Helliwell et al., 2001). *pt* zygotic embryos are first distinguishable from wild type at early heart-stage by a larger distance between the cotyledon primordia giving rise to a broader embryonic SAM. Embryogenic clusters originated from the enlarged SAM of *pt* seedlings germinated in liquid medium containing auxin (Mordhorst et al., 1998). Wild-type seeds treated under the same con-ditions never developed embryogenic clusters. Embryogenic cell lines could also be established from seedlings of other mutants with enlarged SAMs, such as *clavata1* and *clavata3* (*clv*). *pt clv1-4* and *pt clv3-2* double mutants showed additive effects on SAM size and an even higher frequency of seedlings producing embryogenic cell lines. This suggested at that time a positive correlation between SAM size and embryogenic cell formation (Mordhorst et al., 1998).

Interestingly, *pkl* seedlings formed embryogenic clusters that originated from the SAM region as was seen before in the case of *pt*. *ptpkl* double mutants showed additive effects on embryogenic cell formation (Henderson et al., 2004). Because the SAM size of *pkl* and of *ptpkl* double mutants is not increased compared to wild-type and *pt* single-mutant SAMs respectively, the size of the SAM is not decisive for embryogenic cell formation.

For *pt/amp1*, it has been suggested that there could be a delayed switch from the embryonic to the adult SAM program (Conway and Poethig, 1997). This would also fit with the recent interpretation of the *pkl* mutant in derepressing embryogenic traits after germination (Henderson et al., 2004). The additive effect on embryogenic cluster formation of various double mutants compared to the respective single mutants could be explained by the idea that several independent mechanisms contribute in derepression of embryogenic traits, with the effect of an increased number of cells that could become embryogenic upon the right trigger *in vitro*. Interestingly, only cells of the SAM of germinating seeds could be triggered in such a manner.

In *Arabidopsis*, the immature zygotic embryo (IZE), which by definition contains most embryonic cells, has been used with considerable success to initiate embryogenic cultures (Pillon et al., 1996; Luo and Koop, 1997, Mordhorst et al., 1998; 2002). The use of IZEs as explant material is the only system by which embryogenic cell lines of a large number of different *Arabidopsis* wild types ("ecotypes") have been obtained. The use of geminating seeds for establishment of embryogenic cell lines is only possible for mutants such as *pt, pkl, clv1* or *clv3*, whereas wild type seeds do not respond reproducibly. The efficiency of embryogenic cell line establishment depends on the ecotypes used and the developmental stage of the IZE (Luo and Koop 1997; Mordhorst et al., 1998). In contrast to different ecotypes and the above mentioned mutants, using *lec1* IZEs, it was not possible to initiate somatic embryogenesis, confirming the prediction that in *lec1* the ability to produce embryogenic cells is reduced even in immature zygotic embryos (Mordhorst and de Vries, unpublished results). The *lec1* mutation resulted in desiccation-intolerant embryos with overlapping embryonic and seedling characteristics due to an enhanced maturation process (Meinke et al., 1994; West et al., 1993). According to this developmental feature, *lec1* immature zygotic embryos are physiologically more advanced, but still have embryo morphology, a phenotype that is reflected in the *in vitro* culture response. Because the expression of *LEC1* can create the ability to form embryogenic cells and leads to formation of ectopic embryos (Lotan et al., 1998), molecular evidence is now available to suggest that embryogenic competence involves resetting to a state that exists in (some) cells of the IZE.

For wild-type IZEs, it is evident that only a subset of cells is competent to become embryogenic — the region of the embryonic SAM and the cells of the axils of the cotyledons (Mordhorst et al., 2002). After removal of the SAM, formation of somatic embryos from the cotyledon axils was not affected in *STM* (Mordhorst et al., 2002). The embryogenic competence of cells of the cotyledon axil is apparently not influenced by the mutation. Therefore a SAM or SAM activity is not required for embryogenic cell formation in *Arabidopsis*. In addition, it shows that in *Arabidopsis*, IZEs of two different cell types share the competence to initiate somatic embryogenesis. In the wild type, this competence is lost in both cell types after germination, whereas it is maintained in some mutant seedlings (see earlier).

The most likely candidates for cells able to become embryo-competent in the SAM are members of a small stem cell population required to maintain an active meristem status. An indication that this may actually be the case comes from recent experiments where an activation-tagging system was used to identify genes of which overexpression caused somatic embryo formation in all tissues without exogenous hormones. One of these genes was identical to *WUS*, a homeodomain transcriptional regulator (Zuo et al., 2002). The *WUS* gene was previously shown to be involved in specifying stem cell fate in shoot and floral meristems (Laux et al., 1996; Mayer et al., 1998). *WUS* is apparently also able to promote or maintain an embryonic stem cell fate in cells other than those of the SAM.

Besides genes affecting the embryonic to vegetative transition or genes that are part of the controlling pathway to specify and maintain stem cells, several other genes have been found that can confer the ability to promote embryogenic cell formation. Examples are the already mentioned *LEC1* gene, encoding a HAP3 subunit of the CCAAT binding protein (Lotan et al., 1998; Lee et al., 2003), the B3 domain transcription factor *LEC2* (Stone et al., 2001), the AP2 domain transcription factor *BABYBOOM* (*BBM*; Boutilier et al., 2002) and the receptor kinase *AtSERK1* (Hecht et al., 2001). In plants ectopically expressing the first three genes, somatic embryos develop on vegetative organs spontaneously (i.e., without any external hormonal stimulus). *LEC1* was postulated also to be responsible for embryo formation in *pkl* roots, since in *pkl* roots *LEC1* expression is not repressed. The gain of function phenotypes of *LEC1*, *LEC2*, and *BBM* are pleiotrophic. In addition to the development of ectopic somatic embryos, callus development, cotyledon-like and leaf-like structures, and shoot development were described (Stone et al., 2001; Boutilier et al., 2002). None of the target genes of these three seed-expressed transcription factors are yet known. Because they share a common gain-of-function phenotype, they might have overlapping functions during zygotic embryogenesis as well, such as establishing and/or maintaining embryogenic competence (Stone et al., 2001; Boutilier et al., 2002). It is also of interest to note that *BBM* overexpression causes ectopic somatic embryos to develop only from cells normally competent to undergo differentiation, such as cells of the shoot apex or the marginal tissues of young seedling leaves. *AtSERK1* did not induce any morphological phenotype when overexpressed (Hecht et al., 2001). However, ectopic expression of *AtSERK1* facilitated the initiation of embryonic cell clusters from the SAM in germinating seeds, but only after exposure to auxins.

CONCLUSIONS

From the research discussed here, it appears that the change of the cell fate from somatic to embryo competence can be achieved either by overexpression of various genes in the absence of any hormonal stimulus (e.g., *LEC1*, *LEC2*, *BBM*, *WUS*, *AGL15*) or in combination with an auxin treatment (*AtSERK*). On the other hand, recessive mutations in several genes (*pkl*, *pt/amp1*, *clv1*, *clv3*) make embryogenic culture initiation possible using germinating seeds under conditions where it is not possible with wild-type seeds. In the case of *pkl*, spontaneous somatic embryos were also found.

At first sight, there does not appear to be any link between the genes that are able to confer embryogenic potential upon ectopic expression and those that confer increased embryogenic potential when mutated. Interestingly, the overexpression of just one gene may be sufficient to initiate the embryogenesis program. Repression of such genes after completion of embryogenesis might be essential for normal plant development. Mutations in negative regulators of such genes may also result in spontaneous embryo development. Alternatively, the presence of increased numbers of embryonic cells after germination could facilitate *in vitro* induction of embryogenic cells. The data so far available suggest the presence of at least one but probably more complex networks of interacting genes involved in initiating, maintaining, and repressing the embryogenic or totipotent status of plant cells.

Interestingly, mutations in two of the genes (*LEC1* and *WUS*) that upon overexpression result in ectopic somatic embryo formation have been analyzed for the presence of embryogenic competent cells in the IZEs. So far, IZEs are the only wild-type tissue in *Arabidopsis* that contain cells competent to become embryogenic in tissue culture. While *LEC1* eliminated such competent cells completely, they were still present in *WUS*.

The underlying molecular mechanism changing cell fate from somatic into embryo-competent remains to be discovered. It also remains to be determined whether there are multiple independent pathways involved or whether a single integrated network of genes controls embryonic competence. Nevertheless, the genes now identified may have practical implications for improving regeneration of recalcitrant species or genotypes.

LITERATURE CITED AND SUGGESTED READINGS

Barton, M.K. and R.S. Poethig. 1993. Formation of the shoot apical meristem in *Arabidopsis thaliana* — An analysis of development in the wild type and in the shoot meristemless mutant. *Development* 119:823–831.

Blanc, G., N. Michaux-Ferrière, L.Teisson, L. Lardet and M.P. Carron. 1999. Effects of carbohydrate addition on the induction of somatic embryogenesis in *Hevea brasiliensis*. *Plant Cell Tissue Org. Cult.* 59:103–112.

Blanc, G. et al. 2002. Differential carbohydrate metabolism conducts morphogenesis in embryogenic callus of *Hevea brasiliensis* (Müll. Arg.). *J. Exp. Bot.* 53:1453–1462.

Boutilier, K., R. Offringa, V.K. Sharma, H. Kieft, T. Ouellet, L. Zhang, J. Hattori, C.M. Liu, A.A. van Lammeren, B.L. Miki, J.B. Custers, and M.M. van Lookeren Campagne. 2002. Ectopic expression of BABY BOOM triggers a conversion from vegetative to embryonic growth. *Plant Cell* 14:1737–1749.

Carman, J.G. 1990. Embryogenic cells in plant tissue cultures: occurrence and behavior. In Vitro *Cell Dev. Biol.* 26:746–753.

Choi, J.H. and Z.R. Sung 1989. Induction, commitment, and progression of plant embryogenesis. In: *Plant Biotechnology*, Vol. 11. *(*Ed.) S.D. Kung and C.F. Arntzen. Butterworth Publishers, Stoneham, MA. 141–159.

Conway, L.J. and R.S. Poethig. 1997. Mutations of *Arabidopsis thaliana* that transform leaves into cotyledons. *Proc. Natl. Acad. Sci. U.S.A.* 94:10209–10214.

Cordewener, J.H.G., G. Hause, E. Gorgen, R. Busink, B. Hause, H.J.M Dons, A.A.M van Lammeren, M. M. van Lookeren Campagne and P. Pechan. 1995. Changes in synthesis and localization of members of the 70-kDa class of heat-shock proteins accompany the induction of embryogenesis in *Brassica napus* L. microspores. *Planta* 196:747–755.

De Jong, A.J., E.D.L. Schmidt, and S.C. de Vries. 1993. Early events in higher-plant embryogenesis. *Plant Mol. Biol.* 22:367–377.

De Vries, S.C., H. Booij, R. Janssen, R.Vogels, L. Saris, F. LoSchiavo, M. Terzi and A. van Kammen. 1988. Carrot somatic embryogenesis depends on the phytohormone-controlled expression of correctly gly-cosylated extracellular proteins. *Genes Devel.* 2:462–467.

Dong, J.Z. and D.I. Dunstan. 1996. Characterization of three heat-stock-protein genes and their developmental regulation during somatic embryogenesis in white spruce [*Picea glauca* (Moench) Voss]. *Planta* 200:85–91.

Dudits, D., L. Bogre, and J. Gyorgyey. 1993. Molecular and cellular approaches to the analysis of plant embryo development from somatic cells *in vitro*. *J. Cell Sci.* 99:475–484.

Dudits, D., J. Gyorgyey, L. Bogre and Bako L. 1995. Molecular biology of somatic embryogenesis. In: In Vitro *Embryogenesis in Plants*. (Ed.) T.A. Thorpe. Kluwer Academic Publishers, Dordrecht, the Netherlands. 267–308.

Endrizzi, K., B. Moussian, A. Haecker, J.Z. Levin and T. Laux. 1996. The *SHOOT MERISTEMLESS* gene is required for maintenance of undifferentiated cells in *Arabidopsis* shoot and floral meristems and acts at a different regulatory level than the meristem genes *WUSCHEL* and *ZWILLE*. *Plant J.* 10:967–979.

Finstad, K., D.C.W. Brown, and K. Joy. 1993. Characterization of competence during induction of somatic embryogenesis in alfalfa tissue cultures. *Plant Cell Tissue Org. Cult.* 34:125–132.

Fujimura, T., A. Komamine, and H. Matsumoto. 1980. Aspects of DNA, RNA and protein synthesis during somatic embryogenesis in a carrot cell suspension culture. *Physiol. Plant.* 49:255–260.

Gyorgyey, J., A. Gartner, K. Nemeth, H. Hirt, E. Heberle-Bors and D. Dudits. 1991. Alfalfa heat shock genes are differentially expressed during somatic embryogenesis. *Plant Mol. Biol.* 16:999–1007.

Halperin, W. 1966. Alternative morphogenetic events in cell suspensions. *Amer. J. Bot.* 53:443–453.

Hanai, H., T. Matsuno, M. Yamamoto, Y. Matsubayashi, T. Kobayashi, H. Kamada and Y. Sakagami. 2000. A secreted peptide growth factor, phytosulfokine, acting as a stimulatory factor of carrot somatic embryo formation. *Plant Cell Physiol.* 41:27–32.

Harding, E.W.,W. Tang, K.W. Nichols, D.E. Fernandez and S.E. Perry. 2003. Expression and maintenance of embryogenic potential is enhanced through constitutive expression of AGAMOUS-Like 15. *Plant Physiol.* 133:653-663.

Hecht, V. et al. 2001. The *Arabidopsis SOMATIC EMBRYOGENESIS RECEPTOR KINASE 1* gene is expressed in developing ovules and embryos and enhances embryogenic competence in culture. *Plant Physiol.* 127:803–816.

Heck, G.R., S. E. Perry, K.W. Nichols and D.E. Fernandez. 1995. AGL15, a MADS domain protein expressed in developing embryos. *Plant Cell* 7:1271–1282.

Helliwell, C.A., A.N. Chin-Atkins, I.W. Wilson, R. Chapple, E.S. Dennis and A. Chaudhury. 2001. The *Arabidopsis AMP1* gene encodes a putative glutamate carboxypeptidase. *Plant Cell* 13:2115–2125.

Henderson, J., H.-C. Li, S.D. Rider, A.P. Mordhorst, J. Romero-Severson, J.-C. Cheng, J. Robey, Z.R. Sung, S.C. de Vries and J. Ogas. 2004. PICKLE plays a role in GA-dependent responses and acts throughout the plant to repress expression of embryonic traits. *Plant Physiol.* 134:995–1005.

Kitamiya, E. et al. 2000. Isolation of two genes that were induced upon the initiation of somatic embryogenesis on carrot hypocotyls by high concentrations of 2,4-D. *Plant Cell Rep.* 19:551–557.

Kreuger, M. and G.J. van Holst. 1993. Arabinogalactan proteins are essential in somatic embryogenesis of *Daucus carota* L. *Planta* 189:243–248.

Kunitake, H., T. Nakashima, K. Mori and M. Tanaka. 1997. Normalization of Asparagus somatic embryogenesis using a maltose-containing medium. *J. Plant Physiol.* 150:458–461.

Laux, T., K.F. Mayer, J. Berger and G. Jürgens. 1996. The WUSCHEL gene is required for shoot and floral meristem integrity in *Arabidopsis. Development* 122:87–96.

Lee, H., R.L. Fischer, R.B. Goldberg and J.J. Harada. 2003. *Arabidopsis* LEAFY COTYLEDON1 represents a functionally specialized subunit of the CCAAT binding transcription factor. *Proc. Natl. Acad. Sci. U.S.A.* 100:2152–2156.

Long, J.A., E.I. Moan, J.I. Medford and M.K. Barton. 1996. A member of the KNOTTED class of home-odomain proteins encoded by the *STM* gene of *Arabidopsis. Nature* 379:66–69.

Lotan, T., M.A. Ohto, K.M. Yee, M. A.L. West, R. Lo, R.W. Kwong, K.I. Yamagishi, R.L. Fisher, R.B. Goldberg and J. J. Harada. 1998. *Arabidopsis LEAFY COTYLEDON1* is sufficient to induce embryo development in vegetative cells. *Cell* 93:1195–1205.

Luo, Y. and H.U. Koop. 1997. Somatic embryogenesis in cultured immature zygotic embryos and leaf protoplasts of *Arabidopsis thaliana* ecotypes. *Planta* 202:387–396.

Lynn, K., A. Fernandez, M. Aida, J. Sedbrook, M. Tasaka, P. Masson and M.K. Barton. 1999. The *PIN-HEAD/ZWILLE* gene acts pleiotropically in *Arabidopsis* development and has overlapping functions with the *ARGONAUTE1* gene. *Development* 126:469–481.

Magioli, C., R.M. Barroco, C.A. Benicio Rocha, L.D. Santiago-Fernandes, E. Mansur, G. Engler, M. Marghis-Pineiro and G. Sachetto-Martins. 2001. Somatic embryo formation in *Arabidopsis* and eggplant is associated with expression of a glycine-rich protein gene (*Atgrp-5*). *Plant Sci.* 161:559–567.

Mayer, K.F.X., H. Schoof, A. Haecker, M. Lenhard, G. Jürgens and T. Laux. 1998. Role of *WUSCHEL* in regulating stem cell fate in the *Arabidopsis* shoot meristem. *Cell* 95:805–815.

McCabe, P.F., T.A. Valentine, L.S. Forsberg and R.I. Pennell. 1997. Soluble signals from cells identified at the cell wall establish a developmental pathway in carrot. *Plant Cell* 9:2225–2241.

McConnell, J.R. and M.K. Barton. 1995. Effect of mutations in the *PINHEAD* gene of *Arabidopsis* on the formation of shoot apical meristems. *Devel. Genet.* 16:358–366.

Meinke, D.W., L. H. Franzmann, T. C. Nickle and E. C. Yeung. 1994. Leafy cotyledon mutants of *Arabidopsis. Plant Cell* 6:1049–1064.

Mordhorst, A.P., M.A.J. Toonen, and S.C. de Vries.1997. Plant embryogenesis. *Crit. Rev. Plant Sci.* 16:535–576.

Mordhorst, A.P., K.J. Voerman, M.V. Hartog, E.A. Meijer, J. van Went, M. Koornneef and S.C. de Vries. 1998. Somatic embryogenesis in *Arabidopsis thaliana* is facilitated by mutations in genes repressing excess meristematic cell divisions. *Genetics* 149:549–563.

Mordhorst, A.P., M.V. Hartog, M.K., T. El Tamer Laux and S.C. de Vries. 2002. Somatic embryogenesis from *Arabidopsis* shoot apical meristem mutants. *Planta* 214:829–36.

Moussian, B., H. Schoof, A. Haecker, G. Jürgens and T. Laux. 1998. Role of the ZWILLE gene in the regulation of central shoot meristem cell fate during *Arabidopsis* embryogenesis. *EMBO J.* 17:1799–1809.

Ogas, J., J-C. Cheng, Z.R. Sung and C. Somerville. 1997. Cellular differentiation regulated by gibberellin in the *Arabidopsis thaliana pickle* mutant. *Science* 277:91–94.

Ogas, J., S. Kaufmann, J. Henderson and C. Somerville. 1999. *PICKLE* is a CHD3 chromatin-remodeling factor that regulates the transition from embryogenic to vegetative development in *Arabidopsis. Proc. Nat. Acad. Sci. U.S.A.* 96:13839–13844.

Perry, S.E., M.D. Lethi, and D.E. Fernandez. 1999. The MADS-domain protein AGAMOUS-like 15 accumulates in embryonic tissues with diverse origins. *Plant Physiol.* 120:121–130.

Perry, S.E., K.W. Nichols, and D.E. Fernandez. 1996. The MADS domain protein AGL15 localizes to the nucleus during early stages of seed development. *Plant Cell* 8:1977–1989.

Pillon, E., M. Terzi, B. Baldan, P. Mariani and F. Loschiavo. 1996. A protocol for obtaining embryogenic cell lines from *Arabidopsis. Plant J.* 9:573–577.

Rounsley, S.D., G.S. Ditta, and M.F. Yanofsky. 1995. Diverse roles for *MADS box* genes in *Arabidopsis* development. *Plant Cell* 7:1259–1269.

Schmidt, E.D.L., F. Guzzo, M.A. Toonen and S.C. de Vries. 1997. A leucine-rich repeat containing receptor-like kinase marks somatic plant cells competent to form embryos. *Development* 124:2049–2062.

Stone, S.L., L.W. Kwong, K.M. Yee, J. Pelletier, L. Lepiniec, R.L. Fischer, R.B. Goldberg and J.J. Harada. 2001. LEAFY COTYLEDON2 encodes a B3 domain transcription factor that induces embryo development. *Proc. Natl. Acad. Sci. U.S.A.* 98:11806–11811.

Toonen, M.A.J., T. Hendriks, E.D.L. Schmidt, H.A. Verhoeven, A. van Kammen and S.C. de Vries. 1994. Description of somatic-embryo-forming single cells in carrot suspension cultures employing video cell tracking. *Planta* 194:565–572.

Toonen, M.A., E.D.L. Schmidt, A. van Kammen and S.C. de Vries. 1997. Promotive and inhibitory effects of diverse arabinogalactan proteins on *Daucus carota* L. Somatic embryogenesis. *Planta* 203:188–195.

West, M.A.L. and J.J. Harada. 1993. Embryogenesis in higher plants, an overview. *Plant Cell* 5:1361–1369.

Zimmerman, J.L. 1993. Somatic embryogenesis: a model for early development in higher plants. *Plant Cell* 5:1411–1423.

Zuo, J., Q.W. Niu, G. Frugis and N.H. Chua. 2002. The WUSCHEL gene promotes vegetative-to-embryonic transition in *Arabidopsis*. *Plant J.* 30:349–359.

Section V

Crop Improvement Techniques

16 Use of Protoplasts for Plant Improvement

*Richard E. Veilleux, Michael E. Compton,
and James A. Saunders*

CHAPTER 16 CONCEPTS

- Protoplasts are plant cells from which the cell wall has been removed.

- Protoplasts can be obtained from various plant tissues and cultures using hydrolytic enzymes to digest cell wall components and a medium with high osmotic pressure to prevent the protoplasts from bursting.

- Protoplast culture has been used to develop plants with improved agronomic and horticultural characteristics, including disease resistance through recovery of culture-induced variants (somaclones), parasexual hybrids (somatic hybrids), and transgenic plants.

- Protoplast fusion is usually accomplished with either polyethylene glycol (PEG) or electric shock (electrofusion). Hybrid cell fusions can be selected from the nonfused protoplasts by physiological characteristics (e.g., antibiotic resistance), complementary fluorescent stains, and/or molecular (DNA) markers.

- Genetic transformation of single cells can be achieved by co-cultivating protoplasts with *Agrobacterium tumefaciens* or by direct DNA transfer using polycationic chemicals, electroporation, liposomes, microinjection, or sonication.

INTRODUCTION

Plant cells from which the cell wall has been removed are termed protoplasts. Protoplasts are somewhat unique in plant cell culture in that they exist as separate cells without cytoplasmic continuity among neighboring cells. Communication among protoplasts sharing the same culture environment is therefore limited to metabolites that can traverse the plasmalemma into the culture medium and influence the behavior of other cells. This limits the direct pathways of communication through cytoplasmic plasmodesmata generally present with intact plant tissues. However, in protoplast cultures, the barrier imposed by plant cell walls has also been eliminated. The plant cell wall, although critical to plant structure and function, is a major impediment in exploiting direct DNA transfer to individual cells and the production of somatic hybrids by cell fusion. Removal of the cell wall temporarily during protoplast culture can result in viable cells with properties otherwise unknown in plants.

Plant protoplasts were first isolated by Klercker in 1892 by slicing onion bulb scales with a thin knife in a plasmolyzing solution, resulting in the release of protoplasts when cells were cut through the wall (Bhojwani and Razdan, 1983). Protoplast yield was low and the procedure was restricted to highly vacuolated, nonmeristematic cells. Methods to isolate plant protoplasts were improved in the 1960s with the extraction and purification of enzymes that could degrade the plant cell wall. Cocking (1960) discovered that intact protoplasts could easily be obtained by incubating plant tissues in a concentrated cellulase solution prepared from the fungus *Myrothecium verrucaria*. By 1968, commercial preparations of purified cell wall degrading enzymes, such as macerozyme and cellulase, were available that allowed easy isolation of protoplasts from plant tissue.

ISOLATION OF PLANT PROTOPLASTS

Protoplast isolation by enzymatic cell wall digestion involves the use of cellulase, hemicellulase, and/or pectinase, which are extracted from various sources, including fungi and snail and termite gut. These hydrolytic enzymes are available commercially in differing formulations of varying purity. Digestion by a combination of these three enzymes is generally conducted at a pH of 5.5 to 5.8 over a period of 3 to 18 hours. Protoplasts can then be collected and purified using centrifugation to separate broken and damaged cells from intact protoplasts by taking advantage of their differing buoyant densities.

USES FOR PLANT PROTOPLASTS

Plant breeders have used sexual hybridization to improve cultivated crops for centuries. This process is generally limited to plants within a species or to wild species that are closely related to the cultivated crop. When possible, sexual hybrids between distantly related species have been useful for the incorporation of single gene traits such as insect and disease resistance. However, intraspecific and interspecific incompatibility barriers limit the use of sexual hybridization for accessing germplasm distantly related to crop species.

Protoplast culture has been used to develop plants with improved agronomic and horticultural characteristics and improved disease resistance through recovery of culture-induced variants (Chapters 25 and 26), parasexual hybrids, and genetically engineered plants. Regenerated plants from cultured protoplasts have included variants (somaclones) with improved characteristics. Parasexual or somatic hybrids can be obtained by fusing protoplasts of distantly related species. Through genetic engineering (Chapter 19), foreign genes can be inserted into plant protoplasts that can be regenerated into improved plants that express the inserted genes. Improved plants obtained by protoplast manipulation can be used in breeding programs to develop new cultivars.

PROCEDURES FOR REGENERATING PLANTS FROM PROTOPLASTS

Leaf tissue from plants grown in a greenhouse or growth chamber with controlled light (16:8 or 18:6 hour [light:dark] photoperiod, 60–300 $\mu M \times m^{-2} \times s^{-1}$ light intensity) and temperature (20–25°C) is commonly used as a source of plant protoplasts. A variable greenhouse environment may hinder the repeatability of experiments. Cells within the leaf mesophyll are loosely packed, allowing penetration of digestive enzymes that facilitate protoplast release. Typically 10^6 to 5×10^7 protoplasts can be obtained per gram of leaf tissue. Generally, the youngest fully expanded leaves from young plants or seedlings are used. Preconditioning — e.g., placing plants in darkness for 24 to 72 hours before protoplast isolation or giving them a cold treatment (4–10°C) — may improve protoplast yield.

Leaves from greenhouse- or growth-chamber-grown plants require surface disinfection prior to protoplast isolation. *In vitro* plantlets are often used as a source of protoplasts because of the advantages conferred by aseptic conditions and a controlled environment. *In vitro* sources of protoplasts include callus, cotyledons, hypocotyls, embryogenic suspension cultures, leaves, shoots, or somatic embryos.

Before protoplast isolation, plant tissues are cut into small pieces or wounded and placed in an osmotically adjusted solution at 20 to 25°C for 1 to 24 hours. During this step, water moves out of the cells by osmosis, causing the cytoplasmic content to shrink and withdraw from the cell wall (plasmolysis). This allows cells to retain their integrity after the cell wall is removed. Sometimes it is beneficial to conduct this step at a low temperature (4–10°C). Next, the plant material is incubated in digestive enzymes for removal of the cell wall and middle lamellae (Chapter 7). Incubation is conducted in darkness on a shaker (30–50 rpm) for 3 to 18 hours at 25 to 30°C. Incubation time and temperature vary with species and tissue. Plasmolysis and enzyme incubation may be simultaneous. Following enzyme treatment, the digest is gently swirled to release protoplasts and filtered (50–100 μm mesh size) to separate protoplasts from debris. The protoplast suspension is centrifuged at low speed (50 H g) for about 5 to 10 minutes. Protoplasts and small debris collect in the pellet at the bottom of the tube. The supernatant is discarded and protoplasts are resuspended in a high sucrose (flotation) medium, which is overlaid with rinse medium containing mannitol. During centrifugation, viable protoplasts collect at the interface of the two media while debris concentrates in the pellet. Protoplasts are removed from the interface with a pasteur pipet and washed two to three times in rinse medium before transfer to culture medium. There are several variations for purification of protoplasts that combine or separate the various steps outlined in this general procedure.

After the final wash, protoplasts are suspended in a small volume (~1 ml) of liquid culture medium. A sample is removed for cell counting in a haemacytometer. Enough culture medium should be added to adjust the protoplast density to 10^4–10^6 cells per ml. Protoplasts may be cultured in a thin layer of liquid medium embedded in agarose, in liquid medium on top of solid medium, or in agarose droplets suspended in liquid. The type of culture depends on the species and the purpose of the experiment. Liquid medium is often used because cell density and osmotic pressure can be efficiently adjusted. In addition, cells of some species are unable to divide in agar-solidified medium. The development of individual cells can be observed if protoplasts are embedded in agarose.

Protoplast viability may be estimated by staining a sample with fluorescein diacetate (FDA). The stain is mixed with protoplasts and fluorescence is observed under a fluorescent microscope equipped with filters for excitation at 488 nm and emission at 530 nm (green). Viable protoplasts actively absorb the stain and exhibit green fluorescence, whereas nonviable protoplasts do not. Nonviable mesophyll protoplasts fluoresce red due to autofluorescence of chlorophyll (assuming that the filters on the microscope permit observation of wavelengths ranging from green to red). Another viability stain, Evans blue, does not require fluorescence microscopy to detect protoplast viability.

Freshly isolated protoplasts are spherical (Figure 16.1) because they are unrestricted by a cell wall. Under suitable conditions, viable protoplasts regenerate a new cell wall within 48 to 96 hours after isolation. The presence of the cell wall can be determined by a change in shape from spherical to ovoid or by staining cells with Calcofluor White, which detects the presence of cellulosic wall materials. Stained cells or protoplasts with remnant cell wall material or protoplasts that have resynthesized a new cell wall fluoresce bluish-white under UV light. Protoplasts without a cell wall do not fluoresce. Protoplasts that fail to regenerate a wall generally will not divide and will eventually die.

Not all viable protoplasts divide. Therefore, the plating efficiency (PE) (equivalent to the number of dividing protoplasts divided by the total number of protoplasts) is used as an initial estimate of the potential of a culture to grow. PE is usually calculated 1 week after protoplast isolation and varies from 0 to 80%. However, a PE of 20 to 30% is most common. The first cell division

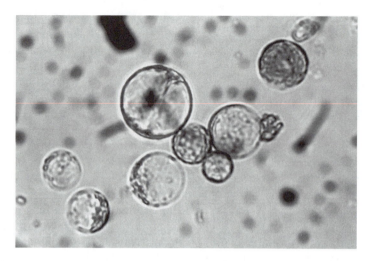

FIGURE 16.1 Freshly isolated protoplasts from *in vitro* plantlets of *Solanum phureja*.

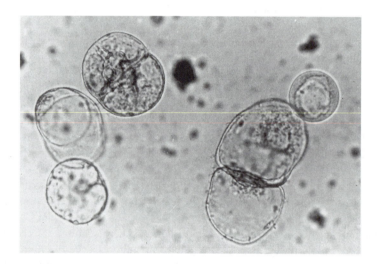

FIGURE 16.2 First mitotic division after protoplast isolation of *Solanum phureja*. Notice that the cells are no longer spherical and are more opaque than freshly isolated protoplasts. (Reprinted from Cheng, J. and Veilleux, R.E., Genetic analysis of protoplast culturability in *Solanum phureja*, *Plant Sci.*, 75: 257–265, ©1991 with permission from Elsevier.)

(Figure 16.2) often occurs within 2 to 10 days after isolation. Division of protoplasts is affected by culture medium, osmoticum, plating density, conditions of incubation, and source tissue.

After protoplasts have regenerated a new wall and divided, they are diluted with culture medium that has lower osmoticum. Multicellular colonies (Figure 16.3) develop within 14 to 21 days, and macroscopic calli (Figure 16.4) within 4 weeks. Plant regeneration occurs primarily by shoot regeneration, or occasionally by somatic embryogenesis. The first shoots or embryos may be seen as early as 1 month after protoplast isolation, but may require much longer incubation periods.

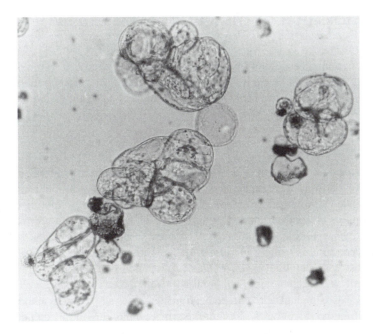

FIGURE 16.3 Multicellular colonies 7 days after isolation of protoplasts of *Solanum phureja*. (Reprinted from Cheng, J. and Veilleux, R.E., Genetic analysis of protoplast culturability in *Solanum phureja, Plant Sci.*, 75: 257–265, ©1991 with permission from Elsevier.)

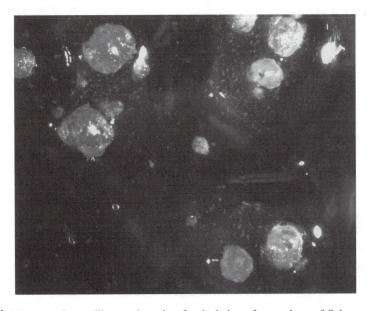

FIGURE 16.4 Macroscopic p-calli several weeks after isolation of protoplasts of *Solanum phureja*.

FACTORS THAT INFLUENCE PROTOPLAST GROWTH AND DEVELOPMENT

Modifications to normal tissue culture procedures often are required due to the delicate nature of protoplasts. Changes often include adjustment to the inorganic salt concentration, the addition of organic components, vitamins, or sugars to control osmolarity, and the addition of plant growth regulators (PGRs) to stimulate cell division.

Culture Media

The best culture medium for protoplasts is often similar to, or slightly modified from, that used for organ regeneration from other explants. Adjustment of ammonium nitrate and calcium is often required to stimulate cell division. Ammonium nitrate is essential for promoting cell division, but is toxic to protoplasts at concentrations (20 mM) used in most tissue culture media. Hence, the ammonium nitrate concentration is lowered to 1/4 to 1/2 of the normal concentration. In contrast, the calcium concentration is usually increased for most protoplast procedures. The calcium concentration of most tissue culture media ranges from 0.5 to 3 mM. At this low concentration, protoplasts typically aggregate and brown rapidly. Raising the calcium concentration (14–40 mM) promotes early cell division and cell synchronization, decreases aggregation, and reduces browning of protoplasts during the early stages of culture. Normal ammonium nitrate and calcium levels are restored after protoplasts have regenerated a new wall and divided.

Organic components such as inositol, nicotinic acid, pyridoxine, thiamine, glycine, folic acid, and biotin are found in many tissue culture media. In addition, casein hydrolysate, D-Ca-pantothenate, choline chloride, cysteine, malic acid, ascorbic acid, adenine sulfate, riboflavin, and glutamine are often added to protoplast media in small amounts (0.01–10 mg/l) to hasten cell wall synthesis and promote cell division. Most of these compounds are not used after protoplasts have regenerated a new cell wall and divided.

Sugars are used in protoplast culture media as osmotic stabilizers and carbon sources. Due to the elimination of wall pressure (i.e., pressure potential), a component of water potential, culture media must be osmotically adjusted to prevent rupture of protoplasts during isolation and early culture. Mannitol, sorbitol, and glucose are used as osmotica at concentrations from 0.3 to 0.7 M. Sucrose and glucose are used as carbon sources at concentrations from 0.2 to 0.6 M. The sugar concentration is gradually reduced after cells synthesize a new wall and divide. Osmotic sugars are usually eliminated by the time macroscopic colonies are visible.

Exogenous PGRs are necessary to promote cell division. Both auxins and cytokinins typically are used. The type and concentration differ with the species and mode of regeneration. Naphthaleneacetic acid (NAA) and 2,4-dichlorophenoxyacetic acid (2,4-D) at 0.45 to 10.7 μM are typical auxins and concentrations. Benzyladenine (BA) and zeatin are the usual cytokinins used at concentrations of 2 to 5 μM. Kinetin and zeatin riboside are sometimes used instead of, or in combination with, BA or zeatin at similar concentrations. Coconut milk has been substituted for cytokinin at concentrations of 20 to 40 ml/l.

Culture Environment

Following isolation, protoplasts are sensitive to light and should be incubated in darkness or exposed only to filtered light (5–30 μ$M \times m^{-2} \times s^{-1}$) until they synthesize a new wall and divide. At this point, they can be exposed to elevated light levels (30–300 μ$M \times m^{-2} \times s^{-1}$). White light is often best as blue or red light alone may inhibit shoot formation. Protoplasts are generally incubated at 20 to 25°C. Optimum environmental conditions vary and should be tested for each species.

SOMACLONAL VARIATION

Spontaneous mutation rates in plants vary depending on the species and DNA sequence. When mutations occur in somatic cells, they generally are of little consequence because single somatic cells harboring a point mutation within mature plant tissue do not divide to give rise to visibly mutated sectors of tissue. Protoplast culture can be used as a means of recovering somatic mutations because the culture environment encourages the division of individual somatic cells. About 10^4 to 10^6 protoplasts are cultured per milliliter of medium with 40,000 to 10,000,000 cells present in a typical 15×60 mm petri dish with 4 to 10 ml of medium. Given that each cell is viable and has a chance of regenerating into a plant, there is a high possibility of recovering somatic variants (Chapter 26), derived either from preexisting mutations harbored in the explant tissue or from novel variation induced during culture.

Somaclonal variants have been recovered from protoplast cultures of many species (Veilleux and Johnson, 1998). Variations include changes in leaf and flower morphology, fertility, improved disease resistance, and variation in secondary product production. For example, protoclones — i.e., protoplast-derived plants of an asexually (clonally) propagated crop — have been examined in potato with the hope of obtaining plants with improved horticultural characteristics. Protoclones that exhibited compact growth habit; variation in tuber characteristics such as maturity date, color, russeting, protein content, size, and number of tubers per plant; changes in flowering response to photoperiod; and improved resistance to early blight (*Alternaria solani*) and late blight (*Phytophthora infestans*) were regenerated from protoplasts of potato (*Solanum tuberosum* L. cvs. Russet Burbank [Shepard et al., 1980] and Bintje [Burg et al., 1989]). Plants with improved resistance to fusarium wilt (*Fusarium oxysporum* f. sp. *lycoperici* race 2) were obtained by challenging tomato (*Lycopersicon esculentum* Mill. cv. UC 82) protoplasts and protoplast-derived calli (p-calli) with fusaric acid (Shahin and Spivey, 1986), and plants with elevated lanatoside C (a glycoside that yields digoxin, glucose, and acetic acid on hydrolysis) content were regenerated from *Digitalis lanata* J.F. Ehrh. protoplasts (Diettrich et al., 1991).

The most frequent form of somaclonal variation observed after regeneration of plants from protoplasts is polyploidy. This occurs through endopolyploidization (doubling of the chromosome number during mitosis without cytokinesis), spontaneous cell fusion, or regeneration from polyploid cells already present in the plant tissue from which protoplasts were isolated. Recovery of aneuploids (plants having a chromosome number that is not a multiple of the basic number) is also possible. Polyploid or aneuploid regenerants may outnumber or even exclude regenerants that express the original chromosome number predominant in the source tissue. However, more often somaclonal variants of this type comprise only a small portion of the total number of regenerants.

PROTOPLAST FUSION AND SOMATIC HYBRIDIZATION

Once purified protoplasts have been obtained from two different plant or tissue sources, various treatments can be applied to induce them to fuse together to form hybrid cell lines. The simplest way to do this is to use spontaneously fusogenic protoplasts. Such cell lines have been described in detail in carrot and other systems (Boss, 1987), but they are not common enough for widespread use in plant protoplast research. Generally, chemical agents or electrical manipulation are necessary to induce membrane instability that leads to protoplast fusion.

POLYETHYLENE GLYCOL-MEDIATED PROTOPLAST FUSION

Several aqueous solutions have been used to induce chemical fusion of plant protoplasts (Saunders and Bates, 1987). These include salt solutions (NaCl, KCl, $NaNO_3$, KNO_3), dextran sulfate, polyvinyl alcohol, lysolethicin and polyethylene glycol (PEG). Of these fusogenic agents, PEG has been most frequently used, in conjunction with alkaline pH and high calcium concentration.

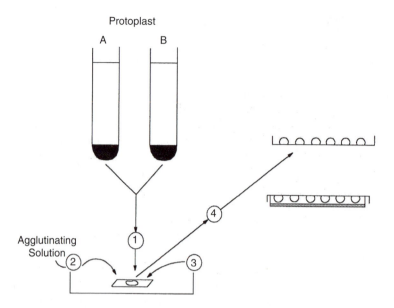

FIGURE 16.5 Typical protocol for protoplast fusion of potato using PEG. (1) Dilute protoplast mixture to 10^4 cells per ml in 0.5 M glucose, 3 mM $CaCl_2$, 0.7 mM KH_2PO_4, pH 5.7. (2) Add 3 to 200 µl volumes of agglutinating solution (0.09 mM PEG 6000, 10 mM $CaCl_2$, 0.7 mM KH_2PO_4, pH 5.5). (3) Add 5 to 100 µl volumes of diluting solution (50 mM $CaCl_2$, 0.5 M glucose, 50 mM glycine buffer, pH 10.0). (4) Wash cells with culture medium and culture in hanging drop suspensions.

There are many steps in the fusion of plant protoplasts using PEG as a chemical facilitator (Figure 16.5). Initially, the cell membranes must be brought into close physical contact (agglutination). As most cell membranes possess a net negative surface charge, adjacent cells with similar charges tend to repel each other. Chemical fusogenic facilitators, such as concanavalin A and immune antisera, promote cell fusion by overcoming the repellent effect of similar net negative surface charges on the protoplast membranes. PEG is a potent cell agglutinator and functions as a membrane modifier. Once cell membranes are in close contact, their surface proteins migrate to create lipid-rich regions. During this period, the dehydrating effect of PEG on the cell membrane and the ability of PEG to bind to phospholipids within the membrane induce cell adhesion between adjacent cells. Subsequently, successive washing with high concentrations of calcium in a buffer with an alkaline pH effectively completes the fusion process.

ELECTROFUSION-MEDIATED PROTOPLAST FUSION

Protoplast fusion can also be induced by the manipulation of cell membranes using electrical currents. Electrofusion of plant protoplasts is often preferred over PEG fusion because it does not employ reagents that are toxic to the cells undergoing fusion. Conditions for electrofusion must be optimized for specific cell types. Typically, a yield of 20% or greater fusion products can be obtained by electrofusion of protoplasts, compared to less than 1% fusion products with PEG.

For effective electrofusion, the protoplasts must be brought closely together (agglutination) and the cell membranes perturbed by an electrical shock to promote membrane fusion. With electrofusion, dielectrophoresis is used to bring protoplasts into close physical contact. The protoplasts are subjected to a low voltage (50–100 V/cm electrode gap) oscillating sine wave current of approximately 1 MHz frequency that polarizes the cells so that each has a positive and negative area. The positive end of one protoplast is drawn toward the negative area of an adjacent protoplast such that "pearl chains" of protoplasts are formed prior to fusion. The fusion itself may be accomplished by a relatively high voltage (0.5–1.5 kV/cm electrode gap) pulse that destabilizes the cell membranes at specific sites on each protoplast. The electrical pulse does not affect the organelles inside the cell because the current only passes along the cell membranes that are generally more conductive than the cytoplasm. It is necessary to supply the voltage of the fusion pulse for some critical duration at a minimal threshold level in order to achieve successful fusion.

SELECTION OF HYBRID CELL LINES

An important feature of any somatic hybridization procedure is the identification and selection of fused cells. Typically, it is desirable to separate hybrid cell lines from other cells in the mixture because cell lines that result from fusion of two or more cells from the same parent (homokaryotic fusions) or from parental cells that did fuse at all may greatly outnumber the heterokaryotic fusion products. Furthermore, there is no guarantee that just because a cell fusion occurred, nuclear fusion will follow. Fusion of the two membrane-bound nuclei within a single cell and coordination of mitosis involving the chromosomes from what were formerly two independent nuclei are required to form a stable hybrid. Unless there is some identification and selection system incorporated into each of the parental cell lines prior to fusion, collection of hybrid cell lines can be very difficult.

Several selection systems have been developed based on specific requirements of the prospective parent cell lines. These selective systems include antibiotic-resistant cell lines, herbicide-resistant cell lines, and cell lines with the ability to grow on specific amino acid analogs, among others. It is most convenient to combine two parental cell lines with differing growth requirements to employ selective screens. Such selective screens prohibit the growth of unfused cells, allowing only fused cells that possess the complementary traits of both parents to thrive.

Unfortunately, most commercially important cultivars of crop plants that are candidates for cell fusion do not have sophisticated cell selection systems. Therefore, selection may be based on growth characteristics; e.g., callus tissue from hybrid fusions may grow more vigorously than that of either parent. Parental cell lines can be given different properties by the use of vital stains (stains that do not adversely affect the viability of cells) that can be recognized by fluorescence detectors. Two such stains, fluorescein isothiocyanate (FITC) and rhodamine isothiocyanate (RITC), have been used successfully on plant protoplasts. FITC stains cells green and RITC stains cells red when viewed with a fluorescence microscope. Fusion products that have both FITC and RITC fluoresce yellow. The fused (yellow) cells can be isolated mechanically using a micromanipulator and cultured individually on an enriched tissue culture medium, supplemented either with nurse cultures or "conditioned" medium. Large numbers of fused cells marked with fluorescent dyes can be isolated more efficiently using a laser activated flow cytometer, also known as a cell sorter. Cells are sorted according to the wavelength of their fluorescent emission, and collected and cultured in enriched medium. Although these procedures have been used successfully, it may be difficult to maintain sterile conditions throughout the lengthy process. The process may also adversely affect the viability of the sorted protoplasts.

Frequently, molecular markers, such as microsatellites, amplified fragment length polymorphism (AFLP), and randomly amplified polymorphic DNA (RAPD) can be used even at the callus stage, but are more often used after putative somatic hybrid plants have been regenerated to identify and select hybrids.

There are several reports of useful somatic hybrid plants that have been produced by protoplast fusion (Johnson and Veilleux, 2001). Two male sterile potato dihaploids were fused to obtain a fertile tetraploid somatic hybrid. Citrus somatic hybrids have been developed and are under evaluation as widely adapted rootstocks. Insect and disease resistance have been observed in somatic hybrids between potato and various unadapted *Solanum* relatives. Chilling tolerance has been observed in somatic hybrids between tomato and its green fruited weedy relative, *L. peruvianum*. Somatic hybrids between crop *Brassica* species and unadapted *Brassica* have been found to be resistant to bacterial and fungal diseases. Intergeneric somatic hybridization has been used to transfer disease resistance in citrus, potato, and *Brassica* species. There is really no limit in the choice of fusion partners, as the processes of electrofusion or chemical fusion are nondiscriminatory. Discrimination occurs during the regeneration process. Perhaps the most remarkable somatic hybridization to date has been between barley, a monocot, and carrot, a dicot. Electrofusion resulted in an aneuploid somatic hybrid that closely resembled carrot, but chloroplast and mitochondrial recombination had occurred.

GENETIC TRANSFORMATION

Genetic transformation (Chapter 19) of single cells has been accomplished by co-cultivating protoplasts with *Agrobacterium tumefaciens* or by direct DNA transfer using polycationic chemicals, electroporation, liposomes, microinjection, or sonication (Sawahel and Cove, 1992; Fisk and Dandekar, 1993). Transformation of plant protoplasts using polycationic chemicals and/or electroporation has been most often reported. Transformed protoplasts offer the possibility of examining factors affecting transient gene expression, regenerating transformed plants where other methods have been difficult or unsuccessful, as in many monocots, or for studies of gene function or protein targeting.

AGROBACTERIUM-MEDIATED TRANSFORMATION

Agrobacterium tumefaciens, the causative agent of crown gall disease, stimulates gall formation by inserting a portion of the tumor-inducing (Ti) plasmid into plant cells (see Chapter 19). For *Agrobacterium*-mediated gene transfer, *in vitro* plant regeneration systems are frequently exploited where cells are transformed prior to regeneration and then regeneration is conducted of selective medium to exclude the participation of untransformed cells. If protoplasts are used as the system of plant regeneration for transformation, they are first isolated and cultured for 2 to 3 days prior to co-cultivation with *Agrobacterium* cells. Protoplasts are mixed with *Agrobacterium* at a rate of 10^4 to 10^6 plant cells to 10^6 to 10^8 bacterial cells per milliliter and incubated together for 24 to 30 hours. After co-cultivation, antibiotics that inhibit growth of the bacterium are added to the culture medium. Treated plant cells are allowed to grow to the multicellular stage before the addition of selective antibiotics (kanamycin, hygromycin, etc.) that inhibit growth of untransformed cells, due to the presence of a resistance gene on the T-DNA, that part of the *Agrobacterium* that inserts into the plant genome. The T-DNA generally comprises some gene of interest to be inserted into the target species and a selective marker, such as antibiotic resistance, both of which are associated with plant promoters to facilitate gene expression in the host. Transformed cells grown on selective medium are able to regenerate plants with the introduced gene. Transformation frequencies usually range from 0.26 to 15.3%. Not all species are amenable to *Agrobacterium*-mediated transformation. In addition, simpler regenerative culture systems are preferred to the use of protoplasts for transformation because of the high frequency of somaclonal variants often found among plants regenerated from protoplasts.

POLYCATIONIC CHEMICAL-MEDIATED TRANSFORMATION

The most commonly used polycationic chemical for direct DNA transformation is polyethylene glycol (PEG). It has been used to transform protoplasts of many plant species. Transformation frequencies vary from 0.8×10^{-6} to 6×10^{-3} depending on the species.

With the PEG transformation method, protoplasts are isolated and adjusted to the proper cell density before adding transforming DNA and PEG. The mixture is incubated for 10 to 30 min before dilution with culture medium. A heat shock treatment (~45°C) prior to the addition of DNA may be used. It is not completely known how PEG promotes direct DNA uptake. The polycationic charge of PEG may interact with negatively charged DNA to form a positively charged complex that binds to the anionic plasmalemma. Absorption of DNA into cells occurs through endocytosis or active uptake. PEG-mediated transformation has low-cell toxicity, promotes high transformation frequencies, increases membrane permeability, protects DNA against degradation by nucleases, is independent of DNA structure (supercoiled, circular, or linear), allows integration of DNA into the host genome without a carrier, and does not have a limited host range. Another polycationic chemical, hexademethrine bromide (Polybrene) has occasionally been used to transform plant protoplasts (Antonelli and Stadler, 1990).

ELECTROPORATION-MEDIATED TRANSFORMATION

In electroporation, an electric pulse is used to induce DNA uptake by plant protoplasts. DNA is thought to enter protoplasts through pores in the cell membrane generated in response to the electric pulse. Transforming DNA is mixed with plant protoplasts prior to the application of one or more electric pulses of 200 to 1000 V/cm and 16 to 1000 μF capacitance. Short pulses (typically in μsec) are typically less damaging to protoplasts than long (msec) pulses. Optimum electroporation parameters vary and should be tested for each new species.

Stable transformation using electroporation either with or without PEG has been obtained in protoplasts of many plant species. Genes incorporated into plant protoplasts using electroporation have included antibiotic resistance, *GUS A*, or phosphoenolpyruvate from *Sorghum bicolor* (L.) Moench. Electroporation is a fast and efficient means of delivering DNA to plant cells and proto-plasts, is not limited by the size or structure of transforming DNA, and does not have a limited host range. Electroporation, if not optimized for individual generators and plant sources, can result in low viability of protoplasts. Other methods of incorporating foreign DNA into plant protoplasts have been described recently and show considerable promise (Rakoczy-Trojanowska, 2002).

CONCLUSIONS AND FUTURE PROSPECTS

Plants with improved characteristics have been obtained through protoplast manipulations, including both somaclonal variants and somatic hybrids. The methods for manipulation of protoplasts and integration of foreign DNA into protoplasts have continually improved over the past decade, expanding the number of species amenable to such techniques to include most major crop plants. However, initial somatic hybrids are generally genetic monstrosities that require several generations of additional breeding to deliver desirable genes into useful agronomic formats. Somaclonal variation is not controllable and is more likely to result in deleterious rather than beneficial changes to agronomic characters. Hence, it is generally suppressed as much as possible in transformation procedures that require a tissue culture step by reducing the tissue culture duration to a minimum. Therefore, protoplast culture is not the technology of choice for this application, but may be used when all else fails. Plant protoplasts are elegant life forms of single-celled versions of higher plants, capable of expressing totipotency under ideal conditions. Their culture is complex, yet they are quite forgiving and ready to regenerate after outrageous prodding and manipulation, making us ponder the adaptability of plants even to harsh laboratory environments.

LITERATURE CITED

Antonelli, N.M. and J. Stadler. 1990. Genomic DNA can be used with cationic methods for highly efficient transformation of maize protoplasts. *Theor. Appl. Genet.* 80:395–401.

Bhojwani, S.S. and M.K. Razdan. 1983. *Plant Tissue Culture: Theory and Practice*. Elsevier, Amsterdam. 237–260.

Boss, W. 1987. Fusion-permissive protoplasts. A plant system for studying cell fusion. In: *Cell Fusion*. (Ed.) A.E. Sowers. Plenum Press, New York. 145–166.

Burg, H.C.J. et al. 1989. Patterns of phenotypic and tuber protein variation in plants derived from protoplasts of potato (*Solanum tuberosum* L. cv. Bintje). *Plant Sci.* 64:113–124.

Cheng, J. and R.E. Veilleux. 1991. Genetic analysis of protoplast culturability in *Solanum phureja*. *Plant Sci.* 75:257–265.

Cocking, E.C. 1960. A method for the isolation of plant protoplasts and vacuoles. *Nature* 187:927–929.

Diettrich, B., V. Schneider, and M. Luckner. 1991. High variation in cardenolide content of plants regenerated from protoplasts of the embryogenic cell strain VII of *Digitalis lanata*. *J. Plant Physiol.* 139:199–204.

Fisk, H.J. and A.M. Dandekar. 1993. The introduction and expression of transgenes in plants. *Scientia Hort.* 55:5–36.

Johnson, A.A.T. and R.E. Veilleux. 2001. Somatic hybridization and application in plant breeding. *Plant Breed. Rev.* 20:167–225.

Rakoczy-Trojanowska, M. 2002. Alternative methods of plant transformation — A short review. *Cell. Mol. Biol. Lett.* 7:849–858.

Saunders, J.A. and G.W. Bates. 1987. Chemically induced fusion of plant protoplasts. In: *Cell Fusion*. (Ed.) A.E. Sowers. Plenum Press, New York. 497–520.

Sawahel, W.A. and D.J. Cove. 1992. Gene transfer strategies in plants. *Biotech. Adv.* 10:393–412.

Shahin, E.A. and R. Spivey. 1986. A single dominant gene for *Fusarium* wilt resistance in protoplast-derived tomato plants. *Theor. Appl. Genet.* 73:164–169.

Shepard, J.F., D. Bidney, and E. Shahin. 1980. Potato protoplasts in crop improvement. *Science* 208:17–24.

Veilleux, R.E. and A.A.T. Johnson. 1998. Somaclonal variation: molecular evidence and utilization in plant breeding. *Plant Breed. Rev.* 16:229–268.

17 Haploid Cultures

Sandra M. Reed

CHAPTER 17 CONCEPTS

- Haploids may be produced from male gametophytes (androgenesis) or female gametophytes (gynogenesis).

- *In vitro* androgenesis involves culture of anthers or isolated microspores.

- Gynogenesis may be induced in unfertilized ovule or ovary culture.

- Success of haploid cultures depends on genotype, media, and cultural conditions.

- Stage of microspore development at time of culture is critical to the success of androgenesis.

- Chromosome doubled haploids (diploids) are useful in breeding projects.

INTRODUCTION

The ability to produce haploid plants is a tremendous asset in genetic and plant breeding studies. Heritability studies are simplified because, with only one set of chromosomes, recessive mutations are easily identified. In addition, doubling the chromosome number of a haploid to produce a doubled haploid results in a completely homozygous plant. Theoretically, the genotypes present among a large group of doubled haploids derived from an F_1 hybrid represent in a fixed form the genotypes expected from an F_2 population. Use of doubled haploids in breeding programs can thus greatly reduce the time required for development of improved cultivars. To be most useful, a large number of haploids from many different genotypes are required.

Haploids have been available for genetic studies for many years. Prior to the 1960s, they were mostly obtained spontaneously following interspecific hybridization or through the use of irradiated pollen, but usually only infrequently and in very small numbers. Haploid methodology took a giant step forward 40 years ago when Guha and Maheshwari (1964) found that haploid plants could be obtained on a regular basis and in relatively large numbers by placing immature anthers of *Datura innoxia* Mill. into culture. This work was rapidly expanded using tobacco (*Nicotiana tabacum* L.), which became the model species for anther culture experiments. To date, androgenic haploids have been produced in more than 170 species; several good reviews provide lists of these species (Maheshwari et al., 1982; Bajaj, 1983; Heberle-Bors, 1985; Dunwell, 1996). While efforts have been more limited, haploids have also been obtained from *in vitro* culture of the female gametophyte in more than 30 species (Keller and Korzun, 1996; Lakshmi Sita, 1997). Gynogenesis has been successfully applied to several species in which androgenesis is generally ineffective, such as sugar beet (*Beta vulgaris* L.), onion *(Allium cepa* L.), and the Gerbera daisy (*Gerbera jamesonii* H. Bolus ex Hook).

Although much of the terminology used in this chapter has been discussed in previous chapters, the *in vitro* induction of haploids involves a few specialized terms. A haploid is a plant with the gametic or *n* number of chromosomes. Doubled haploids, or dihaploids, are chromosome doubled haploids or 2n plants. Androgenesis is the process by which haploid plants develop from the male gametophyte. When anthers are cultured intact, the procedure is called anther culture. Microspore culture involves isolating microspores from anthers before culture and is sometimes referred to as pollen culture. Haploids are derived from the female gametophyte through a process referred to as gynogenesis. *In vitro* gynogenesis involves the culture of unfertilized ovules or ovaries. While both androgenesis and gynogenesis may occur *in vivo*, the usage of the terms in this chapter will refer to the *in vitro* induction of haploids via these two mechanisms.

This chapter will begin with general discussions of androgenesis and gynogenesis, followed by a review of the factors that affect the successful production of androgenic and gynogenic haploids. Finally, some of the basic procedures used for the *in vitro* production of haploids will be summarized. Excellent discussions of *in vitro* haploid production, along with specific protocols for a number of crop species, can be found in Jain et al., 1996a; 1996b; 1996c; 1996d; 1997.

ANDROGENESIS

DEVELOPMENT OF HAPLOIDS

Haploid plants develop from anther culture either directly or indirectly through a callus phase. Direct androgenesis mimics zygotic embryogenesis; however, neither a suspensor nor an endosperm is present. At the globular stage of development, most of the embryos are released from the pollen cell wall (exine). They continue to develop, and after 4 to 8 weeks, the cotyledons unfold and plantlets emerge from the anthers. Direct androgenesis is primarily found among members of the tobacco (Solanaceae) and mustard (Cruciferae) families.

During indirect androgenesis, the early cell division pattern is similar to that found in the zygotic embryogenic and direct androgenic pathways. After the globular stage, irregular and asynchronous divisions occur and callus is formed. This callus must then undergo organogenesis for haploid plants to be recovered. The cereals are among the species that undergo indirect androgenesis.

The early cell divisions that occur in cultured anthers have been studied (for review, see Reynolds, 1990). For species cultured during the uninucleate stage, the microspore either undergoes a normal mitosis and forms a vegetative and a generative nucleus or divides to form two similar-looking nuclei. In those cases where a vegetative and a generative nuclei are formed in culture, or where binucleate microspores are placed into culture, it is usually the vegetative nucleus that participates in androgenesis. The only species in which the generative nucleus has been found to be actively involved in androgenesis is black henbane (*Hyoscyamus niger* L.). When similar looking nuclei are formed, one or both nuclei may undergo further divisions. In some cases, the two nuclei will fuse, producing homozygous diploid plants or callus. Because diploid callus may also arise from somatic tissue associated with the anther, diploids produced from anther culture cannot be assumed to be homozygous. To verify that plants produced from anther culture are haploid, chromosome counts should be made from root tips or other meristematic somatic tissues (see Chapter 4). Because haploids derived from diploid species are expected to be sterile or have greatly reduced fertility, pollen staining, which is much quicker and requires less skill than chromosome counting, can also be used to identify and eliminate potential diploids. However, pollen staining may not distinguish between haploids and plants that have reduced fertility because they have a few extra or missing chromosomes (i.e., aneuploids). Haploids and diploids recovered from anther culture may also be distinguished by comparing size of cells, particularly stomatal guard cells, or through the use of flow cytometry.

PROBLEMS ASSOCIATED WITH ANTHER CULTURE

Problems encountered in plants during or as a result of anther culture range from low yields to genetic instability. Many of the major horticultural and agronomic crops do not yield sufficient haploids to allow them to be useful in breeding programs. In other species, genetic instability has often been observed from plants recovered from anther and microspore cultures.

The term "gametoclonal variation" has been coined to refer to the variation observed among plants regenerated from cultured gametic cells and has been observed in many species. While often negative in nature, some useful traits have been observed among plants recovered from anther and microspore culture. Gametoclonal variation may arise from changes in chromosome number (i.e., polyploidy or aneuploidy) or chromosome structure (e.g., duplications, deletions, translocations, inversions, and so on). In tobacco, gametoclonal variation has been associated with an increase in the amount of nuclear DNA without a concomitant increase in chromosome number (DNA amplification). In many cereals, a high percentage of the plants regenerated from anther culture are albino; changes in cytoplasmic DNA have been associated with this albinism. Good discussions and reviews of gametoclonal variation can be found in several chapters of Jain et al., 1996b.

GYNOGENESIS

As with androgenesis, gynogenic haploids may develop directly or indirectly via regeneration from callus (for review, see Keller and Korzun, 1996). The first cell divisions of gynogenesis are generally similar to those of zygotic embryogenesis. Direct gynogenesis usually involves the egg cell, synergids, or antipodals with organized cell divisions leading first to the formation of proembryos and then to well-differentiated embryos. In indirect gynogenesis, callus may be formed directly from the egg cell, synergids, polar nuclei, or antipodals, or may develop from proembryos. Plants regenerated from callus may be haploid, diploid, or mixoploid. As with plants produced from anther cultures, chromosome counts can be used to identify haploids. Distinguishing between homozygous dihaploids, in which chromosome doubling occurred in culture, and diploids that developed from somatic tissue requires the use of molecular markers.

The major problems affecting the use of gynogenesis are the lack of established protocols for most species, poor yields, and production of diploid or mixoploid plants. Gametoclonal variation among gynogenic haploids has not been widely studied; however, it has been noted that, unlike androgenesis, gynogenesis of cereal species does not result in the production of albino plants.

FACTORS AFFECTING ANDROGENESIS

GENOTYPE

The choice of starting material for an anther or microspore culture project is of utmost importance. In particular, genotype plays a major role in determining the success or failure of an experiment. Haploid plant production via androgenesis has been very limited or nonexistent in many plant species. Furthermore, within a species, differences exist in the ability to produce haploid plants. Even within an amenable species, such as tobacco, some genotypes produce haploids at a much higher rate than others. Because of this genotypic effect, it is important to include as much genetic diversity as possible when developing protocols for producing haploid plants via anther or microspore culture.

CONDITION OF DONOR PLANTS

The age and physiological condition of donor plants often affect the outcome of androgenesis experiments. In most species, the best response usually comes from the first set of flowers produced by a plant. As a general rule, anthers should be cultured from buds collected as early as possible during the course of flowering. Various environmental factors that the donor plants are exposed to may also affect haploid plant production. Light intensity, photoperiod, and temperature have been investigated, and at least for some species, these are found to influence the number of plants produced from anther cultures. Specific optimum growing conditions differ from species to species and are reviewed by Powell (1990). In general, the best results are obtained from healthy, vigorously growing plants.

STAGE OF MICROSPORE DEVELOPMENT

The most critical factor affecting haploid production from anther and microspore culture is the stage of microspore development. For many species, success is achieved only when anthers are collected during the uninucleate stage of pollen development. In contrast, optimum response is obtained in tobacco and *Brassica napus* L. from anthers cultured just before, during, and just after the first pollen mitosis (late uninucleate to early binucleate microspores).

In developing a protocol for anther culture, one anther from each bud is usually set aside and later cytologically observed to determine the stage of microspore development. In many cases, anthers within a bud are sufficiently synchronized to allow this one anther to represent the remaining cultured anthers. Measurements of physical characteristics of the flower, such as calyx and corolla length and anther color, shape, and size, are also recorded. Results of the experiments are analyzed to determine which microspore stage was the most responsive. The physical descriptions of the buds and anthers are then examined to determine if this microspore stage correlates to any easily identified inflorescence, flower or anther characteristics. For example, in tobacco, buds in which the calyx and corolla are almost identical in length usually contain anthers having microspores at or near the first pollen mitosis. A researcher wishing to produce a maximum number of haploid plants of tobacco would collect only buds fitting this physical description.

PRETREATMENT

For some species, a pretreatment following collection of buds, but before surface disinfestation and excision of anthers, has been found to be beneficial. Yields of tobacco haploids are often increased by storing excised buds at 7 to 8°C for 12 days prior to anther excision and culture (Sunderland and Roberts, 1979). For other species, temperatures from 4 to 10°C and durations from 3 days to 3 weeks have been utilized. For any one species, there may be more than one optimum temperature and length of treatment combination. In general, lower temperatures require shorter durations, whereas a longer pretreatment time is indicated for temperatures at the upper end of the cold pretreatment range mentioned above.

MEDIA

Androgenesis can be induced in tobacco and a few other species on a simple medium such as that developed by Nitsch and Nitsch (1969). For most other species, the commonly used media for anther culture include MS (Murashige and Skoog, 1962), N6 (Chu, 1978), or variations on these media. In some cases, complex organic compounds, such as potato extract, coconut milk, and casein hydrolysate, have been added to the media. For many species, 58 to 88 mM (2–3%) sucrose is added to the media; whereas other species, particularly the cereals, have responded better to higher (up to 435 mM or about 15%) concentrations of sucrose. The higher levels of sucrose may fulfill an osmotic rather than a nutritional requirement. Other sugars, such as ribose, maltose, and glucose, have been found to be superior to sucrose for some species.

For a few species, such as tobacco, it is not necessary to add plant growth regulators (PGRs) to the anther culture media. Most species, however, require a low concentration of some form of auxin in the media. Cytokinin is sometimes used in combination with auxin, especially in species in which a callus phase is intermediate in the production of haploid plants.

Anther culture media is often solidified using agar. Because agar may contain compounds inhibitory to the androgenic process in some species, the use of alternative gelling agents has been investigated. Ficoll, Gelrite™ (Merck and Co., Inc., Rahway, NJ), agarose, and starch have proven superior to agar for solidifying anther culture media in various species (Calleberg and Johansson, 1996). The use of liquid medium has been advocated by some researchers as a way to avoid the potentially inhibitory substances in gelling agents. Anthers may be placed on the surface of the medium, forming a so-called "float culture." Alternatively, microspores may be isolated and cultured directly in liquid medium.

TEMPERATURE AND LIGHT

Various cultural conditions, such as temperature and light, may also affect androgenic response. Anther cultures are usually incubated at 24 to 25°C. In some species, an initial incubation at a higher or lower temperature has been beneficial. Haploid plant production was increased in *Brassica campestris* L. by culturing the anthers at 35°C for 1 to 3 days prior to culture at 25°C (Keller and Armstrong, 1979). In contrast, androgenesis was promoted in *Cyclamen persicum* Mill. by incubating cultured anthers at 5°C for the first 2 days of culture (Ishizaka and Uematsu, 1993).

Some species respond best when exposed to alternating periods of light and dark, whereas continuous light or dark cultural conditions have proven beneficial in other species. Other physical cultural factors, such as atmospheric conditions in the culture vessel, anther density, and anther orientation, have been studied and found to affect androgenic response in some species; however, species have varied greatly in their response to these physical factors.

FACTORS AFFECTING GYNOGENESIS

GENOTYPE

Gynogenesis has not been investigated as thoroughly or with as many species as has androgenesis; therefore, less information is available concerning the various factors that contribute to the successful production of haploids from the female than the male gametophyte. However, several studies have identified genotype as a critical factor in determining the success of a gynogenesis experiment. Not only are there differences between species, but also genotypes within individual species have responded differently. As with androgenesis, it is important to include a wide range of genotypes in ovule and ovary culture experiments.

MEDIA

Media has also been identified as an important factor in gynogenesis. The most commonly used basal media for recovering gynogenic haploids are MS, B-5 (Gamborg et al., 1968), Miller's (Miller, 1963), or variations on these media. Sucrose levels have ranged from 58 mM to 348 mM (2–12%). While gynogenic haploids have developed in a few species without the use of growth regulators, most species have required auxins and/or cytokinins in the medium. For those species that undergo indirect gynogenesis, both an induction and a regeneration medium may be required. Most ovule and ovary culture experiments have been conducted using solid medium. A list of specific media components used for gynogenesis in several crop species can be found in Keller and Korzun (1996).

STAGE OF GAMETOPHYTIC DEVELOPMENT

Because the female gametophyte is difficult to handle and observe, determining the optimum stage of gametophytic development for gynogenesis is usually based on other, more easily discerned, characteristics. Performance of ovule and ovary cultures has often been correlated with stage of microspore development. Depending on species, the best results have been obtained when the female gametophyte was cultured from the late uninucleate to trinucleate stage of megaspore development. In other studies, the number of days until anthesis has been used as an indicator of stage of gametophytic development. A few gynogenesis studies that involved direct observations of the female gametophyte have been conducted. For several species, gynogenesis was most successful where cultures were initiated when the embryo sac was mature or almost mature (for review, see Keller and Korzun, 1996).

OTHER FACTORS

Cold pretreatment of flower buds at 4°C for 4 to 5 days has been effective in increasing yields of haploid embryos or callus in a few species, but has not been widely investigated. Seasonal effects have been observed in several species. Many of the other factors that affect androgenesis probably also affect gynogenesis; however, in most cases, insufficient data is available to detect trends in response. These variables should, however, be considered when initiating gynogenesis experiements.

GENERAL ANDROGENESIS PROCEDURES

COLLECTION, DISINFESTATION, EXCISION, AND CULTURE

Floral buds may be collected from plants grown in the field, greenhouse, or growth chamber. Entire inflorescences or individual buds are harvested and kept moist until ready for culturing. If buds are to be pretreated (i.e., kept at low temperature), they should be wrapped in a moistened paper tissue and placed into a small zipper-type plastic bag.

Flower buds are typically disinfested using a 5% sodium or calcium hypochlorite solution for 5 to 10 min, and then are rinsed thoroughly in sterile distilled water. Anthers are aseptically excised in a laminar flow hood, taking care not to cause injury. If the anther is still attached to the filament, the filament is carefully removed.

If a solid medium is used, the anthers are gently pressed onto the surface of the medium (just enough to adhere to the medium), but should not be deeply embedded. When using a liquid medium for intact anthers, the anthers are floated on the surface. Care must be taken when moving float cultures so as not to cause the anthers to sink below the surface.

For most species, disposable petri dishes are utilized for anther cultures. For a species with large anthers, such as tobacco, the anthers from 4 to 5 buds (20–25 anthers) may be cultured together on one 100×15 mm diameter petri dish. For species with smaller anthers, or for certain experimental designs, smaller petri dishes or other containers may be more useful. Petri dishes are usually sealed and placed into an incubator; the specific temperature and light requirements of the incubator depends on the species being cultured.

While many of the steps involved in microspore culture are similar to those of anther culture, microspore culture also requires the separation of the microspores from the surrounding anther tissue. Microspores may be squeezed out of anthers using a pestle or similar device, or a microblending procedure may be used. See Dunwell (1996) for a review of literature pertaining to microspore culture.

DETERMINING STAGE OF MICROSPORE DEVELOPMENT

For most species, stage of microspore development can be determined by "squashing" an entire anther in aceto-carmine or propiono-carmine and then observing the preparation under the low-power objective of a light microscope. The early uninucleate microspore is lightly staining with a centrally located nucleus. As the uninucleate microspore develops, its size increases and a large

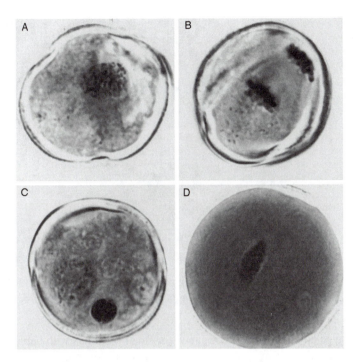

FIGURE 17.1 Microspores of tobacco. (A) Late uninucleate microspores. Vacuole can be seen on right side of cell. (B) Anaphase of first pollen mitosis. (C) Early binucleate microspore. Large diffuse vegetative nucleus and smaller, more darkly staining generative nucleus can be distinguished. (D) Mid-binucleate microspore. Staining is darker, obscuring vegetative nucleus, but the generative nucleus is still visible. (From Trigiano, R.N. and D.J. Gray, (eds.) 2000. *Plant Tissue Culture Concepts and Laboratory Exercises*, 2nd edition. CRC Press, Boca Raton, FL.)

central vacuole is formed. As the microspore nears the first pollen mitosis, the nucleus is pressed up near the periphery of the microspore (Figure 17.1A). Staining will still be fairly light. Pollen mitosis is of short duration, but it may sometimes be observed; it is recognized by the presence of condensed chromosomes (Figure 17.1B). The product of the first pollen mitosis is a binucleate microspore containing a large vegetative and a small generative nucleus. The vegetative nucleus is often difficult to recognize because it is so diffuse and lightly staining. However, this stage may be definitively identified by the presence of the small densely staining generative nucleus (Figure 17.1C). As the binucleate microspore ages, the intensity of the staining increases and starch granules begin to accumulate (Figure 17.1D). Eventually, both nuclei may be hidden by the dark staining starch granules.

HANDLING OF HAPLOID PLANTLETS

For species undergoing direct androgenesis, small plantlets can usually be seen emerging from the anthers 4 to 8 weeks after culture (Figure 17.2). When these get large enough to handle, they should be teased apart using fine-pointed forceps and then either placed on a rooting medium (usually low salt, with a small amount of auxin) or transplanted directly into a small pot filled with soilless potting mixture. The callus produced in species that undergo indirect androgenesis must be removed from the anther and placed onto a regeneration medium containing the appropriate ratio of cytokinin to auxin.

To produce dihaploid plants, it is necessary to double the chromosome number of the haploids, and for many species, a colchicine treatment is used. Published procedures for producing polyploids from diploids can be modified for use with anther culture derived haploids. For example, it may

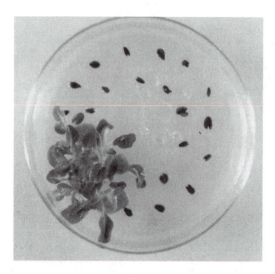

FIGURE 17.2 Haploid tobacco plantlets emerging from an anther. (From Trigiano, R.N. and D.J. Gray, (eds.) 2000. *Plant Tissue Culture Concepts and Laboratory Exercises*, 2nd edition. CRC Press, Boca Raton, FL.)

be possible to use a colchicine treatment designed for small seedlings with haploid plants directly out of anther culture. Alternatively, established procedures using larger plants may be used. In wheat and other cereals, chromosome doubling is induced by initially culturing anthers on a medium containing a low concentration of colchicine. In addition to leading to the direct regeneration of homozygous dihaploids, the inclusion of colchicine in the medium for the first few days of culture caused a decline in the number of albino regenerants (Barnabás et al., 2001).

GENERAL GYNOGENESIS PROCEDURES

Gynogenesis experiments are usually conducted using unfertilized ovules or ovaries, although entire immature flower buds have been cultured in a few species. It is easier to dissect ovaries than ovules without damaging the female gametophyte. However, in polyovulate ovaries, it may be advantageous to excise the ovules so that they can be in direct contact with the culture medium.

Inflorescences must be collected before pollen is shed, unless the species is highly self-incompatible or a male-sterile line is used. In developing a gynogenesis protocol for a species, it may be necessary to collect explants from several days before anthesis to just before anthesis. As discussed earlier, the stage of microspore development is sometimes recorded as an indicator of developmental stage of the female gametophyte. Procedures used for determining the stage of microspore development are described earlier in this chapter.

Disinfestation varies depending on species, growing conditions of explant source, and choice of explant. Woody plant material often requires longer disinfestation times and/or stronger sterilizing agents than herbaceous materials. Tissue from greenhouse-grown plants is usually easier to disinfest than that of field-grown plants. If ovules are to be cultured, a harsh surface sterilization procedure should be applied to ovaries. It should not be necessary to disinfest the ovules, since they are presumed to have been removed from a sterile environment inside the ovary. Commonly used sterilizing agents and disinfestation times are presented in Chapter 3 of this book.

Techniques used for the excision of ovules depend on the arrangement of ovules within the ovary. Care must be taken not to let ovules dry out during excision. A solid medium is typically used for gynogenesis experiments; choice of culture vessel depends on size of explant. Disposable petri dishes work well for culturing ovules of small-seeded polyovulate species, whereas test tubes may be preferable for large ovaries.

Handling procedures for gynogenic haploids are similar to those described for androgenic haploids. As plants emerge from cultured ovules or ovaries, they can be transferred to a rooting medium or transplanted directly to a soilless potting mixture. Colchicine or another mitotic inhibitor is typically used for doubling chromosome number to produce dihaploids.

SUMMARY

Haploids of many plant species can be produced *in vitro*. Anther culture has been the most widely used *in vitro* technique for producing haploids, but androgenic haploids have been obtained in a few species through the culture of isolated microspores. While fewer studies have been conducted involving the induction of haploids from the female gametophyte, gynogenesis has proven successful in several species. Yields of androgenic and gynogenic haploids differ greatly depending on species, and are also affected by cultural conditions, such as media formulation, stage of microspore or embryo sac at time of culture, and use of a low- or high-temperature pretreatment. Both androgenic and gynogenic haploids may arise directly, or may be produced indirectly through a callus intermediate. *In vitro*-derived dihaploids of several important crop species are now produced routinely. Use of dihaploids in breeding programs of these species has shortened cultivar development time. Expansion of this valuable breeding technique to additional species should occur as continued efforts are made to identify factors critical to *in vitro* induction of haploidy.

LITERATURE CITED

Bajaj, Y.P.S. 1983. *In vitro* production of haploids. In: *Handbook of Plant Cell Culture, Vol. 1: Techniques for Propagation and Breeding*. (Ed.) D.A. Evans et al. Macmillan, New York. 228–287.

Barnabás, B. et al. 2001. *In vitro* androgenesis of wheat: from fundamentals to practical application. *Euphytica* 119: 211–216.

Calleberg, E.K. and L.B. Johansson. 1996. Effect of gelling agents on anther culture. In: In Vitro *Haploid Production in Higher Plants, Vol. 1: Fundamental Aspects and Methods*. (Ed.) S.M. Jain, S.K. Sopory, and R.E. Veilleux. Kluwer Academic Publishers, Dordrecht, the Netherlands. 189–203.

Chu, C. 1978. The N6 medium and its applications to anther culture of cereal crops. In: *Proceedings of Symposium on Plant Tissue Culture*. Science Press, Peking, China. 51–56.

Dunwell, J.M. 1996. Microspore culture. In: In Vitro *Haploid Production in Higher Plants, Vol. 1: Fundamental Aspects and Methods*. (Ed.) S.M. Jain, S.K. Sopory, and R.E. Veilleux. Kluwer Academic Publishers, Dordrecht, the Netherlands. 205–216.

Gamborg, O.L., R.A. Miller, and K. Ojima. 1968. Nutrient requirements of suspension cultures of soybean root cells. *Exp. Cell Res.* 50:157–158.

Guha, S. and S.C. Maheshwari. 1964. *In vitro* production of embryos from anthers of *Datura*. *Nature* 204:497.

Heberle-Bors, E. 1985. *In vitro* haploid formation from pollen: A critical review. *Theor. Appl. Genet.* 71:361–374.

Ishizaka, H. and J. Uematsu. 1993. Production of plants from pollen in *Cyclamen persicum* Mill. through anther culture. *Jpn. J. Breed.* 43:207–218.

Jain, S.M., S.K. Sopory, and R.E. Veilleux, (Eds.) 1996a. In Vitro *Haploid Production in Higher Plants, Vol. 1: Fundamental Aspects and Methods*. Kluwer Academic Publishers, Dordrecht, the Netherlands.

Jain, S.M., S.K. Sopory, and R.E. Veilleux, (Eds.) 1996b. In Vitro *Haploid Production in Higher Plants, Vol. 2: Applications*. Kluwer Academic Publishers, Dordrecht, the Netherlands.

Jain, S.M., S.K. Sopory, and R.E. Veilleux, (Eds.) 1996c. In Vitro *Haploid Production in Higher Plants, Vol. 3: Important Selected Plants*. Kluwer Academic Publishers, Dordrecht, the Netherlands.

Jain, S.M., S.K. Sopory, and R.E. Veilleux, (Eds.) 1996d. In Vitro *Haploid Production in Higher Plants, Vol. 4: Cereals*. Kluwer Academic Publishers, Dordrecht, the Netherlands.

Jain, S.M., S.K. Sopory, and R.E. Veilleux, (Eds.) 1997. In Vitro *Haploid Production in Higher Plants, Vol. 5: Oil, Ornamental and Miscellaneous Plants*. Kluwer Academic Publishers, Dordrecht, the Netherlands.

Keller, E.R.J. and L. Korzun. 1996. Ovary and ovule culture for haploid production. In: In Vitro *Haploid Production in Higher Plants, Vol. 1: Fundamental Aspects and Methods*. (Ed.) S.M. Jain, S.K. Sopory, and R.E. Veilleux. Kluwer Academic Publishers, Dordrecht, the Netherlands. 217–235.

Keller, W.A. and K.C. Armstrong. 1979. Stimulation of embryogenesis and haploid production in *Brassica campestris* anther cultures by elevated temperature treatments. *Theor. Appl. Genet.* 55:65–67.

Lakshmi Sita, G. 1997. Gynogenic haploids *in vitro*. In: In Vitro *Haploid Production in Higher Plants, Vol. 5: Oil, Ornamental and Miscellaneous Plants*. (Ed.) S.M. Jain, S.K. Sopory, and R.E. Veilleux. Kluwer Academic Publishers, Dordrecht, the Netherlands. 175–193.

Maheshwari, S.C., A. Rashid, and A.K. Tyagi. 1982. Haploids from pollen grains — retrospect and prospect. *Amer. J. Bot.* 69:865–879.

Miller, C.O. 1963. Kinetin and kinetin-like compounds. In: *Moderne Methoden der Pflanzenanalyse*, Vol. 6. (Ed.) H.F. Liskens and M.V. Tracey. Springer-Verlag, Berlin. 194–202.

Murashige, T. and F. Skoog. 1962. A revised medium for rapid growth and bioassays with tobacco tissue cultures. *Physiol. Plant.* 15:473–497.

Nitsch, J.P. and C. Nitsch. 1969. Haploid plants from pollen grains. *Science* 163:85–87.

Powell, W. 1990. Environmental and genetical aspects of pollen embryogenesis. In: *Biotechnology in Agriculture and Forestry, Vol. 12: Haploids in Crop Improvement*. (Ed.) Y.P.S. Bajaj. Springer-Verlag, Berlin. 45–65.

Reynolds, T.L. 1990. Ultrastructure of pollen embryogenesis. In: *Biotechnology in Agriculture and Forestry, Vol. 12: Haploids in Crop Improvement*. (Ed.) Y.P.S. Bajaj. Springer-Verlag, Berlin. 66–82.

Sunderland, N. and M. Roberts. 1979. Cold-treatment of excised flower buds in float culture of tobacco anthers. *Ann. Bot.* 43:405–414.

18 Embryo Rescue

Sandra M. Reed

CHAPTER 18 CONCEPTS

- Embryo rescue procedures have been widely used for producing interspecific and intergeneric hybrids.

- Depending on the organ cultured, embryo rescue is referred to as embryo, ovule, or ovary culture.

- Ovule and ovary culture are more suitable than embryo culture for small-seeded species or very young embryos.

- Cultures must be initiated before embryo abortion occurs.

- Media requirements depend on the stage of embryo development.

- Young embryos (proembryos) require a medium with a high osmotic potential.

INTRODUCTION

The term "embryo rescue" refers to a number of *in vitro* techniques whose purpose is to promote the development of an immature or weak embryo into a viable plant. Embryo rescue has been widely used for producing plants from hybridizations in which failure of endosperm to properly develop causes embryo abortion. In embryo rescue procedures, the artificial nutrient medium serves as a substitute for the endosperm, thereby allowing the embryo to continue its development. Embryo rescue techniques are among the oldest and most successful *in vitro* procedures.

One of the primary uses of embryo rescue has been to produce interspecific and intergeneric hybrids. While interspecific incompatibility can occur for a wide variety of reasons, one common cause is embryo abortion. The production of small, shrunken seed following wide hybridization is indicative of a cross in which fertilization occurred but seed development was disrupted. Embryo rescue procedures have been very successful in overcoming this barrier to wide hybridization in a wide range of plant materials (Collins and Grosser, 1984). In addition, embryo rescue has been used to recover maternal haploids that have developed as a result of chromosome elimination following interspecific hybridization.

Embryo rescue techniques also have been utilized to obtain progeny from intraspecific hybridizations that do not normally produce viable seed. For example, triploids have been recovered from crosses between diploid and tetraploid members of the same species, and progeny have been obtained from crosses utilizing early-ripening and "seedless," or stenospermacarpic, fruit genotypes as maternal parents. Embryo rescue techniques have also been used in situations in which embryo abortion is not a concern, such as for overcoming seed dormancy and studying seed development and germination. The various applications of embryo rescue to both applied and basic plant research have been reviewed by Bridgen (1994), Collins and Grosser (1984), Ramming (1990) and Sharma et al. (1996).

Depending on the organ cultured, embryo rescue may be referred to as embryo, ovule, or ovary culture. While the disinfestation and explant excision processes differ for these three techniques, many of the factors that contribute to the successful recovery of viable plants are similar. This chapter will begin with a discussion of general factors that should be considered when utilizing embryo rescue and then will turn to techniques specific to each type of embryo rescue procedure.

FACTORS INVOLVED IN EMBRYO RESCUE

MEDIA

Murashige and Skoog (MS) (Murashige and Skoog, 1962) and Gamborg's B-5 (Gamborg et al., 1968) media are the most commonly used basal media for embryo rescue studies (Bridgen, 1994). Types and concentrations of media supplements required depend greatly on the stage of development of the embryo.

Raghavan (1976) identified two phases of embryo development. In the heterotropic phase, the young embryo, which is often referred to as a proembryo, is dependent on the endosperm. Embryos initiated at this stage require a complex medium. Amino acids, particularly glutamine and aspargine, are often added to the medium. Various vitamins may also be included. Natural extracts, such as coconut milk and casein hydrolysate, have sometimes been used instead of specific amino acids. Young embryos require a medium of high osmotic potential. Sucrose often serves both as a carbon source and osmoticum. High osmotic concentration in the medium prevents precocious germination and supports normal embryonic development. For heterotropic embryos, 232 to 352 mM (8–12%) sucrose is commonly used. Other sugars have been successfully used instead of or in addition to sucrose; however, sucrose has been by far the most commonly utilized sugar for embryo rescue.

The second stage of embryo development is the autotrophic phase, which usually begins in the late heart-shaped embryo stage (Raghavan, 1976). At this time the embryo is capable of synthesizing substances required for its growth from salts and sugar. Germination will usually occur on a simple inorganic medium, supplemented with 58 to 88 mM (2–3%) sucrose.

Growth regulators have been extensively used in embryo rescue studies, especially for heterotropic embryos; however, their effects have been highly inconsistent. In general, low concentrations of auxins have promoted normal growth, gibberellic acid has caused embryo enlargement, and cytokinins have inhibited growth (Sharma, 1996). In addition to supplying vitamins and amino acids to the medium, natural extracts often also supply growth regulators.

As stated earlier, media requirements differ depending on the stage of embryo development. For cultures initiated using very young embryos, more than one media formulation may be needed. For example, embryos of *Trifolium* interspecific hybrids were first cultured on a high sucrose medium containing a moderate level of auxin and a low level of cytokinin. After 1 to 2 weeks on this medium, embryos stopped growing. Growth resumed after they were transferred to a medium with a lower sucrose concentration, a low level of auxin, and a moderate level of cytokinin (Collins and Grosser, 1984).

For interspecific hybrids, it may be useful to develop media that can nurture embryos of one or both parental species. While the nutritional needs of the hybrid may be different from the parents, the parental media formulations will serve as a good starting point for the hybrid.

TEMPERATURE AND LIGHT

Temperature and light requirements vary among species. According to Sharma (1996), the growth requirements of embryos often mimic those of their parents, with embryos of cool-season crops requiring lower temperatures than those of warm-season crops. Cultures are often incubated at

25 to 30°C, although considerably lower temperatures are needed for some species. In species that normally exhibit seed dormancy, a cold treatment may be required.

Cultures are usually initially cultured in the dark to prevent precocious germination, but are moved to a lighted environment to allow chlorophyll development after 1 to 2 weeks in the dark.

TIME OF CULTURE

When attempting to rescue embryos of incompatible crosses, it is critical that the cultures be initiated prior to embryo abortion. However, because it is more difficult to rear young embryos than those that have reached the autotrophic phase of development, chances of success are maximized by allowing the embryo to develop *in vivo* as long as possible. Histological examinations can be used to determine the time of endosperm failure and embryo abortion; however, these evaluations can be very laborious. Cultures are often initiated at various intervals following pollination to maximize chances of recovering viable plants. Because an interaction between media and time of culture is expected, it is important to test a range of media ranging from complex with high sucrose to simple with low sucrose at the various culture times.

GENERAL EMBRYO RESCUE PROCEDURES

EMBRYO CULTURE

The most commonly used embryo rescue procedure is embryo culture, in which embryos are excised and placed directly onto culture medium. Fruit from controlled pollination of greenhouse- or field-grown plants is collected prior to the time at which embryo abortion is thought to occur. Since embryos are located in a sterile environment, disinfestation of the embryo itself is not required. In some cases, the entire ovary is surface-sterilized. In other cases, ovules are removed from the ovary under nonaseptic conditions and then disinfested. In either instance, a harsh disinfestation procedure can usually be applied, since the embryo is protected by the surrounding tissue.

Careful excision of the embryo is critical to the success of embryo culture. A stereomicroscope is usually required, and must be placed in the laminar flow hood in such a manner as not to restrict airflow. The best point of incision into the ovule differs among species. In some cases, embryos can be extracted by cutting off the micropylar end of the ovule and then applying gentle pressure at the opposite end of the ovule. This results in the embryo being pushed out through the opening. It is crucial that the embryo be placed directly into culture after its excision so that it does not become dry. For heart-shaped and younger embryos, the embryo should be excised with the suspensor intact (Hu and Wang, 1986). Because of the extreme importance and frequent difficulty of excising embryos without causing damage, it may be helpful to develop and practice an excision technique under nonaseptic conditions.

Embryo culture is sometimes preceded by ovule or ovary culture. One advantage of this technique (sometimes termed ovule-embryo or ovary-embryo culture) is that embryo excision is delayed until the embryo becomes large enough to remove without damage. Also, the presence of the integument during the ovule or ovary culture phase has been found to reduce the possibility of precocious germination (Ramming, 1990). Once excised, the embryo may benefit from being in direct contact with the medium. Also, for those species affected by dormancy, removing the embryo may overcome any inhibitory effects imposed by the surrounding ovular tissues.

Nurse cultures have been used for rescuing embryos (Williams et al., 1982). This technique involves inserting the embryo from an incompatible cross into endosperm removed from a related compatible cross. For example, the embryo of an interspecific hybrid may be inserted into endosperm from an intraspecific cross involving one of the parental species. The embryo and endosperm are then placed into culture together.

Ovule Culture

Embryos are difficult to excise when they are very young or from small-seeded species. To prevent damaging embryos during the excision process, they are sometimes cultured while still inside the ovule. This technique is referred to as ovule culture or *in ovolo* embryo culture. As with embryo culture, ovaries are collected prior to the time at which embryo abortion is thought to occur. The ovary is surface-sterilized and the ovules removed and placed into culture. This step ranges from extremely easy to accomplish for large-seeded species in which only a single ovule is present, to time-consuming and difficult for small-seeded polyovulate species. Excision of the ovules may require the use of a stereomicroscope. Including placental tissue in ovule cultures has been found to be beneficial in some species (Rangan, 1984).

Recent modifications of the standard ovule culture technique have been developed for use in peach (*Prunus persica* [L.] Batsch) (Pinto et al., 1994). One technique, ovule perforation, requires making small holes in each ovule just prior to its placement on the culture medium. These perforations, which should be made with care not to damage embryos, increase water and nutrient uptake. Two types of ovule support systems have been developed. The filter paper support system involves culturing ovules on top of filter paper placed over liquid medium, whereas the vermiculite support technique entails placing ovules micropylar side down into a sterile vermiculite/liquid media mixture (vermiculite support). While the ovule perforation and vermiculite support systems may not be feasible in small-seeded species, ovule size should not pose a limitation for using the filter paper support system in species other than peach.

Ovary Culture

In ovary or pod culture, the entire ovary is placed into culture. Ovaries are collected and any remaining flower parts removed. Disinfestation protocols must remove surface contaminants without damaging the ovary. The ovary is placed into culture so that the cut end of the pedicel is in the medium. At the end of the experiment, seeds are removed from the fruit that develop in culture.

A technique known as ovary-slice culture has been utilized for rescuing *Tulipa* interspecific hybrid embryos (Van Creij et al., 1999). Ovaries were cut transversely into sections and the basal cut end of the sections placed on the culture medium. In *Tulipa*, ovule culture and ovary-slice culture produced similar germination rates; however, the ovary-slice culture procedure was considered to be the superior of the two techniques because it was less time-consuming.

SUMMARY

Embryo rescue procedures have been successfully used for many years for producing interspecific and intergeneric hybrids and progeny of other incompatible crosses. While embryo culture is the most widely used embryo rescue procedure, ovule and ovary culture are more suitable for small-seeded species or very young embryos. For all three procedures, the probability of success increases with maturity of the embryos; however, cultures must be initiated before embryos abort. The type of medium needed for rescuing embryos is strongly dependent on the stage of embryo development. Young embryos require a complex medium with high sucrose concentrations, while more mature embryos can usually develop on a simple medium with low levels of sucrose. Continued investigations into nutritional requirements of young embryos, along with modifications of existing embryo rescue techniques, should lead to successful application of this highly valuable *in vitro* procedure to additional crop species.

LITERATURE CITED

Bridgen, M.P. 1994. A review of plant embryo culture. *HortScience* 29:1243–1246.

Collins, G.B. and J.W. Grosser. 1984. Culture of embryos. In: *Cell Culture and Somatic Cell Genetics of Plants, Vol. 1: Laboratory Procedures and Their Applications*. (Ed.) I.K. Vasil. Academic Press, New York. 241–257.

Gamborg, O.L., R.A. Miller, and K. Ojima. 1968. Nutrient requirements of suspension cultures of soybean root cells. *Exp. Cell Res.* 50:157–158.

Hu, C. and P. Wang. 1986. Embryo culture: Technique and application. In: *Handbook of Plant Cell Culture, Vol. 4*. (Ed.) D.A. Evans, W.R. Sharp, and P.V. Ammirato. Macmillan, New York. 43–96.

Murashige, T. and F. Skoog. 1962. A revised medium for rapid growth and bioassays with tobacco tissue cultures. *Physiol. Plant.* 15:473–497.

Pinto, A.C.Q., S.M.D. Rogers, and D.H. Byrne. 1994. Growth of immature peach embryos in response to media, ovule support method, and ovule perforation. *HortScience* 29:1081–1083.

Raghavan, V. 1976. *Experimental Embryogenesis in Vascular Plants*. Academic Press, London.

Ramming, D.W. 1990. The use of embryo culture in fruit breeding. *HortScience* 25:393–398.

Rangan, T.S. 1984. Culture of ovules. In: *Cell Culture and Somatic Cell Genetics of Plants, Vol. 1: Laboratory Procedures and Their Applications*. (Ed.) I.K. Vasil. Academic Press, New York. 227–231.

Sharma, D.R., R. Kaur, and K. Kumar. 1996. Embryo rescue in plants — a review. *Euphytica* 89:325–337.

Van Creij, M.G.M., D.M.F.J. Kerckhoffs, and J.M. Van Tuyl. 1999. The effect of ovule age on ovary-slice culture and ovule culture in intraspecific and interspecific crosses with *Tulipa gesneriana* L. *Euphytica* 108:21–28.

Williams, E.G., I.M. Verry, and W.M. Williams. 1982. Use of embryo culture in interspecific hybridization. In: *Plant Improvement and Somatic Cell Genetics*. (Ed.) I.K. Vasil, W.R. Scowcroft, and K.J. Frey. Academic Press, New York. 119–128.

19 Genetic Engineering Technologies*

Zhijian T. Li and Dennis J. Gray

CHAPTER 19 CONCEPTS

- Over the past three decades, great advancements in plant transformation methodologies have occurred. Several reliable transformation techniques are available to produce transgenic plants of many crop species.

- Genetic engineering technologies complement and extend conventional breeding efforts by allowing incorporation of foreign DNA and novel characteristics into target plants without the hindrance of biological barriers.

- Transfer of foreign DNA into plant cells can be accomplished by a variety of biological, physical, and chemical means, among which the most commonly used are *Agrobacterium*-mediated transformation, protoplast-mediated transformation, and microprojectile bombardment.

- The incorporation of transformation technologies into contemporary plant improvement programs has yielded new cultivars with improved agronomic traits.

INTRODUCTION

Genetic engineering of crop plants represents a major milestone in modern agricultural science. The advent of recombinant DNA technology in the early 1970s and the subsequent development of DNA transfer techniques provided exciting opportunities for plant scientists to insert foreign genes from both prokaryotic and eukaryotic organisms into the genome of crop plants and achieve transgene expression. Technological advancements in plant tissue culture techniques facilitated introduction of foreign genes into the plant genome to produce transgenic plants. Transgenic plants expressing novel traits now are being widely cultivated for their improved yield, quality, and other value-added characteristics. It should be noted, however, that in most instances genetic engineering techniques provide only an alternative approach to conventional breeding programs. In crop improvement, conventional breeding and hybrid seed production are the mainstay in ongoing efforts directed toward varietal development (Morandini and Salamini, 2003). Nonetheless, modern genetic engineering technologies offer several unique advantages over conventional hybridization approaches. For example, *in vitro* DNA transfer techniques permit introduction of genes and other genetic elements among sexually unrelated organisms, thereby bypassing biological barriers. Such genetic manipulation can be accomplished using a large quantity of plant materials in a relatively small space with a year-round artificially controlled growth environment. Hence, use of genetic engineering techniques complements and expedites conventional breeding programs by increasing

* Florida Agricultural Journal Series No. R-10167

diversity of genetic resources, enhancing efficiency and reducing length of time needed to introgress desirable traits into existing elite crop varieties. Genetic engineering also allows utilization of exotic genes for development of transgenic plants to produce proteins with novel nutritive, pharmaceutical, agrichemical, and industrial characteristics (Fischer and Emans, 2000).

Although several studies reported transfer and expression of prokaryotic genes, the first expression of eukaryotic genes in genetically modified plant cells was demonstrated in transgenic cells of sunflower containing a gene encoding the seed storage protein phaseolin (Murai et al., 1983). In these experiments, transformation was accomplished using the soil-borne phytopathogenic bacterium *Agrobacterium tumefaciens* to transfer desired genes and cause them to be incorporated into the plant genome. This DNA transfer technique was developed as a result of years of extensive studies on crown gall disease and molecular mechanisms that control tumor formation (Ream and Gordon, 1982). *Agrobacterium*-mediated transformation has become the most commonly used method of creating transgenic plants. Desired gene expression was derived from so-called hybrid genes in which target gene coding sequences were operably linked to functional promoter and terminator sequences. In the case of transgenic sunflower, gene transcription and translation were confirmed by accumulation of phaseolin gene-specific mRNA and production of phaseolin protein in transformed tissues. The first intact plants that expressed transgenes were subsequently obtained using tobacco.

Stimulated by these and more recent studies, plant scientists worldwide have continued to develop and refine transformation technologies in order to enable the generation of transgenic plants from numerous species. Today, three major DNA transfer methods are widely utilized. These include (1) *Agrobacterium*-mediated transformation, (2) direct protoplast-mediated DNA transfer, and (3) microprojectile bombardment-mediated transformation. In addition, several alternative DNA delivery systems also have been developed. Successful utilization of these DNA transfer techniques has resulted in examples of transgenic plants among almost all major crop species. In this chapter, emphasis is given to discussion of major transformation methods currently being used to transfer foreign DNA into plant cells.

AGROBACTERIUM-MEDIATED TRANSFORMATION

A. tumefaciens is a Gram-negative, soil-borne, phytopathogenic bacterium responsible for inciting crown gall disease in a large number of gymnosperms and angiosperms. The development of crown gall disease occurs through an intricate interplay between bacterial genetic elements and plant host responses. Early molecular studies revealed that this bacterium was capable of transferring a short piece of DNA (T-DNA) from its tumor-inducing (Ti) plasmid into the genome of susceptible host plant cells. The T-DNA contains genes encoding proteins that are involved in the biosynthesis of phytohormones (oncogenes) and novel conjugates of organic acids and amino acids or sugars called opines (the opine synthesis genes). These phytohormones and bacterial metabolites are necessary for survival and proliferation of bacterial cells in the modified host cell environment. They also stimulate tumor formation (i.e., the "crown gall"). In addition, genes located in the virulence (*vir*) region of the Ti plasmid encode a number of virulence proteins (Vir proteins) that mediate the T-DNA transfer process. Activation of these genes is controlled by plant regulatory factors that are produced by infected host plant cells.

For producing transgenic crop plants, there are several advantages associated with *Agrobacterium*-mediated transformation. *Agrobacterium*-mediated transformation results in transfer of DNA with defined ends and with minimal rearrangement. Relatively large segments of DNA also can be accommodated within the T-DNA region and subsequently transferred. In addition, compared to other gene transfer techniques, only one or a few copies of the transferred genes are generally integrated into plant chromosomes with *Agrobacterium*-mediated transformation. With a minimum degree of gene disruption resulting from genomic integration of transgenes, transgenic plants with normal agronomic performance and fecundity can be obtained.

The T-DNA Transfer Process

The T-DNA transfer process can be separated into the following major steps:

1. Activation of virulence genes
2. T-strand processing and transfer
3. T-DNA integration into the plant cell genome (Figure 19.1)

For successful DNA transfer, *Agrobacterium* detects the presence of low molecular mass phenolic and sugar compounds (e.g., acetosyringone) produced by wounded plant cells. This molecular sensing process is mediated by an inner membrane protein, VirA, which subsequently transduces information to a transcriptional activator, VirG, by a mechanism involving protein phosphorylation. Activated VirG then triggers transcription and expression of the *vir* region contained in the Ti plasmid, leading to production of up to 20 Vir proteins from six operons: *virA, -B, -C, -D, -E,* and *-G.* These proteins perform a variety of functions to ensure the success of T-DNA transfer. Of particular interest is the site-specific endonuclease VirD2. This protein, assisted by VirD1 protein, recognizes the T-DNA border sequences (25 bp direct repeats at its ends), creates a nick site and initiates the formation of a single strand (ss) copy of the T-DNA (T-strand). VirE2 is a sequence-nonspecific ssDNA binding protein. It functions to bind ssDNA regardless of sequence, forming a fully coated VirE2:T-strand complex (T-complex) that effectively protects the T-DNA from nucleolytic degradation during the transfer process. In addition, nuclear localization signals (NLSs) present in VirE2 and VirD2 facilitate the import of T-complex into the plant cell nucleus. During the transfer process, the VirD2 protein covalently binds to the right border of the

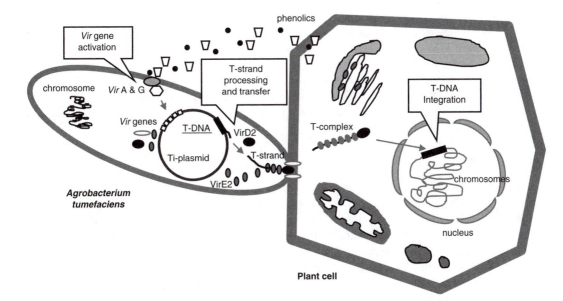

FIGURE 19.1 Schematic representation of the *Agrobacterium* T-DNA transfer process. *Agrobacterium*-mediated T-DNA transfer can be viewed as a three-step process. During the first step, bacterial attachment to plant cells facilitates interactions between plant inducers, including low molecular mass phenolic and sugar compounds and the signal-sensing *virA* product. The latter then activates the transcription activator *virG* product. The activated *virG* product induces transcription of the rest of the *vir* loci located on the Ti plasmid. In the second step, a T-strand containing a single-strand copy of T-DNA is cut out from the Ti plasmid by the *virD2* product and transferred from bacterial cell to plant cell through coordinated action of a variety of other *vir* products, including the single-strand DNA binding protein VirE2. In the final step, T-DNA is introduced into the nucleus and subsequently integrated into the plant cell chromosome.

T-strand and facilitates its transfer into the nucleus and subsequent genomic integration in the plant cell (Figure 19.1). More details on *Agrobacterium*-mediated T-DNA transfer can be found in several excellent reviews (Zambryski, 1992; Tinland, 1996; Gelvin, 2000; Tzfira et al., 2000; Zhu et al., 2000; Zupan et al., 2000).

TRANSFORMATION STRATEGIES

Different strategies have been developed to make use of *Agrobacterium*-mediated T-DNA transfer. It was first essential to modify the native T-DNA to remove tumor-inducing genes responsible for phytohormone biosynthesis so that crown gall disease did not occur. The unwanted genes then could be replaced with target genes encoding products of interest. Ti plasmids from which tumor-inducing genes are removed are called disarmed Ti plasmids.

One early strategy of *Agrobacterium*-mediated DNA transfer utilized the so-called co-integrating Ti plasmid vectors (Rogers et al., 1988). In this method, an intermediate plasmid carrying the right border sequence, a portion of Ti plasmid DNA, and sequences for the origin of replication (replicon) from *Escherichia coli* was used to provide a vehicle to manipulate genes of interest in *E. coli* cells. A disarmed Ti helper plasmid carrying the left border was also prepared in an appropriate *Agrobacterium* strain. In addition, a transfer helper plasmid providing plasmid DNA transfer functions between bacterial cells was maintained in *E. coli* cells. When all three types of bacteria containing these specialized plasmids were mixed together, the intermediate plasmid was transferred from *E. coli* cells into the *Agrobacterium* cells where it was cointegrated in the T-DNA region of the disarmed Ti plasmid through a process called homologous recombination. *Agrobacterium* cells harboring cointegrated Ti plasmids that contain a fully functional T-DNA region with both the right and left borders are identified by selection for antibiotic resistances encoded by the different integration partners and are propagated for subsequent use in transformation experiments.

A second strategy that has received more popular usage involves the use of binary plasmid vectors (Bevan, 1984). This type of plasmid combines both right and left borders of the T-DNA and compatible replicons from both *E. coli* and *Agrobacterium* in a single plasmid molecule, such that it can be propagated and maintained in both hosts for DNA manipulation and subsequent transfer. However, use of the binary vector system requires an *Agrobacterium* host strain that carries a wild-type Ti plasmid or a disarmed Ti plasmid in order to provide functions of the *vir* genes essential for the T-DNA transfer process.

FACTORS AFFECTING *AGROBACTERIUM*-MEDIATED TRANSFORMATION

T-DNA transfer is a complex biological process necessitating activation of a large number of Ti plasmid and related bacterial-encoded genes that involve active molecular interactions between *Agrobacterium* and host plant cells. Accordingly, development of efficient *Agrobacterium*-mediated transformation systems for different crop species can only be accomplished if the essential requirements for improving *vir* gene expression and plant cell responses to *Agrobacterium* infection are met. Among factors that affect the efficacy of DNA transfer into plant cells and the obtainment of transgenic plants are virulence of the *Agrobacterium* strain, plant genotypic response, tissue culture procedures and manipulations, and selection and regeneration of transformants.

In general, there are differences in levels of virulence among *Agrobacterium* strains. The virulence of a particular *Agrobacterium* strain represents its capability to transfer T-DNA through the bacterial cell wall and to introduce it into affected plant cells. Hence, screening different *Agrobacterium* strains for high levels of virulence and infectivity is frequently utilized as a crucial first step in the development of an efficient transformation system for a particular plant species. The degree of virulence is primarily attributed to the level of Vir protein expression from the Ti plasmid. For example, the supervirulence of a wild-type strain A281 was found to be associated with expression of *virG* and *virB* loci of its Ti plasmid pTiBo542. Consequently, use of a disarmed version of this supervirulent Ti plasmid (pEHA101) resulted in virulent strain EHA101, which has been used to transform numerous plant species (Hood et al., 1986). In addition to screening for virulence, several techniques have been developed to enhance the virulence of a chosen *Agrobacterium* strain. For instance, culturing bacteria in the presence of plant phenolic compound inducers, such as acetosyringone, resulted in dramatically enhanced *vir* gene expression and subsequently led to a higher transformation frequency (Engstrom et al., 1987). In another approach, key virulence genes, such as *virG* encoding a transcriptional activator, were modified for constitutive expression in bacterial cells to improve bacterial virulence and host range (Hansen et al., 1994). Use of these modified supervirulent binary vectors ultimately resulted in the successful transformation and recovery of trangenic plants from recalcitrant monocotyledonous species, including rice (Hiei et al., 1997).

Judicious selection of the receptor genotype and explant type and age, along with physiological state and pretreatment of receptor tissues, are also critical factors affecting successful transformation using *Agrobacterium*. While mechanisms still remain unknown, different genotypes of a receptor species often display different levels of susceptibility to *Agrobacterium* infection. Many economically important crop species and elite genotypes of particular species are highly recalcitrant to *Agrobacterium*-mediated transformation. Hence, a few susceptible genotypes (model genotypes) were typically first utilized in order to identify factors that limit the efficacy of T-DNA transfer and to develop optimized procedures for transformation of other genotypes. Cells from various tissues or tissues subjected to various *in vitro* culture conditions or physical or chemical pretreatments prior to transformation may become differentially susceptible to transformation and plant regeneration. In addition, it is important to define tissue culture regeneration systems to maximize the population size of transformable and regenerable cells and to identify effective pretreatments that enhance infection and DNA uptake. Recent efforts to identify plant genes and proteins responsible for resistance to *Agrobacterium* infection may facilitate the eventual transformation of recalcitrant species and increased transformation frequencies in general.

PROTOPLAST-MEDIATED TRANSFORMATION

Protoplasts are cells without cell walls (Chapter 16). In a protoplast system, the only barrier between living protoplasm and the external environment is the plasma membrane. Thus, the semi-fluid nature of the plasma membrane allows direct movement of macromolecules, such as DNA, into protoplasts by using relatively simple physical or chemical treatments. When cultured under suitable conditions, protoplasts are capable of cell wall regeneration and subsequent growth into whole plants. These unique characteristics of protoplasts permit development of efficient protoplast-mediated DNA transfer and transformation systems. Protoplast-mediated transformation has been successfully utilized in numerous molecular and genetic studies of genes and genetic elements to elucidate gene functionality and mechanisms regulating gene expression. Efficient protoplast-mediated transformation systems also have been utilized for transformation of a number of major crop species, including rice (Li et al., 1990).

PROTOPLAST ISOLATION AND REGENERATION

Protoplasts can be isolated from a variety of plant materials, including leaves, callus tissues, suspension cultured cells, somatic embryos, pollen, cotyledons, and other storage organs. Leaf tissue is an excellent source of protoplasts due to the relative abundance of more or less uniform mesophyll cells and a relatively simple anatomical structure that facilitates protoplast release. When leaves of young plants are used, surface sterilization is normally needed, followed by removal of the epidermis, enzymatic digestion, and ultimately the purification of protoplasts. The use of *in vitro* shoot cultures or plantlets can reduce problems of contamination during and subsequent to protoplast isolation. However, suspension cultures are considered to be a better protoplast source material for transformation studies because they provide high quality, actively growing cells, often with synchronized cell division cycles. In addition, protoplasts from suspension cultures may have a higher potential for rapid cell division and a higher competency for DNA uptake and integration. These characteristics result in a better efficiency of transformation and transgenic plant regeneration. (Details of protoplast isolation and culture are provided in Chapter 16.) After protoplast isolation, it is important to clean protoplasts thoroughly to remove cell debris and residual enzymes that may otherwise interfere with DNA transfer or hamper cell wall regeneration during subsequent protoplast culture. The combination of filtration through nylon mesh (40 μm) and simple low-speed centrifugation in an isotonic washing solution proved sufficient to ensure efficient recovery of viable protoplasts for use in transformation studies (Li et al., 1990; 1995).

Regeneration of plants from isolated protoplasts is a prerequisite for the successful utilization of protoplast-mediated transformation for crop improvement. Over the years, a wide range of techniques have been developed for regeneration of plants from protoplasts. In particular, immobilization of protoplasts in a low melting-point agarose gel to facilitate selection and restrict movement of transformed protoplasts during the culture process, and the utilization of nurse cells to enhance protoplast regeneration, have been successfully applied to several important crop species (Li et al., 2000).

PROTOPLAST-MEDIATED DNA TRANSFER

Electroporation and polyethylene glycol (PEG) treatments are the two most widely used methods for delivering DNA into protoplasts. Cationic liposome-mediated transformation of protoplasts is less frequently used. Electroporation treatment provides a technically simple way to introduce DNA into protoplasts through electrically induced membrane pores. Electroporated protoplasts can be immediately subjected to culture treatments facilitating the recovery of transformed cells. Transformation frequency following electroporation is often affected by several factors including electroporation voltage, current and pulse duration, DNA concentration, and buffer type. The development of an efficient electroporation procedure necessitates the optimization of these parameters.

PEG-mediated DNA uptake offers an inexpensive, simple, and efficient approach for introducing DNA into protoplasts (Negrutiu et al., 1987). PEG molecules chemically induce the formation of membrane pores and thus allow the movement of macromolecules into protoplasts. PEG-mediated DNA uptake was up to 100 times higher than that obtained using electroporation (Hayashimoto et al., 1990). PEG concentration is the most important parameter affecting transformation frequency. For example, it was found that 30% (w/v) PEG dramatically reduced protoplast viability, whereas less than 10% (w/v) PEG was ineffective in improving DNA uptake by rice protoplasts (Hayashimoto et al., 1990). High concentrations of PEG can also induce undesirable protoplast fusion during the transformation process. Thus, one has to utilize an appropriate PEG concentration during the transformation process in order to obtain a high transformation efficiency while maximizing protoplast viability after treatment.

MICROPROJECTILE BOMBARDMENT

Microprojectile bombardment is a technique by which solid particles such as tungsten or gold microspheres or microparticles (carriers) are coated with DNA molecules and accelerated by high-pressure helium gas or propulsion of an explosive charge at high velocity into plant cells. These particles have sufficient momentum to penetrate cell walls and move themselves into target cells without causing lethal damage. Once inside the cells, DNA molecules are released from their carriers and integrated into the chromosome by cellular components. Transformed cells are then identified and transgenic plants subsequently recovered through appropriate tissue culture selection and regeneration procedures. As with other direct DNA transfer systems, microprojectile bombardment permits introduction of naked DNA into plant cells, making construction of gene expression units relatively simple because the need for the *Agrobacterium* binary vector system is eliminated. It also allows simultaneous introduction of multiple transgenes or large DNA fragments into plant cells in a single operation. This technique enables the use of a wide range of regenerable plant tissues as transformation target materials for studies of gene expression and obtainment of transgenic plants (Taylor and Fauquet, 2002). Microprojectile bombardment is especially useful for the transformation of plastids (Bogorad, 2000).

PARTICLE BOMBARDMENT TECHNOLOGY

A particle bombardment device was first demonstrated by Sanford and co-workers (Sanford et al., 1987). This device utilized tungsten particles (microcarriers) that were coated with DNA and placed onto the surface of small plastic bullets (macrocarriers). A gunpowder charge was then employed to propel the plastic bullets at high velocity into a stopping plate that effectively stopped the plastic bullets, but allowed the tungsten particles to be released with sufficient speed and force on impact to penetrate the target tissues placed below. Based on a similar mechanism, several modified devices were subsequently developed in efforts to provide a better control for the bombardment process, improve transformation efficiency, and increase operational safety, while reducing the cost.

BioRad (Hercules, CA) markets the Biolistic particle delivery system PDS-100/He. This device utilizes helium gas to propel microcarriers in combination with plastic rupture discs that break at specified pressures ranging from 450 to 2200 psi. A detailed description of the device and operation procedures can be found on the BioRad 2002 website (http://www.bio-rad.com). Parameters that affect DNA delivery, damage to target tissues due to microcarrier impact, and transformation efficiency can be experimentally manipulated and optimized. This device continues to be popular in plant biotechnology laboratories and is capable of efficiently and reproducibly transforming numerous crop species. BioRad has also developed a hand-held, helium-powered device called the Helios Gene Gun for transforming target tissues and organs that are not suitable for placement in a vacuum chamber. This device has been used primarily to study transient gene expression and for the inoculation of plant viral pathogens.

Several variations of the original particle gun have been developed. The Particle Inflow Gun (PIG) device developed by Vain et al. (1993) differs from the BioRad device in that it utilizes a burst of helium to carry microcarriers into the vacuum chamber, thus eliminating the need for rupture discs or macrocarriers and associated parts. The PIG device is relatively inexpensive and easy to operate, and has been used successfully to produce transgenic plants in several major crop species. The PIG device is functionally described in Chapter 21.

Another variant of the particle gun is named "ACCELL technology" and utilizes shock waves generated by high-voltage electrical discharge to propel microcarriers into the target tissues (McCabe and Christou, 1993). Because this device does not rely on gaseous propellants to transfer power to microcarriers, damage to the target tissues can be greatly minimized. In addition, the electrical discharge mechanism provides a high degree of controllability, making it possible to deliver microcarriers into desirable layers of plant cells within target tissues. Transgenic plants containing transgenes with low-copy number integration were produced in several crop species using the ACCELL system.

FACTORS AFFECTING MICROPROJECTILE BOMBARDMENT-MEDIATED TRANSFORMATION

The successful application of microprojectile bombardment technology to plant genetic engineering largely depends on the development of a set of optimized parameters associated with the bombardment process. These parameters include choosing the appropriate type and size of microspheres, coating procedures, DNA concentrations, microcarrier dispersing and loading procedures, vacuum levels, helium gas pressures, electrical discharge settings, and number of repeated bombardments. It is important to make certain that the DNA of interest can be delivered into target cells of selected tissue without causing excessive cellular damage that could otherwise impair the subsequent regeneration process.

The ability to deliver foreign DNA into cells by microprojectile bombardment greatly increases the range of cell and tissue types and the number of elite genotypes that can be used in transformation experiments. However, selection of appropriate target cells and tissues is crucial for achieving high transformation efficiencies and, ultimately, transgenic plants. There are several criteria to be evaluated in choosing an appropriate target tissue. Target tissues should have the following capabilities:

1. Integrating introduced foreign DNA into its genome and germline cells
2. Expressing selectable marker genes that have been engineered into the transferred DNA so that transformants can be successfully selected
3. Proliferating and regenerating into normal transgenic plants at a high frequency

Embryogenic and meristematic tissues are most frequently selected for use as target materials for the production of transgenic plants using microprojectile bombardment technology.

ALTERNATIVE DIRECT DNA TRANSFER

Over the years, the previously discussed transformation methods have been widely utilized in efforts to produce transformed tissues and transgenic plants of numerous crop species. However, there are certain limitations associated with these methods. For instance, the use of *Agrobacterium*-mediated transformation is limited by target tissue availability and low transformation efficiencies associated with poor susceptibility to infection by recalcitrant species and genotypes. Protoplast-mediated transformation often requires an elaborate and time-consuming transformant selection regime and plant regeneration systems that can increase the occurrence of somaclonal variations. Microprojectile bombardment is frequently associated with high costs for acquiring the operating apparatus and associated accessories. To circumvent these problems, several alternative transformation methods based on the use of direct DNA transfer were developed.

Silicon-carbide whisker-mediated transformation is perhaps the best documented alternative method of direct DNA transfer. It involves a brief vortex of a transformation mixture containing needle-like silicon-carbide microbodies, DNA, and target tissues. DNA is transported into target cells through holes generated in cell walls by silicon-carbide whiskers (Keappler et al., 1990). This efficient and inexpensive method has been used successfully to produce fertile transgenic maize plants (Frame et al., 1994). Transformation via DNA infiltration, electroporation of cells and intact tissues, electrophoresis, imbibition of embryos, microprojectile bombardment of pollen tubes, and liposome-mediated transformation also have been attempted. However, the application of these techniques in practical transformation of major crop plants requires further technical development.

CONCLUDING REMARKS

Over the past three decades, great advancements have been made by plant scientists through concerted efforts to develop transformation methodologies. This has resulted in the identification of several reliable transformation techniques that can be used routinely in many crop species to transform plant cells and to produce transgenic plants. Continuous refinement of these techniques and discovery of new approaches will further enhance our ability to genetically modify existing elite genotypes, and expedite efforts to improve the agronomic performance and value of crop plants. Despite these advances, genetic engineering of crop plants continues to face a variety of challenges. For example, factors affecting the level and stability of transgene expression remain to be fully elucidated. Innumerable in-depth studies of transgene expression in transgenic plants have revealed that plants employ sophisticated internal mechanisms to cope with the intrusion of exogenous DNA and associated gene expression components. Molecular strategies that facilitate sustainable transgene expression in transgenic crop plants are needed. In addition, further studies are needed in areas such as the removal of objectionable marker genes from transgenic plants (Zuo et al., 2002), multigene engineering for development of novel biological and metabolic pathways (Daniell and Dhingra, 2002), and the introduction of artificial chromosomes into crop plants. With technological advancements in these and other areas, genetic engineering of plants will continue to play an increasingly important role in global crop improvement efforts.

LITERATURE CITED

Bevan, M.W. 1984. Binary *Agrobacterium* vectors for plant transformation. *Nucl. Acid Res.* 12:8711–8721.

Bogorad, L. 2000. Engineering chloroplasts: an alternative site for foreign genes, proteins, reactions and products. *Trends Biotech.* 18:257–263.

Daniell, H. and A. Dhingra. 2002. Multigene engineering: Dawn of an exciting new era in biotechnology. *Curr. Opin. Biotech.* 13:136–141.

Engstrom, P. et al. 1987. Characterization of *Agrobacterium tumefaciens* virulence proteins induced by the plant factor acetosyringone. *J. Mol. Biol.* 197:635–645.

Fischer, R. and N. Emans. 2000. Molecular farming of pharmaceutical proteins. *Transgenic Res.* 9:279–299.

Frame, B.R. et al. 1994. Production of fertile transgenic maize plants by silicon carbide whisker-mediated transformation. *Plant J.* 6:941–948.

Gelvin, S.B. 2000. *Agrobacterium* and plant genes involved in T-DNA transfer and integration. *Annu. Rev. Plant Physiol. Plant Mol. Biol.* 51:223–256.

Hansen, G., A. Das, and M.D. Chilton. 1994. Constitutive expression of the virulence genes improves the efficiency of plant transformation by *Agrobacterium. Proc. Natl. Acad. Sci. U.S.A.* 91:7603–7607.

Hayashimoto, A., Z. Li, and N. Murai. 1990. A polyethylene glycol-mediated protoplast transformation system for production of fertile transgenic rice plants. *Plant Physiol.* 93:857–863.

Hiei, Y., T. Komari, and T. Kubo. 1997. Transformation of rice mediated by *Agrobacterium tumefaciens. Plant Mol. Biol.* 35:205–218.

Hood, E.E. et al. 1986. The hypervirulence of *Agrobacterium tumefaciens* A281 is encoded in a region of pTiBo542 outside of T-DNA. *J. Bacteriol.* 168:1291–1301.

Kaeppler, H.F. et al. 1990. Silicon carbide fiber-mediated DNA delivery into plant cells. *Plant Cell Rep.* 9:415–418.

Li, Z., M.D. Burow, and N. Murai. 1990. High frequency generation of fertile transgenic rice plants after PEG-mediated protoplast transformation. *Plant Mol. Biol. Rep.* 8:276–291.

Li, Z. et al. 1995. Improved electroporation buffer enhances transient gene expression in *Arachis hypogaea* protoplasts. *Genome* 38:858–863.

Li, Z. et al. 2000. Transgenic peanut (*Arachis hypogaea*). In: *Biotechnology in Agriculture and Forestry, Vol. 46: Transgenic Crops I.* (Ed.) Y.P.S. Bajaj. Springer-Verlag, Berlin. 209–224.

McCabe, D. and P. Christou. 1993. Direct DNA transfer using electrical discharge particle acceleration (ACCELLTM technology). *Plant Cell. Tiss. Org. Cult.* 33:227–236.

Morandini, P. and F. Salamini. 2003. Plant biotechnology and breeding: Allied for years to come. *Trends Plant Sci.* 8:70–75.

Murai, N. et al. 1983. Phaseolin gene from bean is expressed after transfer to sunflower via tumor-inducing plasmid vectors. *Science* 222:476–482.

Negrutiu, I. et al. 1987. Hybrid genes in the analysis of transformation conditions. I. Setting up a simple method for direct gene transfer in plant protoplasts. *Plant Mol. Biol.* 8:363–373.

Ream, L.W. and M.P. Gordon. 1982. Crown gall disease and prospects for genetic manipulation of plants. *Science* 218:854–859.

Rogers, S.G. et al. 1988. Use of cointegrating Ti plasmid vectors. In: *Plant Molecular Biology Manual.* (Ed.) S.B. Gelvin and R.A. Schilperoort. Kluwer Academic Publishers, Dordrecht, the Netherlands. A2/1–12.

Sanford, J. et al. 1987. Delivery of substances into cells and tissues using particle bombardment process. *J. Part. Sci. Tech.* 5:27–37.

Taylor, N.J. and C.M. Fauquet, 2002. Microprojectile bombardment as a tool in plant science and agricultural biotechnology. *DNA Cell Biol.* 21:963–977.

Tinland, B. 1996. The integration of T-DNA into plant genomes. *Trends Plant Sci.* 1:178–184.

Tzfira, T. et al. 2000. Nucleic acid transport in plant-microbe interactions: The molecules that walk through the walls. *Annu. Rev. Microbiol.* 54:187–219.

Vain, P. et al. 1993. Development of the particle inflow gun. *Plant Cell. Tiss. Org. Cult.* 33:237–246.

Zambryski, P.C. 1992. Chronicles from the *Agrobacterium*-plant cell DNA transfer story. *Annu. Rev. Plant Physiol. Plant Mol. Biol.* 43:465–490.

Zhu, J. et al. 2000. The bases of crown gall tumorigenesis. *J. Bacteriol.* 182:3885–3895.

Zuo, J. et al. 2002. Marker-free transformation: increasing transformation frequency by the use of regeneration-promoting genes. *Curr. Opin. Biotech.* 13:173–180.

Zupan, J. et al. 2000. The transfer of DNA from *Agrobacterium tumefaciens* into plants: A feast of fundamental insights. *Plant J.* 23:11–28.

20 Genetically Modified Plant Controversies: Sensational Headlines versus Sensible Research

Harry A. Richards, Laura C. Hudson, Matthew D. Halfhill, and C. Neal Stewart, Jr.

CHAPTER 20 CONCEPTS

- Genetically modified (GM) plants have been widely grown throughout the world and in the United States.

- No measurable harm has been detected from the cultivation of GM plants.

- Popular press and public perceptions often focus on the sensational side.

- Established methods to test food safety of GM plants and products are available.

- Testing ecological interactions between GM plants and wild plants is complicated.

- While today's GM plant products are safe, biotechnology risk assessment research should be continued.

INTRODUCTION

Controversies over scientific issues have often been initiated by sensational media reports of research that capture the imagination of the public. Before the application of biotechnology, the biology of crop plants had never been the focus of controversy and debate. However, with recent attention drawn to plausible harm to foodstuffs and the environment, the previously unheralded subject of agriculture has been transformed into media fodder. We want to illuminate the facts and discuss the research behind the issues by focusing on a few timely plant biotechnology controversies.

BIOTECHNOLOGY IN THE SPOTLIGHT

A remarkable development occurred in the early 1980s: the first genetically engineered plants were produced. For the potential impact that this technology would have on agriculture as we know it, the event was rather unheralded. As biotechnology grew in application, it remained nearly a nontopic for news media and the general populace. There were rumblings from some concerned parties, environmentalist groups for one, but in general, genetically engineered foods were accepted into use with little fanfare or concern. Opponents were thought of as technophobes or extremists.

However, the situation exploded in 1999, and awareness of genetically modified (GM) food sky-rocketed as news media outlets reported that GM crops devastated monarch butterfly populations and could be toxic to men, women, children, and babies.

The question remains to be answered if all this attention is warranted. Clearly, if genetically engineered crops are driving butterflies to extinction or if eating corn chips made from GM corn made people sick, then there would be cause for alarm. However, as is often the case, these reports are not reflective of the scientific experimentation behind them, nor do they indicate the levels of safeguards in place to protect the population from potential risks. Since 2001, the debate has ebbed as other world events and issues have taken priority in the media spotlight. However, the issue has not been settled. The purpose of this chapter is to outline the risks posed by food and agricultural biotechnology and discuss examples of the research that has been conducted to evaluate these risks. The more popular aspects of the GMO controversy will be addressed by outlining the experiments that are the basis for the hysteria and by reporting the research conducted as follow-up to these preliminary studies. Finally, we will provide our perspective on these issues and offer opinions as to the lessons learned and the future direction of plant biotechnology.

THE INCENTIVE

The transfer of genetic materials to create a new plant variety is not a new concept in agriculture. Intensive breeding programs have long been in place to develop crops that perform better in the field. This technology is limited by sexual compatibility and the difficulties in obtaining the desired result from the transfer of thousands of genes from parents to offspring. Irradiation and mutageni-zation have been used to create heritable changes in the genetic structure of crops to generate new traits. These techniques are limited by their lack of predictability and precision. The incentive for the development and use of biotechnology is to overcome these limitations. Because this type of gene transfer does not rely on compatible crosses, it is possible to introduce traits from any species into plants, and the transfer involves a single gene or a handful of genes, which provides more predictable and precise outcomes for the next generation of plants. This technology does not solve all of the problems associated with modern agriculture, but it does offer a number of potential benefits that make it worth pursuing. However, because it is conceptually different from the technologies employed before, we must ask what new complications and risks are posed by biotechnology and how they can be effectively managed to allow its safe use now and in the future.

THE ENVIRONMENT

Risk can be defined as the probability of a potential hazard occurring. One advantage in evaluating environmental risks of biotechnology is that the parameters of the risk equation are similar to those for crops produced via traditional breeding or mutation techniques. When a new variety is created, it is evaluated to determine how its new characteristics will affect its performance in the field. This includes risk analysis, such as increased invasiveness or sexual compatibility with wild relatives. The difference with varieties created through biotechnology is that the potential list of transferred characteristics is greater and more diverse. Also, the traits are more discreet because they are linked to a single gene or a handful of genes. Therefore, genetically engineered crops must be evaluated on a case-by-case basis as the potential hazards will differ for different transgenes. The risk associated with an identified hazard will be defined by the probability of occurrence. So the key to risk assessment of biotechnological products is to identify potential hazards and evaluate the likelihood of that situation occurring.

Volunteerism and the invasiveness of new varieties could lead to economic complications or the persistence of the plants in the environment. These potential hazards are of concern when transgenes are added that increase the fitness of the plants to either agricultural treatments or environmental stresses. Traits such as herbicide resistance could make volunteers (uncollected seeds from the previous crop) difficult to manage or remove. Traits such as drought tolerance may allow crops to be grown in new climates, but could also lead to the escape of those plants into new environments, adversely affecting the native species. Research into these risks involves characterizing the performance of fitness-enhancing transgenes and evaluating management strategies to reduce their occurrence.

HYBRIDIZATION

Potential complications of hybridization with transgenic crops can be divided into two categories: crosses with nontransgenic plants of the same species and crosses with related wild species. The first category could cause problems for farmers who wish to keep crops free of genetically engineered plants, particularly organic farmers that must abide by a low GM plant threshold. The second category encompasses the risk of transgene escape from crops to weedy relatives, the concern being the generation of "superweeds." These hazards are addressed through containment strategies to limit the movement of transgenic material. To define the risk of the creation of a superweed, the effects of the transgenic traits in agricultural or ecological environments must be considered. Plants expressing traits that do not enhance fitness, such as increased carotinoid expression, do not pose this type of risk, but genes such as insect resistance or salt tolerance may need consideration. In addition, hybridization requires sexual compatibility. In the United States, there is no concern that soybeans will hybridize with anything other than other soybeans; therefore the superweed risk is very low. A significant amount of research has been conducted on these issues for crop species that have wild relatives near the place of cultivation (Warwick et al., 2003).

TRANSGENE FLOW FROM GENETICALLY MODIFIED CANOLA

Although maize and soybean have no wild relatives that occur near cultivation in the U.S., this is not to say that all GM crops are free of potential consequences from hybridization. Canola (*Brassica napus* L.) is an emerging crop that could pose potential risks if agricultural acreages expand in geographic regions with naturally occurring wild relatives. Canola is of particular interest as a potential source for transgene escape due to several factors. Approximately 11% of the 25 million hectares of canola produced globally is transgenic, and this percentage is even higher in some countries — e.g., 60% in Canada (Warwick et al., 2003). The crop is predominantly selfing, with outcrossing averaging 30% (Beckie et al., 2003); it forms a persistent seed bank and produces large weedy volunteer or feral populations, particularly in the first year after canola production. The transgenic volunteers are a potential problem for several reasons. The GM volunteers represent a management concern, because they can be a "weedy" problem in the next year's crop and can be difficult to control due to transgenic traits such as herbicide tolerance. More importantly, the volunteers also represent a recurring source of transgenes via hybridization to other canola varieties and wild relatives.

Hybridization from genetically modified canola to other canola varieties has been documented in several scientific studies. In one famous case, volunteer individuals were resistant to three different herbicides as the result of hybridization of three herbicide-tolerant varieties over several seasons (Hall et al., 2000). Therefore, in areas where more than one type of herbicide-tolerant canola variety has been grown, multiple herbicide-tolerant volunteer individuals have arisen as the result of intraspecific hybridization (Hall et al., 2000; Beckie et al., 2003). Also, the unintended presence of transgenes in certified seed stocks has the potential to allow the flow of transgenes in unexpected ways. Beckie et al. (2003) found an example where certified seed of a Roundup Ready™ (glyphosate tolerant) canola variety had a small percentage of Liberty Link™ (glufosinate tolerant) seeds. As a result of the unexpected transgene, a higher quantity of double-resistant volunteers were found in the following field season than would be expected due to hybridization with surrounding canola fields. The agronomic characteristics of canola production combined with the use of multiple GM varieties and the occurrence of unexpected transgenes in certified seed lots demonstrate the necessity to monitor canola as a potential source of transgene flow, at least until the consequences can be determined.

Wild relatives also have the potential to be the recipient of transgenes from GM canola varieties. In many areas, wild relatives such as *Brassica rapa* L. (bird rape, field mustard), *Raphanus raphanistrum* L. (wild radish), and *Sinapis arvensis* L. (wild mustard) occur in or near canola cultivation. Several studies have shown that hybridization between canola and bird rape occurs under field conditions and results in the production of hybrid populations (Warwick et al., 2003). Hybridization occurs at a range of frequencies based on the ratio of parental crop species to the wild relative, and is as high as 93%. Transgenic *Bacillus thuringiensis* (Bt) canola varieties have been shown to hybridize with *B. rapa* under field conditions, and the frequency of hybridization was reported to range from 1 to 17% based on the transgenic variety (Warwick et al., 2003). In a recent report, Warwick et al. (2003) have demonstrated the transfer of an herbicide (glyphosate) tolerance gene from commercial fields of canola to a naturally occurring population of *B. rapa* in Quebec, Canada. With relatively high hybridization frequencies observed in these experiments, transgenic hybrid populations should be expected when canola and naturally occurring wild relatives grow in close proximity.

TRANSGENES DISCOVERED IN MEXICAN CORN

Transgene flow from transgenic crop varieties to wild relatives is of particular interest because of the potential ramifications both inside and outside of agriculture. However, there have been some cases where suspected transgene flow within a crop species has generated both scientific and media attention. A recent finding reported the introgression of the cauliflower mosaic virus (CaMV) 35S promoter into the genome of native Mexican landraces of corn (Quist and Chapela, 2001). The corn sampled was from the state of Oaxaca, which was remotely located from areas where transgenic corn had been grown legally. In fact, the cultivation of GM crops had been illegal in Mexico since 1998. The paper initiated a wave of public and scientific debate, which, after 6 months of scrutiny, ended with the *de facto* retraction of the work by the editor of the journal.

The central conclusion of the paper was that transgenic DNA sequences were present in traditional Mexican corn landraces. In addition, the authors concluded that transgenes had been introgressed into the genome of individuals in this population and that the transgenic material was moving within the genome. These conclusions resulted in a firestorm of public and scientific criticism. The predominant argument was that the methodology used for this research was not appropriate for the conclusions drawn (Metz and Futterer, 2002; Kaplinsky et al., 2002). The author's only reported evidence of the CaMV 35S promoter and not actual transgenes in Mexican corn. In addition, PCR was the method used for detection, which creates the distinct possibility that contamination of trace quantities of the cauliflower mosaic virus itself (ubiquitous in the environment) may have resulted in false positive results.

Critics argued that before conclusions of the presence and introgression of transgenes could be drawn, more stringent and direct tests would be required. For instance, a direct assay for the presence of a transgene, such as Bt or the Roundup Ready™ genes, would have been more indicative of transgenic material. Instead of PCR, a genomic hybridization technique, such as Southern blot analysis, would have greatly reduced the likelihood of false positives and would have been more convincing. The authors published a follow-up article in which they conducted a DNA dot blot assay; however, the result was still based on the CaMV 35S promoter and the data could not indicate insert size, copy number, or integration into the genome. The result was continued debate and criticism.

Two follow-up communications in the journal *Nature* reevaluated the data presented by Quist and Chapela (2001). In the original paper, nested PCR suggested that a few kernels on any given ear of corn contained the CaMV 35S promoter. Metz and Futterer (2002) argued that if this result was accurate, then introgression into the genome could not have occurred because the expected result would have been presence of the transgenic sequences in most or all of the kernels. The original publication also suggested that the CaMV 35S promoter was changing locations within the genome of Mexican corn. This data was collected using inverse PCR, which uses known sequences (in this case the CaMV 35S promoter) to determine the identity of adjacent sequences of DNA. However, Quist and Chapela (2001) did not find transgenes located near these putative transgenic promoters, as critics argued would be expected (Metz and Futterer, 2002; Kaplinsky et al., 2002). Additionally, Kaplinsky et al. (2002) indicated that the article mistakenly reported a gene sequence as one commonly used in plant genetic engineering (*adh1* intron) when sequence similarity actually suggested that the sequences were from naturally present corn genes (*adh1* and *bronze1* genes). The intense criticism of the methodology and conclusions in the paper resulted in its *de facto* retraction, though the authors remained steadfast in their assertion that transgenic material was present in Mexican corn landraces. To date, no other research has been published that supports this claim.

NON-TARGET ORGANISMS

GM crops cover millions of acres across the United States, where large quantities of GM foods are consumed. When transgenic crops are utilized, especially those designed for pest control, the potential exists for them to have adverse effects on species that were not intended to be harmed (Losey et al., 1999). This hazard is similar to the risks associated with the application of chemical pesticides in modern agriculture. In fact, GM crops have been developed to reduce the application of these chemicals, to reduce costs to farmers, and to lessen the impact of pest control agents on the environment. Nevertheless, pesticidal transgenic plants receive the lion's share of the attention and criticism. Research has been conducted on the toxicity of these materials to agriculturally benign species and on their persistence in the environment. Until the monarch butterfly became the subject of significant scientific and media interest regarding non-target effects from biotech crops, there was little public debate. This issue focused on the pollen of Bt corn, which could blow onto milkweed leaves, the exclusive diet of monarch caterpillars.

The monarch butterfly became the focus of attention in 1999 when a published article captured media attention. Bt is a naturally occurring soilborne bacterium, which produces a crystal-like protein (Cry protein) that selectively kills a specific group of insects (Lepodopteran, the caterpillars). It is not harmful to agriculturally beneficial insects, animals, or humans and does not persist in the environment. Liquid and granular formulations of the Bt proteins have been used successfully for 50 years on a variety of crops. Bt corn refers to corn that has been enhanced through biotechnology to produce its own Bt insecticidal proteins.

Regulation for Bt corn and other transgenic crop species falls under the jurisdiction of the Environmental Protection Agency (EPA) in the United States. The EPA is responsible for determining environmental effect. Before the approval of Bt corn in 1995, 1996, and 1998, the EPA concluded that the product does not cause any "unreasonable adverse effects to non-target organisms based on evaluations of toxicity and exposure." These conclusions were based on the facts that corn pollen is heavy and can move only a short distance; that cornfields typically contain a low concentration of weeds; and that monarch larvae exposure would be limited to milkweeds within or in close proximity to cornfields during pollen shed.

In May 1999, a communication in the journal *Nature* (Losey et al., 1999) raised concerns regarding non-target effects of transgenic crops containing Bt genes. A laboratory feeding experiment performed by researchers at Cornell University found monarch butterfly larvae that were given no choice but to feed on milkweed leaves dusted with high levels of pollen from Bt corn had slower growth rates and a higher mortality rate than those larvae consuming leaves with no pollen or with non-Bt pollen. This group of scientists suggested that there was a potential risk to monarch butterfly populations if demonstrated in the natural environment because migratory patterns include the central United States, where the majority of corn is grown.

In order to address public concerns about the monarch, researchers from nine universities, the U.S. Department of Agriculture (USDA), and the Agriculture Extension Service conducted numerous field tests in 1999 and reported their findings at the Monarch Butterfly Research Symposia and later in the *Proceedings of the National Academy of Sciences, U.S.A.* This collaborative work found that monarch larvae have very little exposure to corn pollen under field conditions (Sears et al., 2001) because the majority of corn pollen settles in the immediate vicinity of the field. Pollen movement is limited; however, wind can disperse pollen away from the field, depositing it over a broad area and creating low-pollen concentrations on the surfaces of leaves. Further field studies demonstrated that milkweed leaves captured only 30% of the corn pollen available, and that wind and rain can reduce this amount by 90%. Furthermore, Bt endotoxin degrades in pollen upon exposure to sunlight (Pleasants et al., 2001).

Monarch butterflies may have exposure to small quantities of Bt corn pollen on milkweed but at amounts below the threshold for harming the larvae. Monarch larvae can safely consume milkweed leaves with up to 1100 Bt pollen grains per square centimeter (Hellmich et al., 2001), which is high for field levels. Field surveys in Maryland were conducted in 81 locations, showing that the mean pollen level inside of the cornfields or around their perimeters was 56 pollen grains per square centimeter; only 10 out of 127 leaves examined had higher levels (Pleasants et al., 2001). Hellmich et al. (2001) found that monarch larvae would avoid eating milkweed with Bt or non-Bt corn pollen on the surface if milkweeds free from pollen were available. Field studies conducted in Canada and the United States., sponsored by the USDA and the Agricultural Biotechnology Stewardship Technical Committee, concluded that monarch survival, weight gain, and milkweed consumption were similar for monarch larvae feeding for 5 consecutive days on milkweed plants in Bt and non-Bt cornfields during pollen shed (Sears et al., 2001; Hellmich et al., 2001). These experiments confirmed that milkweed leaves in close proximity to Bt cornfields contained pollen levels too low to affect the normal development of the monarch butterfly larvae.

Now that the question of an immediate significant risk has been answered, studies are underway, according to the USDA, to determine if there are subtle effects on the larvae when exposed to Bt corn pollen for extensive time periods and whether older caterpillars consume anthers that fall onto milkweed leaves unintentionally, since Bt protein levels were found to be higher in these plant organs. Many factors contribute to the fact that fewer than 10% of monarch caterpillars make it to adulthood, and this should be weighed in comparison to rates that are associated with typical mortality. Furthermore, future conclusions addressing environmental impacts on non-target organisms by transgenic crops need to be approached with appropriate scientific methods and assessment procedures.

RESISTANCE

As with any chemical pesticide treatment, the reality exists that individuals and eventually populations of pests will be selected that have resistance to the treatment, a fact that keeps the chemical industry searching for new insecticides and herbicides to meet the demands of modern agriculture. The same hazard exists for transgenic crops. To address this risk, the same types of management strategies used for chemical resistance management are used for genetically engineered crops. Because transgenic crops have been in widespread use for nearly a decade, data are now being collected on the occurrence of resistance. These data will be valuable in determining the long-term effectiveness of specific GM crops and in helping to develop new management strategies.

BT-RESISTANT INSECTS

Concerns exist that the utility of transgenic Bt crops will be short-lived due to pest evolution coinciding with the widespread implementation of transgenic Bt cotton, corn, and potato (Gould, 1998). Although insecticidal crop varieties (mainly Bt) have been widely implemented in the United States, pest control failures have not occurred due to pest resistance to recombinant proteins produced in transgenic plants. This is not to say that Bt-resistance alleles are not present in insect pest populations. Resistance to Bt has already evolved in the diamondback moth (DBM) (*Plutella xylostella*) due to the use, not of transgenic plants, but of Bt foliar sprays under field conditions. Also, Bt-resistant DBM populations have the potential to survive and cause damage to transgenic plants. Resistance alleles have been detected in bollworms (*Helicoverpa zea*) collected near areas of Bt cotton production in North Carolina (Burd et al., 2003). The management of pest resistance to Bt transgenic plants is the essential next step for allowing the continued use of Bt in pest control.

Understanding the genetic mechanism for Bt resistance in insect populations is vitally important for developing strategies that control the resistance. Dominant resistance alleles would be considerably more problematic than recessive alleles because the insect only needs one copy of a dominant allele to be resistant to the Bt protein. Also, at least one half of the offspring of a resistant parent who has a dominant allele will be resistant to the Bt protein, which could allow a rapid population shift to the resistant phenotype. Recessive alleles still must be managed, but the effects of a recessive gene might be slower in a population because individuals must have two copies of the recessive alleles to be resistant to the protein. In both cases, management strategies for the control of resistance alleles in pest populations involve nontransgenic refuges to allow the survival of susceptible individuals.

In the case of DBM, Shelton et al. (2000) have shown that resistance alleles are recessive. They and others have suggested a Bt management strategy in which transgenic plants must produce a high lethal dose combined with a nontransgenic refuge in order to delay the onset of Bt resistance within insect populations. This strategy effectively dilutes recessive resistance genes by providing an abundance of susceptible individuals as mates for rare homozygous resistant individuals. The offspring of this mating would be heterozygous for the resistance alleles and would be susceptible to the Bt toxin. In a recent report, Burd et al. (2003) have found evidence that dominant resistance alleles have been found in bollworm populations. Dominant resistance alleles could be a significant problem to the utility of Bt crops because heterozygous individuals that arise from the mating of resistant and susceptible individuals will be resistant. The control of dominant resistance alleles will require stringent management strategies that have larger refuges to allow the survival of larger populations of susceptible individuals. If not properly managed, the utility of Bt transgenic plants in pest control could be lost, and agricultural systems will have to return to chemical-based sprays to control pest populations.

ROUNDUP-RESISTANT HORSEWEED

A parallel situation to that of Bt resistance seems to exist for herbicide resistance management. Apparently, weeds can be selected for resistance to herbicides. It is interesting that while so many scientists were studying intra- and interspecific transgene flow, herbicide overuse was selecting glyphosate-resistant horseweed (*Conyza canadensis*, Asteraceae). Of course, the case of horseweed has absolutely nothing to do with transgene flow from GM plants to non-GM plants. However, the selection of herbicide-resistant biotypes might be accelerated and exacerbated by the habitual application of a single herbicide and the practice of conservation tillage. Roundup Ready crops permit the use of a single herbicide: Roundup™. Can a plant species that is already a problematic weed become herbicide-tolerant, so that it is no longer killed by spraying herbicide on it?

Millions of acres in the United States are planted every year with herbicide-tolerant soybeans, corn, and cotton. Growing these plants provides a tremendous benefit to farmers. They can treat for weeds by applying a single herbicide, sprayed over the top of the crop. Weeds are killed, but the crop thrives because of the one or two added genes that render them tolerant to an herbicide such as Roundup (glyphosate). In addition, there is a twofold environmental benefit to this practice. The first benefit is that glyphosate is a nontoxic chemical (except to plants) that degrades rapidly in the environment, which is in contrast with less environmentally friendly herbicides. The second, and even larger, benefit to the environment is that farmers can better practice no-till farming. Farmers will commonly remove weeds from fields by tilling the ground. In the absence of GM herbicide-tolerant crops, tillage reduces weed load so that chemical applications are kept at a minimum. The downside of tillage, however, is an increase in topsoil erosion. In the past 10 years, the amount of no-till and reduced-tillage agriculture has increased more than threefold to 52 million acres in the United States. In 2002, Roundup Ready soybean varieties were grown on about the same acreage. Erosion has historically been a problem in places like the Mississippi River watershed, which includes western Tennessee. In western Tennessee, reduced tillage agriculture has been practiced for many years to keep fertile and fragile soils from washing down the Mississippi River to the Delta. Also, with no-till agriculture there are generally greater weed problems. Conservation tillage increases the importance of certain weeds such as horseweed, also known as mare's-tail.

Horseweed has, until recently, been controllable by glyphosate. And, in fact, it was doubted that a glyphosate-resistant broadleaf weed, such as horseweed, would spontaneously emerge because of the herbicide's mode of action. However, in Delaware in 2000 (VanGessel, 2001), and later that same year in western Tennessee, a few glyphosate-resistant horseweed plants were discovered. By the following year, glyphosate-resistant horseweed was found on tens of thousands of acres. In 2002, upward of a half a million acres of crops in Tennessee alone contained Roundup-tolerant horseweed. Presumably, one or two genes inherent in the horseweed species allow it to survive Roundup applications, but prior to Roundup Ready crop cultivation, these resistance genes were very rare in the weed populations. But the use of Roundup™ year after year selected for the rare resistance genes, which spread rapidly across populations of horseweed, just as population genetics would predict. Little is known about the molecular biology of glyphosate resistance in horseweed. However, unlike the predominant paradigm in Bt resistance management in which the resistance is recessive, glyphosate resistance seems to be dominant. In our opinion, the risk of herbicide-resistance management, while not unlike the risk conferred by conventional agriculture, is the greatest risk in biotechnology. Environmental risks seem to be borne out as a result of using an agricultural biotechnology as a silver bullet, the final solution. Most of the risks that find a high profile in the popular media and are promulgated by environmental activists are not great, while Bt and herbicide resistance management do seem to be very important risks that need to be managed hand in glove with biotechnology's deployment.

THE FOOD SUPPLY

Whenever a new product or agricultural commodity enters the food supply, there is concern as to whether the food is safe for consumers. Products of biotechnology should be no different; however, they are much more heavily scrutinized than their traditional counterparts because of public reservations and confusion over recombinant DNA technology. While these precautions may have added costs to consumers and delayed the availability of some products, ultimately the public needs to know that regulatory agencies are working to protect them and the food supply. Thus, genetically engineered foods are evaluated for toxicity and allergenicity before approval for consumption.

Toxicity research follows the same design that is used for testing other food additives or agricultural chemicals. Allergenicity testing poses unique difficulties, in that testing traditional foods for new allergens has not been previously feasible and therefore a well-established system does not yet exist. Furthermore, transgenic crops may contain proteins that have not been consumed by humans before, which limit the available allergenicity research tools. The concept of substantial equivalence may be useful in evaluating GM foods. Essentially, it holds that a product that is similar in composition to its processor may be considered to have the same level of benefit and risk. Basically, if a new soybean variety is created, it is still considered a soybean. Transgenic foods will be nearly identical to their processor except for the changes introduced by biotechnology.

ASSESSING TOXICITY

There is, unfortunately, a lack of peer-reviewed, published data on GM food safety. There may be several reasons for this, the most likely being that the research is expensive and often results in negative data, which is difficult to use for external funding or to have it published. Therefore, much of the research on food safety is conducted by private companies, which are typically reticent to publish on proprietary products. As such, opponents make the argument that insufficient research has been conducted in this area and that the research that has been conducted lacks accountability. To dispute this claim, more research on the matter needs to be published in peer-reviewed journals, which requires more private company publications, more public research funding, and more journals willing to publish this type of data. Due to the current climate toward biotechnology, an environment of openness and accountability must be established to address public concerns.

The research that has been published to date has suggested that GM foods are safe for consumption. Safety evaluations on transgenic plants have focused on several aspects of the crop. For instance, studies have been conducted on the safety of consuming genetically engineered DNA. In one study, mice were fed recombinant GFP DNA for eight generations (Hohlweg and Doerfler, 2001) without deleterious effects on the animals. In general, because transgenic DNA accounts for less than 1/10,000 of the total DNA humans consume on a daily basis, it is difficult to imagine it posing a health risk.

Most of the food safety research has been conducted on the engineered product or on the specific recombinant compound that was added. Assessing the safety of a gene product has followed established toxicological practices. Compounds such as enolpyruylshikimate-3-phosphate synthase (an herbicidal agent based on existing plant enzymes) and Bt (an insecticidal agent used in agriculture for 40 years) have an established safe history of consumption and use, which one would not expect to be altered by transgenic plants (Cockburn, 2002). If the transgene product is novel to the food supply or changes the metabolic profile of the plant, then classical toxicity assessment would be conducted, including acute oral and repeat dose studies (Cockburn, 2002).

The research that has generated more interest and controversy is that which has been conducted on novel crops or on the derived GM foods themselves. Most of these studies have not indicated any evidence of toxicity upon consumption. Examples include Flavr Savr™, tested by tube-feeding mice; Bt corn, fed to broiler chickens; insecticidal peas (expressing α-amylase), fed to rats; and GM potatoes expressing soybean glycinin, also fed to rats. Probably the most-tested GM crop is herbicide-tolerant soybeans (Roundup Ready), which have been fed to broiler chickens, catfish, dairy cows, rats, and mice (Besler et al., 2001). In all of these studies, there were no indications of toxicity or compromised health. The most significant criticism has been that detailed histology of the stomach and intestinal tracts had not been conducted and that the conclusions of safety are overstated. However, these studies were of an acceptable quality for effective safety assessment research.

The study that has generated the most attention among the studies that have been published is the one that produced a positive result (Ewen and Pusztai, 1999). This article is still a cornerstone for opponents in the GM foods debate. The authors found that feeding mice GM potatoes expressing an insecticidal protein (lectin) compromised the intestinal tract. This effect was not observed in mice fed control potatoes or control potatoes and purified lectin protein. The conclusion was that the genetic modification itself was responsible for the observed toxic effect. The publication became a lightning rod for both praise and criticism. The consensus in the scientific community was that the experiments were not conclusive and were insufficiently designed. Whether or not the conclusions were warranted, condemnation of the paper did not dilute its effect on the public or on the debate. The research published by Ewen and Pusztai (1999) raises concerns, and more complete experiments should be conducted either to support or to dispute their findings. To date, 4 years later, no follow-up research has been done.

Ultimately, there are systems in place that allow for the screening of products for toxicity before they are approved for the food supply. The likelihood that a harmful GM product would enter the food supply is remote, and is certainly lower than the chance that other novel food products will cause harm. The question becomes, to what degree of scrutiny should GM foods be screened? Currently testing of GM foods is more rigorous than that of their conventional, irradiated and mutagenized counterparts.

POTENTIAL ALLERGENICITY

The potential introduction of food allergens into the food supply by biotechnology is perhaps the most significant concern that has developed in the public consciousness. This issue is complicated because understanding of the biology of IgE-mediated allergenicity is limited (Ladics et al., 2003). Models to test for and identify food allergens are relatively new and are only now in the process of being validated. As such, potential allergenicity has been a focal issue of biotechnology opponents, despite the fact that all novel food products (regardless of origin) are a potential source of new allergens. However, because genetically engineered crops will possess only a handful of modified traits, it is possible to propose models for systematically testing those novel proteins. One standard for allergenicity assessment is the decision tree format. The tree relies upon evaluating several parameters of the transgene and its product, which include the source organism of the gene (whether or not it is a common allergen), the sequence homology of the gene to known allergens, its resistance to peptic digestion, and studies involving sensitized human serum and skin tests. These data are used to classify a protein's relative risk of allergenicity; this classification is influenced by the level of expression of the novel trait and the results of animal model testing.

Sequence comparison as an assessment tool relies upon the existence of a database of sequenced food allergens (http://www.allergenonline.com) and alignment/comparison programs such as FASTA. Proteins that have a high overall identity to a food allergen (about a 70% match in amino acids) may have similar three-dimensional structures and may have potential cross-reactivity to immunoglobulins. While a lack of amino acid similarity does not indicate that a protein is not a potential allergen, such data can contribute to an overall assessment strategy and provide direction for future evaluation. This methodology is difficult to validate because it has not been used to identify a novel food allergen yet. While the principle seems sound, without a positive control to demonstrate that it can be effective, questions will remain as to its effectiveness. However, sequence comparison is still a valuable assessment tool for characterizing novel proteins, as it provides potential targets to direct future research.

Stability to peptic digestion is an important characteristic in food allergen identification. Food allergens, relative to other food proteins, are stable to pepsin (Besler et al., 2001). This characteristic is thought to be important in increasing the level of allergen exposure to the intestinal immune tissues (Ladics et al., 2003). Therefore, it stands to reason that a novel protein sharing this characteristic would be at an elevated level of allergenicity risk. This parameter should be used in conjunction with other criteria to determine potential exposure, as the assay is not intended to mimic the range of the human digestive process. Another factor contributing to intestinal exposure would be the level of protein expression. For example, if a transgenic protein consisted of 40% of the total protein of a food product, then it would be assumed that a significant amount of that protein would survive digestion regardless of its digestive stability. Conversely, a protein consisting of only 0.01% of the total protein would constitute much less potential exposure. The parameters used during simulated gastric digestion, such as pH, can dramatically affect the results; therefore, it is important that a standard protocol be adopted that maximizes the opportunity to identify stable proteins and allows for data comparison between studies.

Bolstering these analyses would be reliable animal models for allergenicity assessment. Unfortunately, though several are in development, none are yet validated as effective for human allergen identification. One strategy is to expose strongly IgE-expressing mice or rats to intrapenitoneal injections of the target protein, in order to characterize its potential to elicit IgE responses (Kimber et al., 2000). Preliminarily, these studies have demonstrated that these models can detect differences between major food allergens and nonallergens in IgE response. The concern about their appropriateness stems from uncertainty as to how well rodent physiology reflects that of humans (Ladics et al., 2003). Nonrodent models have focused on high IgE-producing dogs and swine. These models would be more reflective of human physiology, but are not well characterized and are considerably more expensive (Ladics et al., 2003). While animal models are likely to be critical for future analysis of allergenicity, currently they are in the developmental stage, which means that regulatory agencies should remain cautious when approving novel proteins for consumption. Proteins raising red flags in existing assessment protocols should be withheld from the food supply until these new models are ready for application. The research in the field of allergenicity assessment has progressed significantly. As research continues, more will be learned about food allergens and the biology of their effects, which should result in improved detection methodologies.

STARLINK® CORN

Starlink corn, commercialized by Aventis CropScience USA, has been modified through well-recognized genetic techniques to produce the protein Cry9C from the bacteria *B. thuringiensis*, due to the protein's known insecticidal properties. Cry9C is a variant of a number of Bt toxins including the commercially used Cry1A, mentioned previously. Starlink corn was registered in 1998 for industrial uses and animal feed with the EPA under the Federal Insecticide, Fungicide, and Rodenticide Act. In 2000, the EPA concluded, after granting the registration of Starlink, that the Cry9C protein met the safety standard for use in field corn for animal feed based on the toxicology data and limited exposure expected with animal feed use, and that there was reasonable certainty that no harm would result in exposure to the human population. The EPA did not extend the exemption to human food because there was concern that the Cry9C protein is resistant to simulated gastric digestion.

On September 18, 2000, a press conference was held announcing that taco shells purchased from a grocery store contained trace amounts of GM corn DNA associated with Starlink. This event was convened by Genetically Modified Food Alert, a consortium of seven consumer organizations based out of Washington, D.C., who notified Kraft Foods of their findings. A few days later Kraft Foods announced its voluntary recall of all Taco Bell Home Originals taco shells and taco dinners sold nationwide, which resulted in the return of 2.5 million boxes of taco shell products. Following Kraft's action, a number of other food manufactures issued recalls for products made from corn, resulting in the eventual recall of nearly 300 food products.

An investigation done by the Centers for Disease Control (CDC) established that 28 people who filed adverse event reports after eating a product containing the Cry9C protein had experienced an allergic reaction. However, their study could not confirm a link between Cry9C and the production of detectable amounts of the Cry9C-specific antibody in blood serum from the patients. The CDC stated that although their results did not provide evidence that the allergic reactions experienced by these people were associated with the Cry9C protein, the possibility could not be completely ruled out, leaving the EPA with the responsibility to decide how to regulate plant containing the Cry9C protein (CDC, 2001). Following this string of events, Aventis stopped sales of Starlink seed and agreed to purchase Starlink corn in order to isolate it from the U.S. food supply. A few months later, Aventis announced that is was canceling the registration of Starlink corn, resulting in the corn being banned for any agricultural purpose.

The Consumers Union of Japan reported that Starlink protein had been detected in cornmeal, and exports to Japan, the largest foreign market for U.S. corn, dropped by about two-thirds, while South Korea, the second largest consumer of U.S. corn, banned it all together. Many proposals for reform surfaced in light of the Starlink recalls. Legislation was introduced in Congress requiring the FDA to review GM foods (S3184: The Genetically Engineered Foods Act of 2000), and the establishment of a single agency responsible for regulating food safety was considered. The issue of food labeling was also raised after this controversy; however, the FDA reaffirmed its decision not to require special labeling on all bioengineered food, instead proposing guidelines for voluntary labeling.

CONCLUSIONS

While media attention over GM foods has declined over the past few years, unresolved issues remain pertaining to the debate over the application of genetically engineered crops. While it is generally believed that biotechnology is or will be a valuable addition to modern agriculture, there is also agreement that the products of biotechnology must be evaluated for safety to the environment, animals, and people. This technology has elicited more concern than previous agricultural technologies in the public arena, and as such, GM products are more heavily scrutinized than other products. For this scrutiny to serve its purpose and increase safety, it must be founded on scientific principles.

Research on the environmental impact of GM crops is steadily increasing. Risk assessment models are being developed based on both previous experience with similar products and on new insights specific to GM plants. High-profile issues, such as the monarch butterfly controversy, are receiving sufficient scientific attention to provide a reasonable and measured understanding of the actual impact. However, potentially more problematic issues such as invasiveness, hybridization, and resistance are not being ignored. In addition, public concern has spurred the development of a rigorous food safety evaluation paradigm that is applied to genetically engineered plants and the resulting food products. This system is more stringent than previous safety evaluation requirements, and involves not only the study of individual components and proteins, but also research on the whole plant or food product.

While these developments are encouraging and reassuring, it is important that this work continue. For this to happen, more data must be published in peer-reviewed journals and more funding must be made available for public scientists to conduct this type of research. Sensational headlines may grab the public's attention and affect policy in the short term, but it is continued diligence in research that will ultimately persuade the populace and set policy. Therefore, follow-up research is critical. The scientific community was very effective in evaluating the monarch butterfly issue, but not nearly as successful in food safety. Dedicated research and quality data will allow for biotechnology to be safely put into practice, and will consequently allow for its acceptance by the public.

LITERATURE CITED

Beckie, H.J. et al. 2003. Gene flow in commercial fields of herbicide-resistant canola (*Brassica napus*). *Ecol. Appl.* 13:1276–1294.

Besler, M., H. Steinhart and A. Paschke. 2001. Stability of food allergens and allergenicity of processed foods. *J. Chromatog. Biomed. Sci. Appl.* 756:207.

Burd, A.D. et al. 2003. Estimated frequency of nonrecessive *Bt* resistance genes in bollworm, *Helicoverpa zea* (Boddie) (Lepidoptera: Noctuidae) in eastern North Carolina. *J. Econ. Entomol.* 96:137–142.

Centers for Disease Control. 2001. Investigation of human health effects associated with potential exposure to genetically modified corn. A report to the U.S. Food and Drug Administration from the Center for Disease Control and Prevention.

Cockburn, A. 2002. Assuring the safety of genetically modified (GM) foods: the importance of an holistic, integrative approach. *J. Biotech.* 98:79–106.

EPA. 2001.

Ewen, S.W.B. and A. Pusztai. 1999. Effects of diets containing genetically modified potatoes expressing *Galanthus nivalis* lectin on rat small intestines. *Lancet* 354:1353–1355.

FAO/WHO. 2001. Evaluation of allergenicity of genetically modified foods. Report of the joint FAO/WHO expert consultation on allergenicity of foods derived from biotechnology. Food and Agriculture Organization of the UN/WHO, Rome.

Fu, T.J., U.R. Abbott, and C. Hatzos. 2002. Digestibility of food allergens and non-allergenic proteins in simulated gastric fluid and simulated intestinal fluid — A comparative study. *J. Agric. Food Chem.* 50:7154.

Gould, F. 1998. Sustainability of transgenic insecticidal cultivars: Integrating pest genetics and ecology. *Annu. Rev. Entomol.* 43:701–726.

Hall, L. et al. 2000. Pollen flow between herbicide-resistant *Brassica napus* is the cause of mutiple-resistant *B. napus* volunteers. *Weed Sci.* 48:688–694.

Helm, R.M. et al. 2002. A neonatal swine model for peanut allergy. *J. Allergy Clin. Immunol.* 109:135–142.

Hellmich, R.L. et al. 2001. Monarch larvae sensitivity to *Bacillus thuringiensis* purified proteins and pollen. *Proc. Natl. Acad. Sci. U.S.A.* 98:11925–11930.

Hohlweg, U. and W. Doerfler. 2001. On the fate of plant or other foreign genes upon the uptake in food or after intramuscular injection in mice. *Mol. Gen. Genet.* 265:225–233.

Kaplinsky, N. et al. 2002. Maize transgene results in Mexico are artifacts. *Nature* 416:601.

Kimber, I. et al. 2000. Predictive methods for food allergenicity: Perspectives and current status. *Toxicology* 147:147–150.

Ladics, G.S. et al. 2003. Approaches to the assessment of the allergenic potential of food from genetically modified crops. In: *Proceedings of the 41st Annual Meeting of the Society of Toxicology*, Nashville, TN.

Losey, J.O., L. Rainer, and M. Carter. 1999. Transgenic pollen harms monarch larvae. *Nature* 399:214.

Metz, M. and J. Fütterer. 2002. Suspect evidence of transgenic contamination. *Nature* 416:600–601.

Pleasants, J.M. et al. 2001. Corn pollen deposition on milkweeds in and near cornfields. *Proc. Natl. Acad. Sci. U.S.A.* 98: 11919–11924.

Quist, D. and I.H. Chapela. 2001. Transgenic DNA introgressed into traditional maize landraces in Oaxaca, Mexico. *Nature* 414:541–543.

Sears, M.K. et al. 2001. Impact of *Bt* corn pollen on monarch butterfly populations: A risk assessment. *Proc. Natl. Acad. Sci. U.S.A.* 98: 11937–11942.

Shelton, A.M. et al. 2000. Field tests on managing resistance to *Bt*-engineered plants. *Nat. Biotech.* 18:339–342.

VanGessel, M.J. 2001. Glyphosate-resistant horseweed from Delaware. *Weed Sci.* 49:703–705.

Warwick, S.I. et al. 2003. Hybridization between transgenic *Brassica napus* L. and its wild relatives: *B. rapa* L., *Raphanus raphanistrum* L., *Sinapis arvensis* L., and *Erucastrum gallicum* (Willd.) O.E. Schulz. *Theor. Appl. Genet.* 107:528–539.

21 Construction and Use of a Simple Gene Gun for Particle Bombardment*

Dennis J. Gray, Michael E. Compton, Ernest Hiebert, Chia-Min Lin, and Victor P. Gaba

CHAPTER 21 CONCEPTS

- Particle bombardment is a widely used method to introduce modified DNA into plant cells, allowing rapid evaluation of gene function and, in some instances, recovery of transgenic plants.

- The technique utilizes gold or tungsten particles which are coated with DNA. The particles are explosively discharged into plant tissue. Some of the particles penetrate living cells, and gene expression occurs when the DNA becomes active within nuclei.

- Several devices with which to accomplish particle bombardment have been developed. These include machines that utilize explosive properties of gunpowder, electricity, or pressurized gas to drive the discharge of particles. The particle inflow gun, one of the pressurized gas devices, is perhaps the simplest to construct and use.

- Directions and procedures for constructing and utilizing a simple form of a particle inflow gun are presented.

INTRODUCTION

As described in Chapter 19, several techniques have been utilized successfully to introduce DNA into plant cells. Of these methods, particle bombardment, wherein microscopic metal particles coated with genetically engineered DNA are explosively accelerated into plant cells, has become the second most widely used vehicle for plant genetic transformation, after *Agrobacterium*-mediated transformation (Gray and Finer, 1993). Several distinct "particle guns" have been described, including the Biolistic PDS 1000/He (Kikkert, 1993), which is the only commercially available device. The most attractive of the noncommercial devices is the particle inflow gun (PIG) (Finer et al., 1992), based on a flowing helium device described by Takeuchi et al. (1992), which can be fabricated from steel plate with readily available parts and offers performance on par with the Biolistic PDS 1000/He (Brown et al., 1994).

This chapter is a modified version of an article published by Gray et al. in *Plant Cell Tissue and Organ Culture* (1994) and is arranged into two parts. In the first part, complete directions for constructing a simplified version of the particle inflow gun are provided. The device is termed a "Plastic PIG," because, although it is based on the PIG, the steel specimen chamber of the PIG is

* Florida Agricultural Journal Series No. R-10168

replaced with a plastic vacuum jar and other parts, which dramatically simplifies construction (Gray et al., 1994). In the second part, directions for operating the Plastic PIG are provided, including preparation of the DNA-particle mixture and methods for obtaining transient expression of the β-*glucuronidase* (*GUS*) gene, a convenient reporter gene, in plant tissues. Fabrication of this particle gun is an ideal laboratory exercise or demonstration for the entire class because it can be assembled in less than 40 minutes using only simple hand tools from parts that are readily available from technical equipment supply companies and hardware stores. The only drawback to its construction is that the components will cost $400 to $500. However, because it offers performance on par with other devices and is durable, once constructed it can be used in subsequent class exercises or research programs.

GENERAL CONSIDERATIONS

CONSTRUCTION OF THE PLASTIC PIG

For convenience, specific brands and suppliers for the following required components are noted (Table 21.1); this does not, however, constitute an endorsement of these brands or suppliers, since others may be as suitable. Although 115 VAC versions of electrical parts are specified here, 220 VAC valves and timers are available through the same suppliers. Similarly, the closest metric-equivalent fittings

TABLE 21.1
Parts List for the Plastic PIG and Specimen Holders

Part	Catalog No.	Supplier
Vacuum chamber	5305-0910	Nalge Co., Rochester, NY
Solenoid valve	S24C-4V, NC 115/60V, $1/4$"	Atkomatic Valve Co., Inc., Indianapolis, IN
Interval timer	G-08683-90	Cole-Parmer Instrument Co., Niles, IL
Apparatus positioner	G-08056-10	Cole-Parmer Instrument Co., Niles, IL
Vacuum pump	G-07055-04	Cole-Parmer Instrument Co., Niles, IL
Gas pressure regulator	G-98200-70	Cole-Parmer Instrument Co., Niles, IL
Barbed fitting	G-06362-40	Cole-Parmer Instrument Co., Niles, IL
Tee connector	G-06455-15	Cole-Parmer Instrument Co., Niles, IL
Tubing clamp	G-06833-00	Cole-Parmer Instrument Co., Niles, IL
13 mm plastic filter holder	4312	Gelman Sciences, Inc., Ann Arbor, MI
0.2 μm air filter	4464	Gelman Sciences, Inc., Ann Arbor, MI
Plastic tubing	01T252PE-RDT	Ark-Plas Products, Inc., Flippin, AR
Helium tank, lab grade	—	Local supplier
Miscellaneous hardware	—	Local hardware supply
Metal hose clamps	—	Local hardware supply
2-prong electrical appliance cord	—	Local hardware supply
Brass close nipple with flange, $1/4$" - 18 SAE	—	Local hardware supply
Brass cap, $1/4$" - 18 SAE	—	Local hardware supply
Brass bushing, $1/4$" × $1/8$" NPT	—	Local hardware supply
Steel jam nut, $9/16$" SAE	—	Local hardware supply
Teflon and electrical tape, rosin-core solder, silicone sealing glue	—	Local hardware supply
Specimen Holders:		
PC petri dishes	5502-0010	Nalge Co., Rochester, NY
Spectra/mesh PP screen	08-670-185	Fisher Scientific, Pittsburgh, PA
Magnets, 3 mm diameter (6)	—	Radio Shack, Inc.
Permatex high-temp RTV silicone gasket maker	—	Local auto parts store

---

TABLE 21.2
Tools Required to Assemble Plastic PIG
and Specimen Holders

Electric hand-held drill
$^7/_{64}$" and $^1/_4$" drill bits for metal
$^7/_{16}$" drill bit for masonry
$^5/_{16}$" - 24 NF and $^1/_4$" - 18 NPT thread taps
$^5/_{16}$" - 24 NF thread die
Adjustable hole saw
Hobby vise
Adjustable wrench
Pliers
Soldering iron

---

may be substituted for the English measurements given below. Electrical connection to supply current should be through a ground fault protected circuit. The setup must be examined and approved by a qualified electrician before it is plugged into a supply current source. The tools required to accomplish fabrication are listed in Table 21.2. Figures 21.1, 21.2, and 21.3 detail construction and use of the Plastic PIG, and the assembly steps are provided in Procedure 21.1.

Procedure 21.1 Assembly of the Plastic PIG	
Step	Instructions and Comments
1	Drill pilot hole in center top of vacuum jar. Enlarge hole with $^7/_{16}$" masonry bit and thread the hole with $^1/_4$" - 18 NPT tap.
2	Drill pilot hole in center of brass cap, then enlarge to $^1/_4$". Use $^5/_{16}$" - 24 NF tap to cut threads into hole.
3	Screw jam nut tightly onto close nipple, add teflon tape to threads on short end of nipple and screw tightly into the #2 port of the solenoid valve. Screw the brass bushing, then the barbed fitting into the #1 port of the valve, using teflon tape to seal.
4	Add a small amount of silicone glue to the exposed threads of the close nipple and screw the valve assembly snugly into the top of the vacuum jar. Add more silicone glue to the exposed threads protruding into the jar before screwing the brass cap tightly onto the nipple (see Figure 21.2).
5	Use the $^5/_{16}$" - 24 NF die to cut threads carefully over the luerlock end of the plastic filter holder. Add teflon tape to threaded end and screw the filter holder into the brass cap. Allow silicone glue on the assembled valve-vacuum jar unit to cure overnight before subjecting to vacuum.
6	Solder the two-prong electrical appliance cord onto exposed wires of valve and wrap securely with electrical tape. Plug cord into interval meter.
7	Clamp one end of plastic tubing to the barbed fitting and the other end to the helium regulator.
8	Attach one end of vacuum-vent tubing to vacuum plate as shown in Figures 21.1 and 21.3 and the other end to the vacuum pump.
9	Place the apparatus positioner in the center of the vacuum plate.

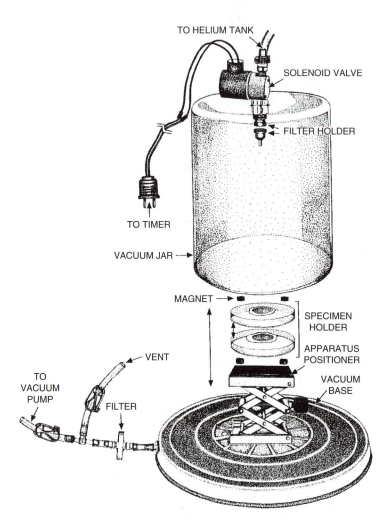

FIGURE 21.1 Construction and setup of the Plastic PIG. (From Gray, D.J. et al., 1994. Simplified construction and performance of a device for particle bombardment. *Plant Cell Tiss. Org. Cult.* 37:179–184. Used with permission.) Exploded view of Plastic PIG and specimen holders.

CONSTRUCTION OF SPECIMEN HOLDERS

Specimen holders consist of the top and bottom of autoclavable plastic petri dishes, in which holes have been drilled to accept screens (see Figure 21.1 and Table 21.1 for parts list). Follow the instructions provided in Procedure 21.2 to construct the specimen holders.

For bombardment, specimens are held between screens by nesting the top piece of the dish over the inverted bottom, both of which are then clamped together with two to three pairs of magnets (Figure 21.1).

PREPARATION OF THE PLASMID-PARTICLE MIXTURE

The plasmid used is pBI221 (Clontech Laboratories, Inc., CA), which consists of a 3kb HindII-EcoRI fragment of pBI121 containing the CaMV 35S promoter, *GUS* gene, and NOS terminator (Jefferson et al., 1987) cloned into pUC19. The plasmid can be maintained in DH5alpha bacterial host and purified according to the QIAGEN plasmid purification kit. Note that any plasmid that

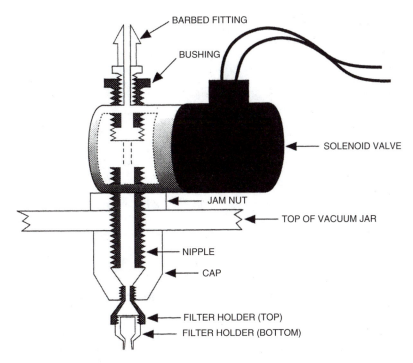

FIGURE 21.2 Detail of solenoid valve assembly. (From Gray, D.J. et al., 1994. Simplified construction and performance of a device for particle bombardment. *Plant Cell Tiss. Org. Cult.* 37:179–184. Used with permission.)

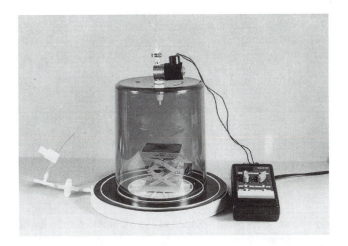

FIGURE 21.3 Setup of plastic PIG and timer. (From Gray, D.J. et al. 1994. Simplified construction and performance of a device for particle bombardment. *Plant Cell Tiss. Org. Cult.* 37:179–184. Used with permission.)

expresses the *GUS* gene in dicotyledonous plants may be substituted. Because instruction on plasmid purification and scale-up are beyond the scope of this book, we recommend that instructors who are unfamiliar with these procedures obtain purified plasmids, which are becoming increasingly available, from colleagues.

Follow the protocols in Procedures 21.3 and 21.4 to complete this part of the exercise.

	Procedure 21.2 **Construction of Specimen Holders**
Step	Instructions and Comments
1	Use hole saw on drill to cut 4 cm diameter hole in center of petri dish tops and bottoms.
2	Autoclave screen material to preshrink. Cut into 5 cm diameter discs. Use high-temperature silicone to attach screen to the *inside surface* of the bottom and the *outside surface* of the top. The petri dish surface under the screen should be roughened with sandpaper prior to attachment.
3	Cover glued surface with plastic wrap and add a weight to hold screen tightly in place. Allow to cure overnight.

	Procedure 21.3 **Preparation of Plasmid/Particle Mixture (as Modified from Finer et al., 1992)**
Step	Instructions and Comments
1	Autoclave M17 tungsten particles (Bio-Rad Laboratories, Inc., Hercules, CA) and place 50 mg in 0.5 ml 95% ethanol. Incubate 20 min, then vortex well.
2	Centrifuge at 10,000 r/min for 5 min and wash with 0.5 ml sterile water. Repeat 5 times.
3	Vortex, then pipette 25 µl of tungsten stock into eppendorf tube.
4	Add 5 µl DNA (1 µg/µl) and mix well. Add 25 µL of 2.5 M $CaCl_2$, then quickly add 10 µl of 100 µm spermidine (free base) and vortex. Incubate on wet ice (important to maintain cold temperature) for 5 min.
5	Without centrifugation, remove about 50 µl of solution, leaving the particles. Use immediately.

	Procedure 21.4 **Operation Cycle of the Plastic PIG**
Step	Instructions and Comments
1	Clamp tissue in a specimen holder using magnets. Center on the positioner and adjust to the desired height. It is imperative that tissue not be "wet," i.e., not covered with a film of liquid, in order for the procedure to be successful.
2	Place a 2 µl drop of DNA/particle mixture in the middle of a plastic filter holder screen and screw tightly into the vacuum jar.
3	Place vacuum jar on base and evacuate the chamber to 90 kPa (27" Hg) vacuum.
4	Set timer to 0.1 sec and fire valve, then vent the chamber and replace specimens on culture medium.

Experiment 1: Comparison of Helium Pressures on Transient Expression of the *GUS* Gene

Particle bombardment is affected by biological factors (e.g., plasmid construct and type and physiological state of target tissue) and physical factors (e.g., particle size, vacuum level, and helium pressure). In this experiment, the effects of two tissue types and three helium pressures will be compared by assessing differences in transient *GUS* reaction. Since *GUS* reaction can be assessed as soon as 48 hours after bombardment, this experiment can be completed within 1 to 2 weeks.

Materials

The following materials are required for each student or team:

- Explant tissue, consisting of the following:
  - Approximately 100 basal cotyledonary explants of melon, cultured for 4 days on embryo induction medium (Murashige-Skoog with sucrose and agar (Sigma Chemical Co.) plus 5 mg/l 2,4-D and 0.1 mg/l thidiazuron)
  - The innermost leaves from fresh cabbage, purchased from a supermarket (approximately 1/2 cabbage for each student or team)
- DNA/particle mixture, prepared continuously, immediately before every four bombardments
- Eight sterile specimen holders
- Three sterile moist chambers; e.g., empty 100×15 mm petri dishes containing moist filter paper (to incubate cabbage tissue after bombardment)
- 10 ml of X-Gluc reagent (40 ml of *GUS*-assay buffer, 0.1 M phosphate buffer with pH 7.0, 5 mM postassium ferricyanide, 5 mM potassium ferrocyanide, and 10 mM EDTA; 40 mg of X-Gluc [5-bromo-4-chloro-3-indolyl-D-glucuronic acid; 400 μl N,N-dimethylformamide])

The experimental design compares three helium pressures (410 kPa [60 psi], 620 kPa [90 psi], and 830 kPa [120 psi]) and two explant tissues (melon cotyledons and cabbage leaves). Each treatment (pressure-explant combination) is replicated four times. Follow the protocol outlined in Procedure 21.5 to complete this experiment.

ANTICIPATED RESULTS

The *GUS* reaction seen in this experiment is due to the direct gene activity in bombarded cells. For such a response to occur, at least one intact DNA molecule must enter a cell and the cell must survive the impact. However, this response does not denote stable transformation (i.e., integration of the *GUS* gene into the cell's chromosomes or genome). Dramatic differences in the number of blue spots will be seen due to different pressure treatments, with the highest pressure typically producing the greatest response in melon cotyledons; however, different tissues would be expected to respond differently. Differences in *GUS* reaction due to tissue type are more difficult to predict due to wide variation in available cabbage tissue. However, when successful, bombarded cabbage leaf tissue demonstrates the "blast" pattern very well. Experiments comparing the other parameters mentioned above can be conducted in a similar manner.

While transient expression of *GUS* is relatively simple to achieve, obtaining stable expression is a more difficult task that involves careful selection over time. Selection usually is accomplished by incorporating a selectable marker gene, such as *neomycin phosphotransferase II (NPT II)*, which confers resistance to the antibiotics kanamycin and gentamycin.

Procedure 21.5
Comparison of Helium Pressures on Transient Expression of the *GUS* Gene

Step	Instructions and Comments
1	Set the particle holder-to-specimen gap at 15 cm, by adjusting the apparatus positioner.
2	Arrange eight cotyledon explants as close together as possible, but not overlapping, in the center of a specimen holder, and carefully center the specimen on the apparatus positioner relative to the bombardment pattern. (For properly "aiming" the gun, the actual particle pattern can be determined by first bombarding the filter paper and viewing the resulting pattern.)
3	As rapidly as possible, perform a bombardment, as described above, and place the explants back onto the culture medium. Repeat four times.
4	For cabbage, dissect young leaves into pieces approximately 2 cm in diameter and bombard. Place the four replicate leaves from one treatment into sterile moist chambers.
5	Incubate explants for 48 hours and then place into X-gluc reagent in the dark at 35°C for 12 to 24 hours.
6	Count the number of distinct blue spots per replicate (specimens can be preserved indefinitely in cold 70% ethanol).
7	Statistically analyze the data as a 2 × 3 factorial and perform a mean separation test, as described in Chapter 6.

LITERATURE CITED AND SUGGESTED READINGS

Brown, D.C.W. et al. 1994. Development of a simple particle bombardment device for gene transfer into plant cells. *Plant Cell Tiss. Org. Cult.* 37:47–53.

Finer, J.J. et al. 1992. Development of the particle inflow gun for DNA delivery to plant cells. *Plant Cell Rep.* 11:323–328.

Gray, D.J. and J.J. Finer. 1993. Editorial introduction. Special section: development and operation of five particle guns for introduction of DNA into plant cells. *Plant Cell Tiss. Org. Cult.* 33:219.

Gray, D.J. et al. 1994. Simplified construction and performance of a device for particle bombardment. *Plant Cell Tiss. Org. Cult.* 37:179–184.

Jefferson, R.A., T.A. Kavanagh, and M.W. Bevan. 1987. GUS fusions: -glucuronidase as a sensitive and versatile gene fusion marker in higher plants. *EMBO J.* 6:3901–3907.

Kikkert, J.R. 1993. The Biolistic PDS-1000/He device. *Plant Cell Tiss. Org. Cult.* 33:221–226.

Takeuchi, Y., M. Dotson, and N.T. Keen. 1992. Plant transformation: a simple particle bombardment device based on flowing helium. *Plant Molec. Biol.* 18:835–839.

22 A Simple Illumination System for Visualizing Green Fluorescent Protein*

Dennis J. Gray, Subramanian Jayasankar, and Zhijian T. Li

CHAPTER 22 CONCEPTS

- The GFP gene provides a convenient visual marker for selecting transgenic cells in plants, due to the protein's low toxicity and ease of obtaining expression.

- Visualizing GFP requires use of an illuminator to supply blue light at 480 nm. However, in order to visualize GFP emission, the blue light must be blocked by a barrier filter.

- A fiber-optic illuminator can be modified to provide the requisite blue light; this is a low-cost alternative to dedicated GFP illumination systems.

INTRODUCTION

Plant cells and tissues grown *in vitro* constitute the basic system used for introducing new traits by transgenic technology, and many commercial products already are being produced from transgenic plants (see Chapters 2 and 20). However, even today, the successful incorporation of a functional gene into a cell and its development into a useful plant are very rare events. This is because gene insertion generally occurs randomly and the probability of obtaining a cell with a functional insertion is low. Coupled with randomness and the rarity of gene insertion, another significant obstacle concerns the difficulty of separating transgenic cells from nontransgenic cells.

Molecular biologists have developed several methods to cope with these obstacles. For example, selection/marker genes such as those for antibiotic resistance commonly are attached to genes that code for the desired traits in order to allow selection of resistant transgenic cells from nontransgenic cells. Thus, with cultures grown on medium containing antibiotic, only transgenic cells survive, divide, and increase in mass, whereas those without functional gene insertions die; this "needle in the haystack" approach allows recovery of transgenic events. In addition to antibiotic resistance, a number of other genes permitting use of chemical selection agents have been developed.

As an adjunct to the use of chemical selection agents, genes that code for visible traits, primarily color changes, also have been developed. The best-known is the β-*glucuronidase* (*GUS*) gene assay (Jefferson et al., 1987). The *GUS* gene was isolated from *E. coli* and produces an enzyme that cleaves a colorless substrate, X-Gluc (5-bromo-4-chloro-3-indolyl glucuronide), into a dark blue product. Thus, cells expressing *GUS* become blue when exposed to the substrate, facilitating sensitive detection of gene function. In fact, entire transgenic plants will become dark blue when soaked in substrate. However, a drawback to the *GUS* assay is that it is destructive — plant tissue

---

* Florida Agricultural Journal Series No. R-10169

is destroyed during processing. This means that the development of a transformed cell or tissue into a plant cannot be followed. An alternative to destructive *GUS* assay is a nondestructive visible marker system employing the luciferase gene, which was obtained from the North American firefly, *Photinus pyralis* (Gould and Subramani, 1988). Cells expressing luciferase fluoresce yellow when provided with the substrate luciferin. Although the assay is nondestructive, luciferin is somewhat cost-prohibitive, and supplying it to tissue can be problematic.

The green fluorescent protein (GFP) gene, isolated from the Pacific jellyfish, *Aequoria victoria*, provides a solution to both the problems of destructive assay procedures and expensive substrates (de Ruijter et al. 2003). GFP is nontoxic to plant and animal cells and requires no substrate, only blue light, to emit green fluorescence (Stewart, 2001). Thus, transgenic cells can be identified as soon as protein is produced and their development into tissues and plants can be followed in "real time" (Mantis and Tague, 2000). The GFP gene has been demonstrated in animals (a variety of fish, monkeys, rabbits, rats, and others) as well as a plethora of plant species. In fact, GFP-positive transgenic zebrafish recently have been commercialized as novelty pets for aquarium lovers.

Visualizing GFP emission in transgenic plants is technically quite simple. Tissue is illuminated with blue light at the 480 nm wavelength and, if *GFP* is present, green light is emitted. However, the green light emitted tends to be cloaked by blue light; therefore a blue-blocking filter is used so that only green emission can be visualized (Gray et al., 2000). Common fluorescent compound microscopes can be fitted with the requisite emission and barrier filters and stereomicroscopes have been developed specifically to view GFP; however, such stereomicroscopes typically cost over $15,000. In this chapter, we describe the construction and operation of a simple system for visualizing GFP that employs a commercially available fiber optic illuminator and filters (Figure 22.1). The illuminator is suitable for visualizing GFP emission both micro- and macro-scopically. Most tissue culturalists have access to a common stereomicroscope and fiber optic illuminator, and their adaptation into a simple illumination system would greatly facilitate the use of GFP technology by reducing cost.

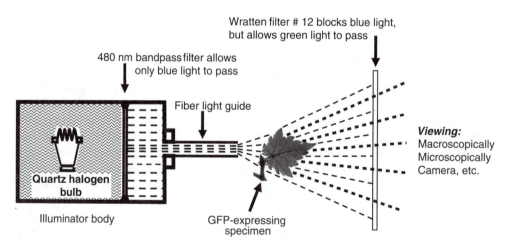

FIGURE 22.1 Diagram of equipment for GFP illumination. An illuminator is fitted with a 480 nm filter that results in blue light transmission through a fiber light guide. The blue light causes excitation of *GFP* molecules, resulting in green light emission. Blue light is eliminated by a barrier filter so that only green GFP emission is visualized.

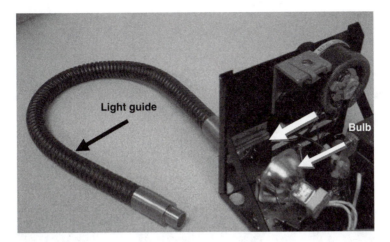

FIGURE 22.2 Modified illuminator showing light guide, quartz halogen bulb, and position of 480 nm filter (arrow) held in place with tape.

COMPONENTS FOR CONSTRUCTING THE *GFP* ILLUMINATOR

For simplicity, specific components are recommended; however, other similar components from different suppliers may be substituted and adapted. For the specified illuminator, all components were purchased from Edmund Industrial Optics. (Send mail to Edmund Industrial Optics, Sales Department W028, 101 East Gloucester Pike, Barrington, NJ 08007-1380. Web address: http://www.edmundoptics.com. Telephone: 800-363-1992.)

Specific components required to assemble the illuminator are as follows:

1. Dolan-Jenner Illuminator Model 180, Cat. #NT38-939
2. Light guide (24" × 1/2"), Cat. #NT39-369
3. 480 nm blue light transmission filter (50.8 mm^2), Cat. #NT43-167
4. Green light transmission filter (Kodak Wratten Filter, Yellow #12), Cat. #NT54-467

Assembly of the components is simple. The fiber optic illuminator is fitted with a 480 nm blue light transmission filter to excite GFP. The microscope or camera is fitted with a Kodak #12 Wratten Filter to block blue but transmit the resulting green fluorescence (Figure 22.1). Alternatively, the filter may be hand-held for macroscopic observation. The specified light pipe provides about 2× the illumination of the bifurcated light pipes generally used with fiber optic illuminators. In the Dolan-Jenner Illuminator Model 180, the glass 480 nm blue light transmission filter fits snugly between light bulb and light guide and is held against the inner housing of the illuminator with a small piece of clear tape placed at the top of the filter (Figure 22.2). Place the filter *with the mirror side facing backward* toward the light bulb to reduce heat transmission; *this will keep the filter from fracturing.* The #12 Wratten filter is yellow and blocks all blue light transmission, but allows green light emission from GFP to pass. It is thin and flexible, and can be easily cut to fit under or over the objective lens of a stereomicroscope or a camera lens; it can even be hand-held to view GFP macroscopically. Exact placement of the Wratten filter is dependent on the optical system and must be determined by the user on a case-by-case basis.

VIEWING *GFP* EMISSION

GFP emission is detected by placing the output side of the light guide as close as possible to cells or tissues that harbor the GFP gene. The tissue becomes flooded with blue light when the illuminator

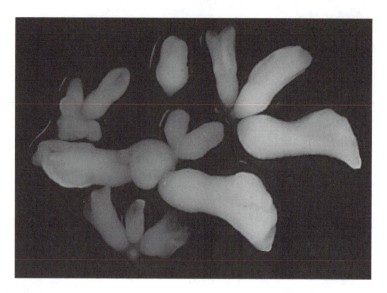

FIGURE 22.3 Digital micrograph of transgenic grape somatic embryos showing expression of GFP, which was obtained using the illumination system. This image is recreated in color on the back cover.

is turned to its highest setting. When the illuminated area is viewed through a #12 Wratten filter, blue light is completely eliminated, and any GFP emission, indicative of protein production by a functional *GFP* gene, is clearly seen as a pale green color emanating directly from the tissue (Figure 22.3). Critical placement of the light guide facilitates the best emission, and hence viewing, of *GFP*. Photographic documentation is best accomplished with a relatively high-speed color film, such as Kodak Ektachrome EPP 135, or with a digital camera (see Chapter 5). For example, the GFP-positive somatic embryos shown in Figure 22.3 were photographed with a digital camera attached to a stereomicrosope (see color version of this image on the back cover of this book). The light guide was positioned to provide the most illumination possible, and the #12 Wratten filter was positioned below the objective lens so that only green light could be visualized with the microscope.

LITERATURE CITED

de Ruijter, N.C.A. et al. 2003. Evaluation and comparison of the GUS, LUC and GFP reporter system for gene expression studies in plants. *Plant Biol.* 5:103–115.

Gould, S. J. and S. Subramani. 1988. Firefly luciferase as a tool in molecular and cell biology. *Anal. Biochem.* 175:5–13.

Gray, D.J. et al. 2000. Visualizing green fluorescent protein. In Vitro *Cell. Dev. Biol.* 36:43A.

Jefferson, R.A., T.A. Kavanaugh, and M.W. Bevan. 1987. GUS fusions: glucuronidase as a sensitive and versatile gene fusion marker in higher plants. *EMBO J.* 6:3901–3907.

Mantis, J. and B.W. Tague. 2000. Comparing the utility of beta-glucuronidase and green fluorescent protein for detection of weak promoter activity in *Arabidopsis thaliana*. *Plant Molec. Biol. Rep.* 18:319–330.

Stewart, C.N. 2001. The utility of green fluorescent protein in transgenic plants. *Plant Cell. Rep.* 20:376–382.

23 Germplasm Preservation

Leigh E. Towill

CHAPTER 23 CONCEPTS

- Germplasm, as defined for a crop, represents all the different genotypes that could be used for improvement, and is valuable because it contains the diversity of genotypes that are needed to develop new and improved lines.

- A key theme for safe preservation of diversity is the use of two sites. The first site, the active collection or genebank, has responsibilities for obtaining diversity, propagation, distribution to users, and characterization. The second site, the base collection or genebank, provides long-term preservation and utilizes the best storage systems available. This site provides the crucial function of being a safety net should materials be lost at the active collection.

- *In vitro* technologies are becoming important in maintaining the diversity of crops and other plants. The greatest use of tissue culture in germplasm preservation is usually for species that are vegetatively propagated.

- Cryopreservation is desired for the base collection (see above) and is applied to axillary or apical shoot tips from *ex vitro* or *in vitro* plants. Cryogenic storage provides long-term storage at a reasonable cost with a low labor input.

- The term "vitrification" refers to the transformation of a liquid to a glass during cooling and procedures that promote vitrification used in cryopreservation. Vitrification also refers to plants that have an abnormal, water-soaked, or glassy appearance, but "hyperhydricity" is now the preferred term for this appearance.

- When recovering plants form cryostorage, rapid warming minimizes any destructive ice crystal growth or devitrification in cells and tissues.

INTRODUCTION

The concept of germplasm is inherently simple, yet extremely complex in scope. Germplasm, as defined for a crop, represents all the different genotypes that could be used for improvement. Improvement is accomplished traditionally by making sexual crosses between individuals and then selecting plants with the desired characteristics, such as disease or stress tolerance or resistance. With the advent of more exotic techniques, such as transformation and protoplast fusion, virtually any species may be modified by incorporating a gene or genes from an unrelated species. Thus, germplasm for a crop becomes more expansive. Nevertheless, we usually consider germplasm more narrowly to be the gene pool from those individuals that are related and may intercross. Germplasm is valuable because it contains the diversity of genotypes that are needed to develop new and improved lines.

The need for preservation of germplasm extends beyond application to crops (Falk and Holsinger, 1991; Stuessy and Sohmer, 1996). Native and endangered species also benefit from preservation strategies, whether it is for reintroduction in habitats or for revegetation, for example on rangeland or mine sites. The mechanics of preservation are similar for all groups, but this chapter will use examples from crop species.

Diversity must be protected from loss to ensure availability for future use. Preservation may be either *in situ* or *ex situ*. The former refers to preservation in the native environment or ecosystem, which is beneficial for evolutionary processes but difficult to manage with the pressures of expanding human populations. The latter refers to preservation not in the wild, but in what are termed repositories or genebanks. Populations (as seed) are maintained through controlled crosses, and individuals (as clones) are maintained in the field or greenhouse.

Diversity has been and is still being collected from centers of crop origin and from landraces developed over many years. Acquisition by field collection or exchange is an ongoing effort. The uniqueness of the individual or population must be maintained. Janick's book (1989) describing the U.S. National Plant Germplasm System (NPGS) (http://www.ars-grin.gov/npgs) is an excellent source for those interested in the details of a comprehensive system. Worldwide, the International Plant Genetic Resources Institute (IPGRI) emphasizes the conservation, improvement, and use of plant genetic resources (http://www.ipgri.cgiar.org).

A key theme for safe preservation of diversity is use of two sites. One site, often termed the active collection or active genebank, has responsibilities for obtaining diversity (either by collection or by exchange with other genebanks worldwide), propagation, distribution to users, and characterization. The second site, the base collection or base genebank, provides long-term preservation and utilizes the best storage systems available. This site provides the crucial function of being a safety net should materials be lost at the active collection. It is important to have the distinction between these two sites firmly in mind in discussing the issue of preservation and in deciding what methods are useful for each collection.

This chapter describes germplasm preservation and emphasizes where and how *in vitro* culture and cryopreservation are used for efficient and effective management at both an active collection and a base collection. Germplasm preservation for most higher plants does not involve *in vitro* culture, but instead utilizes seeds, and a brief discussion of seed preservation is included to provide a complete picture.

PRESERVATION OF SEED-PROPAGATED SPECIES

Seeds are classified as either being desiccation tolerant (capable of retaining viability after being dried to a low-moisture content [~1% moisture, fresh weight basis] — also termed "orthodox" seed) or desiccation sensitive (seeds losing viability after being dried below a species-specific critical limit, usually about 12–30% moisture — also termed "recalcitrant" seed). This classification is a great simplification, and intermediate types exist. Most temperate crop species have desiccation-tolerant seeds, and many studies have shown the importance of adjusting seed moisture and storage temperature for retaining viability (Roos, 1989). Longevity is increased with reduced moisture and with lower temperatures. Recent studies suggest that the optimum seed moisture content for obtaining the greatest longevity depends upon storage temperature used (Walters, 1998). If seeds are dried too much, longevity is reduced. Practically, seeds can be equilibrated with a 25% relative humidity at 5°C and then stored at between –5 and –20°C. Longevities under these conditions vary considerably, ranging from a few years to a couple of centuries depending upon species. In active genebanks, a +2 to –20°C range is often used for storage of desiccation-tolerant seed. A base collection should use lower temperatures (–20°C or cryogenic storage) to increase longevity.

Most desiccation-tolerant seed survives cryogenic exposure, although some physical damage (seed coat or cotyledon cracking) occurs in some species (Stanwood, 1985). Temperatures less than about –130°C are desired for cryopreservation because of very low molecular kinetic energies, the absence of liquid water, and extremely slow diffusion. Thus, many reactions leading to deterioration are minimized and longevities are postulated to be extremely long (measured in centuries) and limited only by the buildup of genetic lesions resulting from background irradiation (Ashwood-Smith and Grant, 1977). Because cryopreservation has only recently been employed, data to support the contention of extreme longevity do not exist. In practice, cryogenic storage is in liquid nitrogen (LN) at –196°C, or in the vapor above LN at ~–150 to –180°C, because this is a relatively cheap and available cryogen.

Desiccation-sensitive seeds are produced by aquatic species, large-seeded species, some species native to tropical areas, and some temperate zone species of trees. Coconut, cocoa, mango, nutmeg, and rubber plants are examples of economically important species with recalcitrant seed. To retain viability, recalcitrant seeds are stored at as low a temperature as possible, under conditions that retain relatively high-seed moisture levels and assure a supply of oxygen for respiration. These seeds have short life spans, ranging from a few weeks to several months. This is too short for adequate preservation, and so these species are usually stored in an active collection as vegetative plants.

Desiccation tolerance and sensitivity are clearly very important for the success of preservation protocols, not only for seeds but also for vegetative propagules (see later). The mechanisms of damage due to desiccation are diverse and reasons for differences among species are not well defined (Walters et al., 2002). Clearly, an understanding of the processes may allow manipulations that will enhance preservation.

In vitro technologies are becoming important in maintaining crops with desiccation-sensitive seed. *In vitro* plants may be maintained for active collections. Embryo excision and culture (Chapter 18) may be used in preservation strategies of desiccation-sensitive seeds. Isolated embryonic axes (with cotyledons removed) from mature seeds may survive partial desiccation and cooling to cryogenic temperatures. The level of desiccation that may be tolerated, and the question of whether this level is sufficiently low to avoid damage from ice formation during cooling, must be defined for each species. Upon retrieval from LN, axes are cultured and can develop into plants. Treatments involving application of cryoprotectants to excised embryos may also prove to be useful and are actively being researched, but so far no procedure has been routinely used for practical, reproducible cryopreservation of truly recalcitrant seeds. Axillary shoot tip or bud cryopreservation (see later) from *in vivo* or *in vitro* plants developing from the seed also is useful for base-collection storage of these species.

PRESERVATION OF POLLEN

Pollen may also be used to store genes from crops (Hanna and Towill, 1995). The major use of germplasm is for improvement of plants by sexual processes, and thus it requires making controlled crosses between desired individuals. Supplying pollen to the user facilitates this process; the user does not have to grow the plant or wait for the flowering of the male parent. This is of obvious benefit for those species which often take many months to flower if grown from seed, and, therefore, is of potential value for some species that are vegetatively propagated. Pollen preservation thus could be integrated into the active genebank, where pollen would be collected, stored, and distributed upon request. Pollen from many species is desiccation tolerant and exhibits storage characteristics similar to desiccation-tolerant seed. Longevity is increased with storage at lower temperatures, and two to several years' storage is feasible at –20°C. Survival with cryogenic storage has been shown for many species.

Pollen from some species, for example within the *Poaceae*, is desiccation sensitive and short-lived, however. Cryogenic storage of this type of pollen is feasible, but only within a narrow range of moisture contents which must be determined for each species. Overall, pollen preservation, whether of tolerant or sensitive types, is a supplement to seed or clone storage, not a substitute for them.

PRESERVATION OF VEGETATIVELY PROPAGATED SPECIES

Some species either have an extremely long juvenile phase (no flowers), produce seed with extremely short longevities, do not easily produce seed, or do not produce viable seed (Towill, 1988). The vegetative plant for these species is stored either in the field or in greenhouses and is regularly propagated asexually to rejuvenate the line and prevent loss. These species are extremely diverse, comprising herbaceous and woody growth habits from tropical to temperate zones (Table 23.1). These species also differ greatly in stress tolerances to temperature, salt, moisture, and other factors, as well as to disease.

ACTIVE COLLECTIONS

It is impossible to grow all species at one location, and thus several repositories, termed clonal repositories, exist throughout the world. The vast number of different clones held precludes holding very many replicates, especially under field conditions. Maintenance is expensive and loss could easily occur due to disease or catastrophe.

The greatest use of tissue culture in germplasm preservation usually is for species that are vegetatively propagated. *In vitro* culture of plants can be established and used to gain more efficient preservation (Withers, 1991; Lynch, 1999). A number of advantages for using *in vitro* plants can be listed, including the small space needed for storage, ease of shipping, year-round availability, the potential for eliminating disease and for maintaining disease-free plants, rapid propagation, and labor and cost savings (Towill, 1988). Many species can now be propagated in culture, but some species are still problematic and methods are often genotype specific, requiring considerable modification. All *in vitro* lines held in one location are also susceptible to loss from catastrophe. Individual lines could be lost due to insect infestation.

The advantages of maintaining *in vitro* versus *in vivo* plants often relate to what is desired by the user community. For example, if crosses are to be made between apple lines, apple budwood from *ex vitro* sources is preferred, since a grafted bud can lead to a flowering shoot within about 18 months. An *in vitro* apple plant, by comparison, might take more than 3 years to flower after transplanting. However, for potato, flowering times from planting a tuber or transplanting an *in vitro* plant are similar. The advantage here for the *in vitro* system is ease in maintaining disease-free status (Chapter 28).

In vitro and *in vivo* systems are not mutually exclusive. To guard against loss, *in vitro* plants may serve as a backup to field or greenhouse lines at the same or another location. Recall that replication for clones *in vivo* is minimal: often only two individuals are held for a given line.

TABLE 23.1
Major Crops Maintained Vegetatively at Different Repositories within the United States National Plant Germplasm System

Location	Species Maintained Clonally
Corvallis, OR	Pear, filbert, mint, strawberry, raspberry, blackberry, currant, hops
Davis, CA	*Prunus* spp., *Vitis vinifera* and related species
Geneva, NY	Apple, *Vitis* spp.
Brownwood, TX	Pecan, walnut
Hilo, HI	Guava, passionfruit, papaya, pineapple, macadamia, lychee, carambola
Griffin, GA	Sweet potato
Miami, FL/Mayaguez, PR	Avocado, banana, plantain, coffee, cacao, mango, sugarcane
Riverside, CA	*Citrus* spp. and related species, date palm
Orlando, FL	*Citrus* spp. and related species

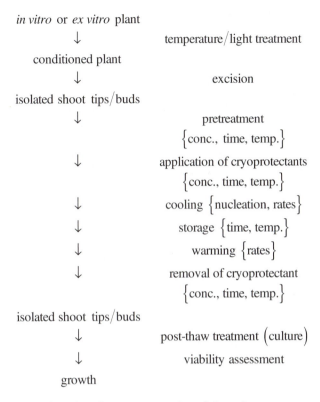

in vitro or *ex vitro* plant
↓   temperature/light treatment
conditioned plant
↓   excision
isolated shoot tips/buds
↓   pretreatment
{conc., time, temp.}
↓   application of cryoprotectants
{conc., time, temp.}
↓   cooling {nucleation, rates}
↓   storage {time, temp.}
↓   warming {rates}
↓   removal of cryoprotectant
{conc., time, temp.}
isolated shoot tips/buds
↓   post-thaw treatment (culture)
↓   viability assessment
growth

FIGURE 23.1 Generalized flow chart for cryopreservation of shoot tips.

Maintenance of *in vitro* cultures in an active collection is often conducted under normal growth conditions. If material is requested by the user community, it can be quickly propagated and distributed. Backup materials at the active collection site may be maintained under conditions that minimize growth (Withers, 1991). This usually means storing the *in vitro* plant at lower temperatures — for example, potato at 10°C or mint at 4°C. Growth is retarded and cultures may be held for one to a few years, providing precautions are taken to avoid extensive water loss. Elevated osmotic pressures and other growth inhibitors may also be used to reduce the frequency of transfer needed to maintain a viable culture. Empirical studies are required to apply this method to different species, and thus longevities are currently difficult to predict.

BASE COLLECTIONS: CRYOPRESERVATION

It is prohibitively expensive to maintain complete duplicate collections for *ex vitro* plants at a second location. A second site for *in vitro* plants may be less costly but still requires considerable maintenance. Hence, cryopreservation is desired for the base collection, the backup whereby lines are held in case loss occurs in the active collection, and is applied to axillary or apical shoot tips from either *ex vitro* or *in vitro* plants. Storage of vegetative propagules at low temperatures, such as −20°C or even −80°C, results in very short longevities, for reasons that will not be discussed here. Cryogenic storage provides long-term storage at a reasonable cost with a low-labor input. As with seed, storage is either in LN or in the vapor phase over LN.

It should be emphasized that, whereas cryopreservation for germplasm purposes uses shoot tips or buds to maintain the integrity of the line (and to avoid undue chances for selecting somaclonal variants), cryopreservation of protoplasts, suspension cells, callus, embryogenic suspensions, and somatic embryos is also needed for many tissue culture operations to guard against loss, to ensure line purity and performance, and to avoid costly, frequent transfers. Although shoot tips are now employed for cryopreservation, the production of synthetic seed from somatic embryos and their use for cryopreservation may be feasible once induction from all genotypes is feasible, off-type production is minimized, and maturation efficiency is improved.

Shoot tips have a higher water content than seeds and cannot withstand extreme desiccation. In the hydrated state, exposure to low temperatures leads to damage due either to intracellular ice or to consequences of desiccation during extracellular ice formation. For these reasons, a more detailed protocol (Figure 23.1) is needed. Virtually all cryopreservation protocols follow these steps, but details vary with species. It is usually necessary to use cryoprotectants and to control cooling and warming rates. A theoretical treatment of rate phenomena is beyond the scope of this article; excellent reviews address this and related topics (Mazur, 1984; Fahy et al., 1987). Recent books provide technical information on application to cells, protoplasts, and shoot tips from many different plant species (Bajaj, 1995; Engelmann and Takagi, 2000; Towill and Bajaj, 2002).

PRETREATMENT OF THE PLANT AND ISOLATED PROPAGULE

The physiological condition of the plant used for cryopreservation is important. Shoot tips isolated from plants that have been acclimated to cold survive the cryogenic protocol in higher percentages than those not acclimated. This is useful only for those species that have the genetic capability to acclimate to cold. Survival of excised shoot tips also can be enhanced by exposure to elevated sucrose levels. The exact effect of treatments on the plant or of the excised shoot tip is uncertain, but they probably alter the intracellular contents of critical metabolites (particularly sugars), cell water content, and physiological characteristics of the cell (e.g., membrane permeability).

CRYOPROTECTANT APPLICATION AND COOLING RATE

After the pretreatment phase, shoot tips are exposed to cryoprotectants. The concentration of cryoprotectants used and the time and temperature of exposure vary with the cooling strategy employed. Two cooling regimens are generally used with shoot tips. The first is termed two-step, slow, or optimum-rate cooling, and employs cooling of the cryoprotected shoot tip at about 0.25 to 1.0°C/min to about –35°C prior to immersion in LN. For this method, shoot tips are incubated in 1 to 2 M concentrations of a suitable permeating cryoprotectant, such as dimethylsulfoxide (DMSO) or ethylene glycol. The second regimen uses rapid cooling and is often termed a vitrification method. Here, shoot tips are exposed to much more concentrated solutions of cryoprotectants and then are immersed directly from about 22°C or 0°C into –196°C.

Research with plant systems initially emphasized two-step cooling techniques to attain cryopreservation. Useful levels of survival were obtained for some species, but not for others, for reasons that have not been elucidated. Summaries of this method can be found in recent reviews (Sakai, 1995; Towill, 2002). Attention has now focused on vitrification for cryopreservation.

The term "vitrification" refers to transformation of a liquid to a glass during cooling. Tissue culture literature has used the term vitrification in an entirely different sense to refer to plants that have an abnormal, water-soaked, or glassy appearance. However, "hyperhydricity" is now the preferred term for this abnormal appearance of plants in tissue culture.

The advantages of vitrification are that it is a simple technique that does not require an expensive apparatus, that it may avoid the damaging consequences of ice formation, and that it may be applicable to larger pieces of tissue than are usually used for two-step cooling methods. Survival after cryogenic exposure has been reported with several different vitrification procedures for protoplasts, suspension cultured cells, embryogenic suspensions, callus, somatic embryos, embryonic axes, and shoot tips.

There are increasing numbers of theoretical and practical studies examining vitrification in animal and plant systems (Fahy et al., 1987; Steponkus et al., 1992; Mehl, 1996). Vitrification requires application of suitable concentrations of compounds, such that, with usually rapid cooling, the system forms a glass (i.e., the solution vitrifies). Information about the physical aspects of the process is given by MacFarlane et al. (1992). Although it is difficult to prove that the cell itself vitrifies, circumstantial evidence suggests that this happens under defined conditions. Indeed, vitrification may occur naturally in extremely cold-hardy plants exposed to very low temperatures.

The general strategy for vitrification is to expose the shoot tips (or protoplasts, cells, or somatic embryos) to low concentrations of cryoprotectants such that permeable components have sufficient time to penetrate (the loading phase). Next, the shoot tips are exposed to a vitrification solution (the dehydration phase). Because water permeability is much greater than solute permeability, the shoot tip dehydrates in the very concentrated solution and endogenous solutes and any permeating cryoprotectants are concentrated. Shoot tips are then rapidly cooled to facilitate glass formation.

It has become apparent that loading of very permeable components is not necessary for shoot tips from some species, and indeed may be detrimental (Steponkus et al., 1992). The cell contents alone, when sufficiently dehydrated, apparently vitrify when cooled at a rapid rate. However, for many species preculture for 1 to 3 days in sucrose solutions (0.3–1.0 M) prior to dehydration enhances desiccation tolerance and is necessary to achieve survival after vitrification. This exposure probably allows for the uptake of sugars and is conceptually a loading process. Short-term (20 min to 3 h) exposure to 2 M glycerol plus 0.4 to 1.6 M sucrose is often applied after the preculture step and enhances survival. This induces plasmolysis, but the mechanism of protection is not well understood.

A major variation of vitrification uses an encapsulation/dehydration sequence (Fabre and Dereuddre, 1990). Shoot tips are encapsulated in an alginate gel, cultured in sucrose solutions, dehydrated in an air flow, and then, usually, rapidly cooled. Encapsulated shoot tips vitrify under the conditions employed. Survival has been found in shoot tips from several unrelated species such as potato, grape, carnation, and pear. Encapsulation followed by exposure to vitrification solutions rather than air desiccation is a further variation that has been useful with tropical species.

WARMING RATE AND RECOVERY

Rapid warming from –196°C to about 22°C is generally beneficial for obtaining high levels of survival for samples cooled either by two-step cooling or by vitrification. Rapid warming minimizes any destructive ice crystal growth or devitrification as the cellular milieu gains more thermal motion. The application of cryoprotectants and the excursion of shoot tips to and from very low temperatures lead to sublethal damage which, in concept, can be repaired by cellular processes and can be influenced by environmental conditions. The subsequent culture of the retrieved shoot tip may require conditions that control shoot tips do not require. Treated shoot tips are usually cultured in the dark or dim light for a few days to minimize any oxidative stress that might occur due to damaged mitochondria or chloroplasts. Brief culture on elevated osmotica may be beneficial, as may use of low ammonium levels. Sometimes different levels of plant growth regulators (PGRs) are required. Eventually the shoot tip can be placed back in normal growing conditions.

CONCLUSION

Seeds and plants are used to preserve the germplasm of most crops, to ensure that materials are available for future breeding and to produce crops with specified characteristics. Whereas desiccation-tolerant seeds are simply stored, desiccation-sensitive seeds present challenges. Preservation of clones is also challenging due to the high cost of plant maintenance and to difficulties in using cryopreservation. Increasingly, plant culture is allowing for efficient and effective preservation of the latter two groups. Many species now may be propagated and maintained in culture. Embryonic axes and shoot tips may be regrown in culture after cryogenic treatment. Some species are more problematic, but with advances in developing cryopreservation methods, such as vitrification, practical use is now feasible (Reed, 2002).

LITERATURE CITED

Ashwood-Smith, M.J. and E. Grant. 1977. Genetic stability in cellular systems stored in the frozen state. In: *Freezing of Mammalian Embryos*. (Eds.) K. Elliot and J. Whelan. Ciba Foundation Symposium No. 52. Elsevier, Amsterdam. 251–272.

Bajaj, Y.P.S., Ed. 1995. *Cryopreservation of Plant Germplasm I*. Biotechnology in Agriculture and Forestry Series No.32. Springer-Verlag, Berlin.

Engelmann, F. and H. Takagi, (Eds.) 2000. *Cryopreservation of Tropical Plant Germplasm: Current Research Progress and Application*. International Plant Genetic Resources Institute, Rome.

Fabre, J. and J. Dereuddre. 1990. Encapsulation-dehydration: A new approach to cryopreservation of *Solanum* shoot tips. *Cryo-letters* 11:413–426.

Fahy, G.M., D.I. Levy, and S.E. Ali. 1987. Some emerging principles underlying the physical properties, biological actions, and utility of vitrification solutions. *Cryobiology* 24:196–213.

Falk, D.A. and K.E. Holsinger, (Eds.) 1991. *Genetics and Conservation of Rare Plants*. Oxford University Press, Oxford.

Hanna, W.W. and L.E. Towill. 1995. Long-term pollen storage. *Plant Breeding Rev.* 13:179–207.

Janick, J., Ed. 1989. *The National Plant Germplasm System of the United States*. Plant Breeding Reviews, Vol. 7. Timber Press, Portland, OR.

Lynch, P.T. 1999. Tissue culture techniques in *in vitro* plant conservation. In: *Plant Conservation Biotechnology*. (Ed.) E.E. Benson. Taylor & Francis, London. 42–61.

MacFarlane, D.R., M. Forsyth, and C.A. Barton. 1992. Vitrification and devitrification in cryopreservation. In: *Advances in Low-Temperature Biology*, Vol. 1. (Ed.) P.L. Steponkus. JAI Press, Greenwich, CT. 221–278.

Mazur, P. 1984. Freezing of living cells: Mechanisms and implications. *Amer. J. Physiol.* 247:c125–c142.

Mehl, P. 1996. Crystallization and vitrification in aqueous glass-forming solutions. In: *Advances in Low-Temperature Biology*, Vol. 3. (Ed.) P.L. Steponkus. JAI Press, Greenwich, CT. 185–255.

Reed, B. 2002. Implementing cryopreservation for long-term germplasm preservation in vegetatively propagated species. In: Cryopreservation of Plant Germplasm II. Biotechnology in Agriculture and Forestry Series No.50, (Eds.) Towill, L.E. and Y.P.S. Bajaj Springer, Berlin. 22–33.

Roos, E.E. 1989. Long-term seed storage. *Plant Breed. Rev.* 7:129-158.

Sakai, A. 1995. Cryopreservation of germplasm of woody plants. pp. 53-69. In: Cryopreservation of Plant Germplasm I. (Ed.) Bajaj, Y.P.S, Springer-Verlag, Berlin.

Stanwood, P.C. 1985. Cryopreservation of seed germplasm for genetic conservation. In: *Cryopreservation of Plant Cells and Organs*, (Ed.) Kartha, K.K., CRC Press, Inc. Boca Raton. 199-226.

Steponkus, P.L., R. Langis and S. Fujikawa. 1992. Cryopreservation of plant tissues by vitrification. In: *Advances in Low-Temperature Biology*, Vol 1. (Ed.) Steponkus, P.L., JAI Press, Greenwich, CN. 1–62.

Stuessy, T.F. and S.H. Sohmer (Eds.) 1996. *Sampling the Green World. Innovative Concepts of Collection, Preservation and Storage of Plant Diversity*, Columbia University Press, New York. 289

Towill, L.E. 1988. Genetic considerations for germplasm preservation of clonal materials. *HortScience* 23:91–97.

Towill, L.E. 2002. Cryopreservation of plant germplasm: Introduction and some observations. p. 3–21. In: *Cryopreservation of Plant Germplasm II*. (Ed.) L.E. Towill and Y.P.S. Bajaj. Biotechnology in Agriculture and Forestry Series No. 50. Springer-Verlag, Berlin.

Towill, L.E. and Y.P.S. Bajaj (Eds.) 2002. *Cryopreservation of Plant Germplasm II*. Biotechnology in Agriculture and Forestry Series No. 50. Springer-Verlag, Berlin.

Walters, C. 1998. Understanding the mechanisms and kinetics of seed aging. *Seed Sci. Res.* 8:223–244.

Walters, C. et al. 2002. Desiccation stress and damage. In: *Desiccation and Survival in Plants: Drying Without Dying*. (Ed.) M. Black and H.W. Pritchard. CAB Publishing, Wallingford, UK.

Withers, L.A. 1991. In-vitro conservation. *Biol. J. Linnean Soc.* 43:31–42.

24 Valuable Secondary Products from *In Vitro* Culture

Mary Ann Lila

CHAPTER 24 CONCEPTS

- A wide diversity of natural products can be induced in plant tissue cultures, including edible, medicinal, and industrial compounds.

- Selection and microenvironmental stimulus regulate productivity *in vitro*.

- Cell culture systems facilitate in-depth investigation of metabolism.

INTRODUCTION

The capacity for plant cell, tissue, and organ cultures to produce and accumulate many of the same valuable chemical compounds as the parent plant in nature has been recognized almost since the inception of *in vitro* technology. The strong and growing demand in today's marketplace for natural, renewable products has refocused attention on *in vitro* plant materials as potential factories for secondary phytochemical products, and has paved the way for new research exploring secondary product expression *in vitro*. However, commercial significance alone does not drive the research initiatives. The deliberate stimulation of defined chemical products within carefully regulated *in vitro* cultures provides an excellent forum for in-depth investigation of biochemical and metabolic pathways, under highly controlled microenvironmental regimes.

Plant-produced secondary compounds have been incorporated into a wide range of commercial and industrial applications, and fortuitously, in many cases, rigorously controlled plant *in vitro* cultures can generate the same valuable natural products. Plants and plant cell cultures have served as resources for flavors, aromas and fragrances, biobased fuels and plastics, enzymes, preservatives, cosmetics (cosmeceuticals), natural pigments, and bioactive compounds.

A series of distinct advantages exist to producing a valuable secondary product in plant cell culture, rather than *in vivo* in the whole crop plant. These include the following:

- Production can be more reliable, simpler, and more predictable
- Isolation of the phytochemical can be rapid and efficient, as compared to extraction from complex whole plants
- Compounds produced *in vitro* can directly parallel compounds in the whole plant
- Interfering compounds that occur in the field-grown plant can be avoided in cell cultures
- Cell cultures can yield a source of defined standard phytochemicals in large volumes
- Cell cultures are a superb model to test elicitation
- Cell cultures can be radio labeled, such that the accumulated secondary products, when provided as feed to laboratory animals, can be traced metabolically

Secondary products in plant cell culture can be generated on a continuous year-round basis; there are no seasonal constraints. Production is reliable, predictable, and independent of ambient weather. At least in some cases, the yield per gram fresh weight may exceed that which is found in nature. Disagreeable odors or flavors associated with the crop plant can be modified or eliminated *in vitro*. Plant cell culture eliminates potential political boundaries or geographic barriers to the production of a crop, such as the restriction of natural rubber production to the tropics or anthocyanin pigment production to climates with high light intensity. When a valuable product is found in a wild or scarce plant species, intensive cell culture is a practical alternative to wild collection of fruits or other plant materials. Extraction from the *in vitro* tissues is much simpler than extraction from organized, complex tissues of a plant. Plant tissue culture techniques offer the rare opportunity to tailor the chemical profile of a phytochemical product, by manipulation of the chemical or physical microenvironment, to produce a compound of potentially more value for human use.

While research to date has succeeded in producing a wide range of valuable secondary phytochemicals in unorganized callus or suspension cultures, in other cases production requires more differentiated microplant or organ cultures (Dörnenberg and Knorr, 1997). This situation often occurs when the metabolite of interest is only produced in specialized plant tissues or glands in the parent plant. A prime example is ginseng (*Panax ginseng* C.A. Meyer). Because saponin and other valuable metabolites are specifically produced in ginseng roots, root culture is required *in vitro*. Similarly, herbal plants such as *Hypericum perforatum* (St. John's wort), which accumulates the anti-depressant hypericins and hyperforins in foliar glands, have not demonstrated the ability to accumulate phytochemicals in undifferentiated cells (Smith et al., 2002). As another example, biosynthesis of lysine to anabasine occurs in tobacco (*Nicotiana tabaccum* L.) roots, followed by the conversion of anabasine to nicotine in leaves. Callus and shoot cultures of tobacco can produce only trace amounts of nicotine because they lack the organ-specific compound anabasine. In other cases, at least some degree of differentiation in a cell culture must occur before a product can be synthesized (e.g., vincristine or vinblastine from *Catharanthus roseus* [L.] G. Don, Madagascar periwinkle). Reliance of a plant on a specialized structure for production of a secondary metabolite, in some cases, is a mechanism for keeping a potentially toxic compound sequestered.

The three long-standing, classic examples of commercially viable production of a secondary metabolite *in vitro* — ginseng saponins, shikonin, and berberine — each feature products that have diversified uses, including medicinal applications. Ginseng is produced in large-scale root cultures, whereas the other two products are produced in highly colored cell cultures. A tremendous research and development effort has advanced a number of other *in vitro*-derived secondary products to semicommercial status, including vanillin and taxol production in cell cultures. In a myriad of other cases, the *in vitro* processes for secondary metabolite production have fallen far short of expectations and have never approached commercial status. Still, the arena of secondary product formation in cell cultures remains as an industrial pursuit. Engineers and biologists are currently joining forces on a global scale to develop new strategies for streamlining the critical bioprocesses. Research efforts on a broad range of plant cell culture-derived extracts can be cited in each of these major product categories: flavors (onion and garlic, peppermint and spearmint, fruit flavors, chocolate aroma, seaweed flavors, vanilla, celery, coffee, spice, sweeteners, and so on); edible colors for foods and medicines (mainly betalains and anthocyanins); nonfood pigments for cosmetics and textiles (shikonin, berberine, and various other products); several examples of fragrances and essential oils; and bioactive natural insecticides and phytoalexins useful in current integrated pest management programs. Of course, intensive activity has centered on production of natural drugs or chemoprotective compounds from plant cell culture. Some of the most prominent pharmaceutical products in this latter category include ajmalicine (a drug for circulatory problems) from *C. roseus*, and taxol (a phytochemical effective in treatment of ovarian cancer) from *Taxus* species.

SELECTION OF PRODUCTIVE CELL LINES AND STIMULUS OF PRODUCTIVITY

Because different levels of secondary metabolite production can be found within a cell line (e.g., sectoring of colored and uncolored callus in a culture), significant headway in the productivity of plant cell cultures has followed on the heels of intensive selection for high-producing cell lines. Plant cell lines can be recurrently selected to amplify the productivity of the cell culture. Shikonin production *in vitro* at levels more than 800 times what is available from plant roots is a case in point.

Similar amplification of yield has followed selection in highly pigmented cell culture lines. Natural pigments including anthocyanins, betalains, shikonin, and other pigmented phytochemicals are conspicuously accumulated in some cell cultures. Some of these products are of interest commercially as replacements for synthetic food dyes that have fallen "under the gun" recently as adverse health and safety concerns have emerged. Negative reports and new medical studies in the popular press have made synthetic additives a target for contention and have created a strong market demand for safe, natural ingredients in foods. As an added bonus, significant chemoprotective benefits (including cardioprotective and anticarcinogenic properties) have been attributed to natural anthocyanin pigments and related flavonoid phytochemicals. These health-enhancing properties are unrelated to the nutritive value of the plants. Even natural nonfood colorants are sought as replacements for environmentally damaging synthetic dyes fabricated with toxic solvents and heavy metals. The myriad of microenvironmental control tools available to *in vitro* production makes this alternative more promising than harvest from the intact plant in nature.

When *in vitro* manipulation results in accumulation of pigments in cell cultures, the result is a superb model to investigate control over anthocyanin, betalain, or other production by introducing gene constructs using regulatory genes to the cell culture tissues. For example, when excessively high anthocyanin levels interfere with cell culture growth, gene constructs may use an inducible promotor system. In this case, the bright red pigment is an obvious marker for research on manipulating flavonoid production and modifications in the simple cell culture system. These lines of research would be complex and unwieldy in whole plant models due to the restrictions of organ-specific product accumulation and interactions with other plant tissues. In many ways, natural plant pigments are ideal target compounds for laboratory classroom or demonstration experiments on metabolite production because their accrual *in vitro* is quick, visible, easy to detect, and fairly simple to quantify. In practice, and for validation of research results, pigment accumulation is typically measured by extraction, then quantification and separation with thin layer chromatography (TLC), high performance liquid chromatography (HPLC), nuclear magnetic resonance (NMR), or absorbance (spectroscopy). However, in a classroom, when traditional methods of pigment quantification are prohibited by the lack of sophisticated laboratory analytical equipment and insufficient student expertise, the production of a pigment *in vitro* can still be very easily detected and quantified by visual means alone or by coupling visual observation with very simple absorbance measurements using a spectrophotometer.

Many of the best-known *in vitro* pigment production systems are actually difficult to demonstrate in a classroom setting because they require cultures derived from specific, hard-to-initiate explants, or because the high yielding models described in research reports have only been achieved after years of laborious repetitive subculture and selection of only highly pigmented sectors. One exception to the general rule is the *Ajuga in vitro* pigment production system, which requires no selection and is readily adaptable to classroom research. *Ajuga reptans* N. Tayl. (bugleweed) and the related species *A. pyramidalis* L. "Metallica Crispa" are both widely distributed groundcover plants that grow rapidly and spread via stolons. *Ajuga* tolerates adverse conditions so well that it is considered an intractable weed when it oversteps its boundaries in the landscape, and stock plants can be maintained long term in a greenhouse with minimal care. *Ajuga* cultivars are typically propagated by division of mother plants. A few select cultivars of *A. reptans*, which feature purple or bronze foliage, include "Purpurea," "Giant Bronze," and "Burgundy Glow." The latter cultivar has a characteristic chimeral green, creamy white, and dark pink variegated foliage, but produces unicolor bronze, albino, or green variants from adventitious buds *in vitro*. *A. pyramidalis* "Metallica Crispa," which is frequently confused with purple forms of *A. reptans* in the trade, is characterized by savoy-type, glossy, bronze-purple foliage.

For a number of reasons, selections in the genus *Ajuga* are particularly well suited for use in plant tissue culture exercises. Like many members of the mint family (*Labiatae*), microplants flourish in culture and may be held in cold storage successfully for up to a year, which helps cut back on the need for repetitive subculture maintenance between semesters. *Ajuga reptans* proliferates readily *in vitro* from axillary or adventitious buds; exhibits classic, predictable responses to gradient increases of cytokinin concentration; and is a good, rapidly growing model plant on which to illustrate stages in an *in vitro* production cycle through acclimatization within a single semester.

The purple or bronze leaves that distinguish some varieties of *Ajuga* are also readily expressed in microcultured plants, and even undifferentiated callus can exhibit pigment expression. These suspensions produced anthocyanin pigments in the dark and pigment expression responded to changes in the light regime and to chemical medium treatments. In our alternative approach, vegetative disks are directly explanted from *Ajuga* foliage and induced to produce callus, which rapidly acquires a vivid purple hue when exposed to illumination (Madhavi et al., 1996). Intense, uniform pigment expression is routinely achieved without selection, and friable callus colonies can be explanted into liquid suspension culture. Cell division and pigment accumulation are not mutually exclusive processes; both cell biomass and pigment accumulation increase steadily throughout each subculture cycle. The growth and pigment expression are so predictable and consistent that this system was used as a model to test the application of video image analysis to regulate bioprocesses by visual characterization of cell size, shape, aggregation, and color intensity.

One often cited drawback of reliance on secondary product accumulation in plants — whether in the field or in an *in vitro* production system — is an observed lack of reproducibility. This problem is a consequence of the variable nature of plant secondary metabolism, since the enzymatic pathways for synthesis of secondary compounds are highly inducible. For example, an alkaloid or bioflavonoid product extracted from plant cells may exhibit a sharp decline in potency or level of accumulation over time. Various stress factors in nature related to climate or pathogen insult or physical stress clearly have an impact on the qualitative and quantitative accumulation of valuable secondary products in nature, and similarly, deliberate introduction of stress agents (elicitation) is a strategy that can regulate secondary product recovery from cell cultures and allow for development of reliable, predictable production systems. Elicitors (compounds that "stress" cells, leading to formation of a secondary product) from biotic and abiotic sources can be added directly to culture media to stimulate production. Fungal filtrates, methyl jasmonate, chitosan, sodium acetate, beta glucan, metal ions (iron and copper, for example), and a wide range of other biotic and abiotic additives have effectively stimulated and intensified productivity in cell culture regimes. Another related approach is biotransformation — the deliberate feeding of metabolic precursors to a cell culture that already contains the necessary enzymes to convert them to product.

Another target for research has been the development of improved instrumentation for cultivation of plant cells *en masse*. For example, improved bioreactors have been specifically designed to permit light to reach plant cells during division and product accumulation. In many cases, the infiltration of light at optimum intensities and spectral qualities is prerequisite to successful synthesis and maximal expression of secondary metabolites. Modifications in aeration, agitation, and nutrient supply have similarly led to enhanced plant product yields in recent years.

UTILITY OF CELL CULTURE SYSTEMS FOR IN-DEPTH INVESTIGATION OF METABOLISM

One of the most important categories of plant secondary compounds, and one that has captured the attention of the general public, is the category of bioactive, medicinally important plant phytochemicals. Throughout history, plants have been formulated in potions and powders and routinely used to cure, to diagnose, and to prevent diseases (as well as related conditions like sleeplessness, impotence, infertility, hangovers, and the like). Current consumers have readily embraced the concept of being proactive about maintenance of their own health. Instead of being dependent on a physician's recommendations and synthetic drugs, they are receptive to having natural plant-derived products available in the marketplace. These natural additives have strong consumer and political support, and face substantially fewer international trade restrictions. As noted earlier, these biologically active compounds can be extracted from field-grown plants or from their cell culture counterparts, the latter usually with greater ease and reproducibility.

However, one of the drawbacks to the use of "natural" plant-derived medicinal substances is that, in fact, relatively little is known about their mechanisms of action or metabolism in the human body. Almost all evidence for their efficacy is based on anecdotal or epidemiological evidence. Two reasons contribute to this: first, plant compounds most commonly work in active mixtures, not as single compounds, which makes it very difficult to track their transport after ingestion; and second, plant compounds break down metabolically in the human body, which further complicates our ability to trace progress in the body or determine organ accumulation or absorption of certain metabolically active plant substances.

Because plant cell cultures can be induced to accumulate complex bioactive compounds and mixtures of compounds in predictable cell cultures using defined substrates, they can also provide a partial solution to the investigative challenge cited above. Recently, radio labeled ^{14}C substrates have been introduced to plant cultures, resulting in the incorporation of traceable labels into the bioactive compounds of interest. These compounds, once introduced to experimental animal models, will soon allow scientists to conclusively determine the mode of action, timing, and metabolic clearance of valuable, medicinally important plant secondary metabolites, which will be a further step toward more effective diagnosis and recommendations for dosage and treatment. These natural additives have strong consumer and political support, and face substantially fewer international trade restrictions.

SUMMARY

Experimental models that result in obvious, measurable accumulation of secondary metabolites are highly amenable to individualized, student-directed experimentation with a vast range of physical or chemical microenvironmental parameters (treatments) that can be deliberately varied to determine the effects on metabolite accumulation. The experimental results also help students appreciate the value of chemical analytical methods for measuring metabolite production as an adjunct to strict visual gauges for pigment production in a callus or suspension culture.

LITERATURE CITED

Dörnenberg, H. and D. Knorr. 1997. Challenges and opportunities for metabolite production from plant cell and tissue cultures. *Food Technol.* 51:47–54.

Smith, M.A.L. et al. 2002. An *in vitro* approach to investigate medicinal chemical synthesis by three herbal plants. *Plant Cell Tiss. Org. Cult.* 70:105–111.

Madhavi, D.L. et al. 1996. Characterization of anthocyanins from *Ajaga-pyramidalis* 'Metallica Crispa' cell cultures. *J. Agric. Food Chem.* 44:1170–1176.

Section VI

Special Topics

25 *In Vitro* Plant Pathology*

Subramanian Jayasankar and Dennis J. Gray

CHAPTER 25 CONCEPTS

- Plants can be regenerated from single cells. Thus, a large number of single cells with the potential to become plants can be grown in a container and can be subjected to intense screening with appropriate selection agents. Surviving cells are regenerated to produce resistant phenotypes. This process is called *in vitro* selection.

- To determine if resistance is actually induced, immediately after the selection process, a number of small tests using the surviving cell or callus culture and the selection agent (or the organism that produces the selection agent) can be performed under laboratory conditions. These "bioassays" are very effective, especially when working with plants that have long regeneration cycles.

- When plant cells are subjected to intensive *in vitro* selection, a number of defense genes are induced; however, only those cells whose defense mechanism is activated quickly and in a sustained manner remain viable and survive recurrent *in vitro* selection. Plants regenerated from such cells exhibit certain native defense genes in a constitutive manner and thus they are more resistant than their parental material.

- This approach of activating innate resistance in the plant using *in vitro* selection constitutes a viable, uncontroversial biotechnological approach of generating resistant plants.

INTRODUCTION

In vitro culture and selection of plant cells and tissues has been used effectively as a tool for developing novel, disease-resistant genotypes. This technique was first demonstrated by producing wildfire-resistant tobacco plants through *in vitro* selection, using methionine sulfoximine, a structural analog of wildfire toxin. Since then, plant cells have been successfully selected against an array of pathogenic microorganisms and regenerated into plants with enhanced disease resistance. In addition, plant cell and tissue culture has become an important tool in the study of plant–pathogen interactions at the cellular and molecular levels. Plant cells react to certain biotic and abiotic stresses in a manner similar to that of an intact plant. This makes such cell cultures ideal candidates for understanding the resistance responses and the changes that occur at the cellular and subcellular levels when infection with pathogenic organisms occurs. Some of the common responses that are well documented include changes in the permeability of plasma membrane and triggering the synthesis of new biochemical compounds, especially defense-related enzymes (such as the pathogenesis-related proteins). Plant cell cultures provide an ideal population of homogeneous genetic material. A single flask of embryogenic cell suspension culture theoretically represents millions of plants that can be effectively screened. For instance, a suspension culture of grape proembryogenic

---

* Florida Agricultural Journal Series No. R-10170

FIGURE 25.1 Suspension culture of grapevine (*Vitis vinifera*). These suspension cultures contain highly embryogenic cell aggregates called proembryogenic masses, which are ideal for *in vitro* selection against plant pathogens.

masses (Figure 25.1) contains enough totipotent cells to regenerate plants for hundreds of acres of vineyard. Furthermore, they are very useful for performing a number of genetic tests such as bioassays, and they can also be useful in the culture of biotrophic pathogens. This chapter will address how *in vitro* culture can best be utilized to study and understand plant pathogen interactions.

DEVELOPING DISEASE-RESISTANT GENOTYPES OF CROP PLANTS

Though cell cultures of most species are homogeneous in nature, subtle differences at the subcellular level do exist among the population. In order to isolate cells with slight genetic differences within a huge population, we have to devise a method for selecting or screening the population with a selection agent. Sometimes such selection agents can also cause minor but specific mutations, thereby altering the genetic makeup of the cells. Plants regenerated from such mutant cells exhibit altered phenotypes. Among the most common selection agents that are used to generate disease-resistant phenotypes are metabolites, such as phytotoxin, produced by the pathogen, crude culture filtrate, and rarely, the pathogen itself (Daub, 1984; 1986). A list of important work done in this field is summarized in Table 25.1.

The use of phytotoxins as selection agents has received the most attention mainly because of the ease in exposing a population of cells to controlled, sublethal doses of the isolated phytotoxin. In addition, toxin selection is easier to handle than pathogen selection because it is a purified chemical compound that can be incorporated into the culture medium at measurable doses. In several instances, cells that were selected against the toxin retained significant levels of resistance to the toxin as well as the pathogen when regenerated into plants. In a few studies, such resistance following selection against the toxin was transmitted to the progeny as well. However, this approach has a number of pitfalls. First, toxin selection has worked well only against host-specific toxins such as the HMT-toxin produced by *Dreschlera maydis* race T (syn. *Helminthosporium maydis*). Second, there are only a handful of characterized phytotoxins, whereas the number of pathogenic microbes is vast. Since selection is targeted against a particular pathogen, the regenerated plants

TABLE 25.1
***In Vitro* Selection For Disease Resistance**

Crop	Pathogen	Selection Agent	Results
Alfalfa *Medicago sativa*	*Colletotrichum gloeosporioides*	Culture filtrate	Plants with increased disease resistance
Asparagus *Asparagus offcinalis* L.	*Fusarium oxysporium* f. sp. *asparagi*	Pathogen inoculation	Plants with increased disease resistance
Barley *Avena sativa* L.	*Helminthosporium sativum*	Partially purified culture filtrate	Plants with increased disease resistance; transmitted to progeny
Chinese cabbage *Brassica campestris* ssp. *pekinensis*	*Erwinia carotovora*	Culture filtrate combined with UV irradiation	Plants with increased disease resistance; transmitted to progeny
Coffee *Coffea arabica* L.	*Colletotrichum kahawae*	Partially purified culture filtrate	Plants with increased disease resistance
Eggplant *Solanum melangena* L.	*Verticillium dahliae*	Culture filtrate	Plants with increased disease resistance
Grapevine *Vitis vinifera* L.	*Elsinoe ampelina*	Culture filtrate	Plants with increased disease resistance
Grapevine *Vitis vinifera* L.	*Colletotrichum gloeosporioides*	Culture filtrate	Plants with increased disease resistance
Maize *Zea mays*	*Helminthosporium maydis*	T-Toxin	Phytotoxin-resistant cells and plants
Mango *Mangifera indica*	*Colletotrichum gloeosporioides*	Colletotrichin and culture filtrate	Resistant embryogenic cultures
Peach *Prunus persica* L.	*Xanthomonas campestris* pv. *pruni*	Culture filtrate	Disease-resistant clones
Potato *Solanum tuberosum*	*Phytophthora infestans*	Culture filtrate	Phytotoxin-resistant plants
Rice *Oryza sativa*	*Xanthomonas oryzae*	Culture filtrate	Filtrate-resistant plants
Strawberry *Fragaria* sp.	*Fusarium oxysporum* f. sp. *fragariae*	Fusaric acid	Resistant shoots
Sugarcane *Saccharum officinarum* L.	*Helminthosporium sacchari*	Culture filtrate	Disease-resistant clones
Tobacco *Nicotiana tabaccum*	*Pseudomonas tabaci*	Methionine sulfoximine	Phytotoxin-resistant plants
Tomato *Lycopersicon esculentum* Mill.	*Fusarium oxysporum*	Fusaric acid	Plants with increased disease resistance

often exhibit resistance only to the particular pathogen in question. When these plants are planted in the field, they are susceptible to other pathogens, thus making evaluation very difficult.

These pitfalls perhaps can be overcome by the use of crude culture filtrates, instead of purified phytotoxin, obtained by growing the pathogen in nutritient broth solutions. Culture filtrates of both pathogenic bacteria and fungi have been successfully used in establishing resistant cultures and regenerating resistant plants (Table 25.1). The use of culture filtrates, though not considered as the best approach by many (Daub, 1986), has its own advantages. The culture filtrate approach more closely approximates aspects of the host–pathogen interaction than toxin selection, since culture filtrates have the entire spectrum of compounds (including the toxin) produced by the pathogen. In addition, recent studies have shown that some compounds (e.g., proteins) produced by pathogens in culture that are not a critical factor in the disease process are often implicated in evoking a general disease-resistance response in the host plant (Strobel et al., 1996). Similarly, culture filtrate selection has also evoked broad-spectrum resistance in grapevine (Jayasankar et al., 2000).

Whether it is purified toxin or crude culture filtrate, the mode of selection is very important. In some studies, solid medium (medium that was gelled with agar or similar compounds) was used for effecting such selections, incorporating measured doses of toxin or culture filtrate in the medium. Plants regenerated from such selection schemes quite often showed epigenetic responses that faded away with time. The probable reason is that in a solid medium, the cultured explants are not in full contact with the medium and often a gradient is established from the top to the bottom of the explant (Litz and Gray, 1992). This results in acclimatization of those cells at the top of the gradient to the selection agent, and the plants regenerated from such cells are generally false positives, exhibiting epigenetic resistance. It is possible to circumvent this gradient factor by plating and growing the cells as a thin layer, but several plant species require a mass of small cells rather than single cells for optimum regeneration. In a suspension culture system of selection, cell masses are completely and rapidly immersed in medium containing the toxin. This not only helps to avoid the gradient factor, but also means that the cell masses do not have sufficient time to elicit an epigenetic response. Hence, suspension cultures are better for such selections.

PROBLEMS ENCOUNTERED IN SELECTION

Selection against toxin or culture filtrate is not possible under certain circumstances and against certain diseases. To give a few examples, it is very difficult to select against fungal pathogens that cause powdery mildew, since none of these fungi grow well in culture. Some bacterial pathogens such as the xylem-limited bacteria (e.g., *Xylella fastidiosa* of grapevine, which causes Pierce's Disease) can kill the plant by physical means rather than chemical action. These bacteria are so restricted in their growth that isolation and culture of these bacteria and addressing resistance through selection is often futile. To date, successful selection is yet to be carried out against viral diseases, since the only way possible to select against virus is to grow the pathogens on their hosts and look for resistance. To impart resistance under such conditions, one has to use gene or genome transfer techniques.

BIOASSAYS

In vitro bioassays are very useful tools in determining the level of resistance in a breeding program or for screening a population for sensitivity to a pathogen or pathogen-derived metabolites. Assays using intact plants (seedlings) or plant parts (leaves or shoots) against pathogens or metabolites are being used routinely for screening purposes. More recently, callus or cell cultures are also used extensively for such screening. Barring a very few exceptions, *in vitro* response exhibits a direct correlation to *en planta* response. Hence, these assays, when combined with a selection program (as described above) are very effective, especially in perennial crops where the regeneration cycles are long. Another added advantage of using cell or callus cultures for screening against pathogen or pathogen-derived metabolites is that this facilitates studying the underlying subcellular mechanisms in those interactions (detailed below).

Germination of seeds in a toxin solution is a very easy and common method of testing phytotoxicity. Both host-specific and nonspecific toxins can be assayed in this manner. Typically, a wide range of concentration is used to determine the LD 50 levels and sublethal doses of toxin, based on germination inhibition. Additional parameters such as inhibition of root growth (especially for soil-borne pathogen), malformation of cotyledons and chlorosis in the emerging leaves of intact plants can also be used to determine toxicity. The most common way to test live pathogen is to spray a measured quantity of spore or bacterial suspension on clean leaves and let it incubate under ideal conditions for 24 to 72 hours. Alternatively, a drop of spore or bacterial suspension can be placed over the leaf lamina, which, after sufficient soaking, are incubated and observed for disease development. In some cases, vacuum infiltration or a prick with a needle may be necessary to facilitate disease development. It is very important to include appropriate controls in these studies. The easiest method involves soaking the same leaf, or a similar leaf from the same plant that is used for pathogen infection, in water or buffer in which the spores are suspended. These tests are equally effective when a pathogen-derived metabolite is tested (see Jayasankar et al., 1999 for details). For assaying "wilt toxins" (toxins that are involved in wilt diseases) or pathogens that cause wilt diseases, the best method is to place a live and clean cutting of the plant into a solution containing the metabolite or the pathogen for a predetermined time and subsequently incubating the cuttings in sterile water for disease symptom development. Parameters such as inhibition of shoot growth, chlorosis or necrosis of leaves, and time required for the plant organ to wilt as compared to the appropriate untreated control will serve to determine the resistance or susceptibility of the plant in question.

Callus cultures are routinely used for *in vitro* screening of multiple genotypes to assess their sensitivity to a pathogen. For instance, Nyange et al. (1995) screened nine genotypes of coffee against *C. kahawae*, the fungal pathogen causing coffee berry disease. Early studies (before plant tissue culture was very common) involving plant tissue cultures aimed only at assessing the effect of phytotoxin. However, our current knowledge of plant tissue culture and plant–pathogen interactions at the molecular level has provided a powerful tool in understanding the disease process. A list of crop plants where tissue-culture-based assays were used to study plant–pathogen interactions is provided in Table 25.2. An effective variant among the tissue culture based bioassays is the dual culture assay. While other screening procedures are based on the growth of pathogen, dual culture assays are based on inhibition of the pathogen by plant cell culture. This assay is an effective tool, where the objective is to induce disease resistance using *in vitro* selection, as shown in some perennial crops such as lemon, grapevine, and mango (Jayasankar et al., 2000).

MOLECULAR STUDIES

In vitro plant–pathogen interaction studies provide an ideally controlled environment in which to study events occurring at the molecular level. Synthesis and accumulation of defense-related proteins have been observed following a pathogen infection, and in several species genes encoding these proteins have been identified, cloned, and characterized. Such studies led to the identification of an important group of proteins, termed pathogenesis-related (PR) proteins (Linthorst, 1991). Examples of PR proteins include chitinases, glucanases, osmotin- and thaumatin-like proteins. To date, several PR proteins have been identified and grouped into 11 "families," based on their serological affinity.

One of the earliest responses that plant cells exhibit following pathogen attack is the rapid increase in reactive oxygen species (ROS), which is also referred to as the oxidative burst. This initial oxidative burst can be detected within a few minutes after infection occurs, regardless of the host's resistance or susceptibility. In a resistance reaction, a second oxidative burst develops a few hours later and is sustained for longer periods. Such elevated levels of ROS can directly or indirectly inhibit the invading pathogen and can also serve as intermediates in the activation of other defense responses (Baker and Orlandi, 1995).

Several days after these initial responses, other defense mechanisms are activated. Among these are the hypersensitive responses (HRs) that result in the rapid and localized death of a small group of plant cells surrounding the site of pathogen infection in a programmed manner. As this programmed cell death proceeds around the infection site, the invading pathogen is killed, thereby preventing any further spread of necrosis (Hammond-Kosack and Jones, 1996). As these cells prepare to die, they also set forth a series of other defense responses, which include the stimulation of genes encoding PR proteins. These responses, culminating in the expression of PR proteins, are

TABLE 25.2
In Vitro **Culture-Based Bioassays Involving Plant–Pathogen Interaction**

Plant Species	Pathogen	Type of Bioassay	Purpose
Alfalfa *Medicago sativa*	*Verticillium albo-atrum*	Viability staining of cell cultures	Evaluation of phytotoxicity of culture filtrate
Alfalfa *Medicago sativa*	*Fusarium* spp.	Intact seedlings in hydroponic culture containing culture filtrate	Evaluation of phytotoxicity of culture filtrate
Coffee *Coffea arabica*	*Colletotrichum kahawae*	Growth of fungus on callus cultures	Screening of genotypes for resistance
Dogwood *Cornus florida*	*Discula destructiva*	Growth of callus on medium containing culture filtrate	Screening of genotypes for resistance
Grapevine *Vitis vinifera*	*Elsinoe ampelina*	Dual culture using proembryogenic mass	Assessing resistance after *in vitro* selection
Lemon *Citrus limon*	*Phoma tracheiphila*	Dual culture using callus	Assessing resistance after *in vitro* selection

Mango
  Mangifera indica
	*Colletotrichum
	  gloeosporioides*
		Dual culture using
		 proembryogenic mass
			Assessing resistance after
			 in vitro selection

Pine
  Pinus taeda
	Cronartium quercuum f.
	  sp. *fusiforme*
		Inoculation of embryos with
		 live fungus
			Assessing *in vitro*
			 resistance of embryos

Soybean
  Glycine max
	Fusarium solani f. sp.
	  glycines
		Viability staining of cell
		 cultures; stem-cutting assay
			Evaluation of phytotoxicity
			 of culture filtrate

Tomato
  *Lycopersicon
  esculentum*
	Alternaria alternata f. sp.
	  lycopersici
		Detached leaf bioassay	Evaluation of phytotoxicity
			 of toxin

Wheat *Triticum aestivum*	*Microdochium nivale*	Detached leaf and young seedlings	Screening of cultivars for resistance

also seen in other uninfected tissues of the plant, and are termed systemic acquired resistance (SAR) (Ward et al., 1991).

In the past ten years SAR has become one of the most widely researched areas in plant–pathogen interactions. These studies led to the finding of several interesting secondary signals, such as ethylene, jasmonates, and salicylic acid (SA), which have crucial roles in mediating the induction of plant defenses. Although it has been more than 20 years since White (1979) showed that application of SA or (acetyl) salicylic acid (aspirin) to tobacco leaves increased its resistance to TMV infection, molecular studies elucidating the role of SA in plant defense have been done only recently. SA has been shown to be an endogenous plant signal that has a central role in plant defense against a variety of pathogens, including viruses (Dempsey et al., 1999). SA treatment also triggers the same set of PR proteins that are expressed as a result of SAR. Plants treated with SA and those exhibiting SAR have heightened resistance to virus as well, whereas none of the PR proteins themselves have been shown to exhibit antiviral activity.

CONCLUSION

It is clear that *in vitro* plant pathology has come a long way over the past 20 years or so. What was once designed to study the plant–pathogen interaction in a controlled environment has slowly evolved into a decisive tool in generating disease-resistant crop plants and, in turn, greatly enhanced our understanding of these interactions at the molecular level. It is crucial that we understand the mechanisms by which plants defend against pathogen attacks so that it will be possible for us in the future to activate their own defenses, instead of using transgenes to confer resistance. Induced resistance conferred by native genes will be more stable and will also help us generate environmentally friendly genotypes of crop plants.

ACKNOWLEDGMENTS

We sincerely thank Dr. Richard Litz, TREC, University of Florida, Homestead, for his encouragement and critical suggestions in the preparation of this chapter. Florida Agricultural Experiment Station Journal series No. R-08485.

LITERATURE CITED

Baker, C.J. and E.W. Orlandi. 1995. Active oxygen in plant pathogenesis. *Annu. Rev. Phytopathol.* 33:299–321.

Daub, M.E. 1984. A cell culture approach for the development of disease resistance: Studies on the phytotoxin cercosporin. *HortScience* 19:382–387.

Daub, M.E. 1986. Tissue culture and the selection of resistance to pathogens. *Annu. Rev. Phytopathol.* 24:159–186.

Dempsey, D.A., J. Shah, and D.F. Klessig. 1999. Salicylic acid and disease resistance in plants. *Crit. Rev. Plant Sci.* 18:547–575.

Hammond-Kosack, K.E. and J.D.G. Jones. 1996. Resistance gene-dependent plant defense responses. *Plant Cell* 8:1773–1791.

Jayasankar, S. 2000. Variation in tissue culture. In: *Plant Tissue Culture Concepts and Laboratory Exercises*, 2nd edition. (Ed.) R.N. Trigiano and D.J. Gray. CRC Press, Boca Raton, FL. 386–395.

Jayasankar, S., Z. Li, and D.J. Gray. 2000. *In vitro* selection of *Vitis vinifera* 'Chardonnay' with *Elsinoe ampelina* culture filtrate is accompanied by fungal resistance and enhanced secretion of chitinase. *Planta* 211:200–208.

Jayasankar, S. et al. 1999. Responses of embryogenic mango cultures and seedling bioassays to a partially purified phytotoxin produced by a mango leaf isolate of *Colletotrichum gloeosporioides* Penz. In Vitro *Cell. Dev. Biol. Plant.* 35:475–479.

Laemmli, U.K. 1970. Cleavage of structural proteins during the assembly of the head of bacteriophage T4. *Nature* 227:680–685.

Linthorst, H.J.M. 1991. Pathogenesis-related proteins of plants. *Crit. Rev. Plant Sci.* 10:123–150.

Litz, R.E. and D.J. Gray. 1992. Organogenesis and somatic embryogenesis. In: *Biotechnology of Perennial Fruit Crops.* (Ed.) F.A. Hammerschalg and R.E. Litz. CAB International, Wallingford, UK. 3–34.

Nyange, N.E. et al. 1995. *In vitro* screening of coffee genotypes for resistance to coffee berry disease (*Colletotrichum kahawae*). *Ann. Appl. Biol.* 27:251–261.

Sambrook, J., E.F. Fritsch, and T. Maniatis. 1989. *Molecular Cloning: A Laboratory Manual.* Cold Spring Harbor Laboratory Press. New York.

Strobel, N.E. et al. 1996. Induction of acquired systemic resistance in cucumber by *Pseudomonas syringae* pv. *syringae* 61HpZpss protein. *Plant J.* 9:431–439.

Ward, E.R. et al. 1991. Coordinate gene activity in response to agents that induce systemic acquired resistance. *Plant Cell* 3:1085–1094.

White, R.F. 1979. Acetylsalicylic acid (aspirin) induces resistance to tobacco mosaic virus in tobacco. *Virology* 99:410–412.

26 Variation in Tissue Culture*

Subramanian Jayasankar

CHAPTER 26 CONCEPTS

- Somaclonal variation is variation or difference from the parent displayed by plants grown in tissue culture.

- Factors that contribute to somaclonal variation fall into three major categories: physiological, genetic, and biochemical.

- Cell culture systems offer plant breeders a well-defined environment where selection pressures can be imposed on thousands of genetically uniform single cells, each capable of growing into a whole plant.

INTRODUCTION

As discussed in Chapter 19 ("Genetic Engineering Technologies"), plant cell culture is an important technique for plant genetic improvement. Historically, and as implied throughout this book, plant cell culture has been viewed by most to be a method for rapid cloning. In essence, it was seen as a method of sophisticated asexual propagation, rather than a technique to add new variability to the existing population. For example, it was believed that all plants arising from such tissue culture were exact clones of the parent, such that terms like "calliclone," "mericlone," and "protoclone" were used to describe the regenerants from callus, meristems and protoplasts, respectively. Although phenotypic variants were observed among these regenerants, often they were considered artifacts of tissue culture. Such variation was thought to be due to "epigenetic" factors such as exposure to plant growth regulators (PGRs) and prolonged culture time. As more and more species were subjected to tissue culture, however, reports of variation among regenerants increased. In a historically significant review, Larkin and Scowcroft (1981) proposed the more general term "somaclones" for the regenerants coming out of tissue culture, irrespective of the explant used. Variation displayed by such regenerants from tissue culture would then be somaclonal variation. Tissue culture studies in the 1970s and early 1980s started to focus their attention on this type of variation, and it was soon recognized that somaclonal variation exists for almost all the phenotypic characters.

To a plant scientist, somaclonal variation is perhaps the best route for studying somatic cell genetics. In contrast to the earlier view of "true to type" regeneration among plants derived from tissue culture, the frequency of genetic variation may actually be quite high. In some species, such as oil palm and banana, variation among tissue culture derived progenies is higher than one would expect to occur *in vivo*. In perennial crops that are asexually propagated, somaclonal variation offers an excellent opportunity to add new genotypes to the gene pool. In such cases it is important to understand and identify the causal mechanism behind the variations, so that we can effectively control them to our advantage.

* Florida Agricultural Journal Series No. R-10170

It is important to differentiate between variation and mutation at this point. In the literature, "variant" is rather loosely used to depict any type of phenotypic change that is present either in cell culture or in the plants that are eventually regenerated. Often, these descriptions are not accompanied by clear evidence of the cause, and most of them are non-Mendelian in their inheritance. "Mutation" on the other hand should be used only for cases where clear evidence is presented of genetic alteration. Mutants need not necessarily present an altered phenotype, as there are point mutations that are not expressed phenotypically. Jacobs et al. (1987) list several criteria for a mutant, among which sexual transmission of the trait to the offspring and molecular evidence for the alteration are noteworthy.

MAJOR CAUSES OF SOMACLONAL VARIATION

Over the years considerable research on trying to identify the causes of variation in culture has been conducted. Seemingly, every possible factor that could result in a genetic change has been accounted for as a cause for somaclonal variation. For convenience, we can pool the factors that contribute to somaclonal variation into three major categories: physiological, genetic, and biochemical.

PHYSIOLOGICAL CAUSES OF VARIATION

Variations induced by physiological factors can be identified quite early, often without the aid of any tools. Classic examples of such variation are those induced by habituation to PGRs in culture and culture conditions. Often such variations are epigenetic and may not be inherited in a Mendelian fashion. Prolonged exposure of explanted tissue to powerful auxins such as phenoxyacetic acids (e.g., 2,4-D or 2,4,5-T) often results in variation among the regenerants. For example, oil palm plants (*Elaeis guineensis* Jacq.) generated from long-term callus cultures in the presence of 2,4-D showed significant amounts of variability in the field. In grapevine (*Vitis vinifera* L.), embryogenic cells that have been maintained in culture for several years gradually lose their ability to differentiate and regenerate into plants over time.

GENETIC CAUSES OF VARIATION

Genetic variation occurs among tissue culture regenerants as a result of alterations at the chromosomal level. Although the explanted tissue may be phenotypically similar, plants often have tissue comprised of diverse cell types. In other words, there are cytological variations among the cell types within an explanted tissue. Such a preexisting condition often results in plants regenerated from tissues that are dissimilar. These species are referred as "polysomatic" species. Species such as barley (*Hordeum vulgare* L.) and tobacco (*Nicotiana tabacum* L.) have been documented to possess such polysomatic tissues.

Chromosomal rearrangements such as deletion, duplication, and somatic recombination are the chief sources of the genetic variations exhibited by somaclones. Extensive studies have been conducted using cultured cells of both plant and animal species to demonstrate chromosomal rearrangements. Lee and Phillips (1988) have described in detail the possible mechanisms that lead to these chromosomal changes. They point out that late-replicating heterochromatin is the primary cause of somaclonal variation in maize (*Zea mays* L.) and broad beans (*Vicia faba* L.). Transposable elements are activated during culture in explanted tissue and this results in altered genotypes among the regenerated plants. The well-known transposable element complex of *Ac–Ds* in maize has been shown to be activated following *in vitro* culture.

BIOCHEMICAL CAUSES OF VARIATION

Biochemical deviations are the predominant type of variation in tissue culture. Many of them are barely noticeable, unless a specific test is performed. To date, several biochemical variations have been identified in various crop plants in tissue culture. While some of these variants show Mendelian inheritance, many may be epigenetic and may be lost in the plants regenerated. Examples of such biochemical variations include alterations in carbon metabolism leading to lack of photosynthetic ability ("albinos" in cereals such as rice), starch biosynthesis, carotenoid pathway, nitrogen metabolism, and antibiotic resistance. In contemporary plant science, any variation in antibiotic resistance or susceptibility has significant implications because antibiotic resistance is a vital marker in transformation studies.

Genomic DNA exhibits normal methylation patterns. Methylation is a process where a particular nucleotide — usually adenine (A) or cytosine (C) — has a methyl group (CH_3) attached to it. Prolonged exposure of plant tissues to *in vitro* culture has resulted in the alteration of normal methylation patterns. When such methylation occurs in a region of DNA that encodes an active gene, it prevents the gene from further processing and the gene is silenced. Results of such gene silencing due to methylation may not be noticed phenotypically. Although methylation due to tissue culture has been shown in several species, such as maize, potato (*Solanum tuberosum* L.), and grapevine, at present we do not know why this process happens.

INDUCED (OR DIRECTED) CAUSES OF VARIATION

In conventional plant improvement, a plant breeder will have to screen an extensive number of plants in the greenhouse or field if selecting for a particular trait of interest. For instance, if the goal is to develop a salt tolerant line of a particular species, a large number of individual plants must be grown and subjected to various doses of salt to eventually identify the plants that can withstand the screening process. In doing so, the amount of material that can be tested will be limited by availability of space and time. In addition, environmental factors will also interfere with the selection process. Cell culture systems provide the breeder with the ability to select from a very large amount of genetically uniform material and to conduct the screening quickly in a few petri dishes or flasks. This provides much greater control over the selection process. As cell and tissue culture systems were developed for various species, the potential of *in vitro* selection was quickly recognized.

IN VITRO MUTATION

Certain crop plants, such as bananas and plantains (*Musa* spp.), do not have a large genetic base and have been obligately propagated by asexual means for thousands of years. Genetic improvement in these species is cumbersome as they seldom produce fertile seeds. In such cases, the only available options are to look for natural somatic mutations, which do occur at an extremely low frequency, or to induce mutations. In many vegetatively propagated crops, even inducing mutations is difficult, because their propagules are quite large, as in the case of banana suckers. Attempts to irradiate such large vegetative propagules results either in mosaics or fatalities. Development of a cell regeneration system such as somatic embryogenesis provides an opportunity to expose a large number of regenerative cells to gamma irradiation or to a chemical mutagen, such as ethylmethylsulfonate, in a very controlled manner, and thus widen the existing germplasm base (Novak, 1992).

USE OF SELECTION AGENTS

Methotrexate is a toxin produced by *Pseudomonas tabaci*, and the causal bacterium of tobacco wildfire disease. Since the first report of successful regeneration of methotrexate-resistant tobacco plants from protoplast culture by Carlson (1973), the use of phytotoxins for developing disease resistance in plants from tissue culture has been pursued. Fungal and bacterial toxins have been used to select disease resistant lines of rice (*Oryza sativa* L.), maize, tobacco, and alfalfa (*Medicago sativa* L.). Identification and isolation of phytotoxins from plant pathogenic bacteria and fungi during the 1970s greatly aided resistance breeding through tissue culture. Selection was carried out at the protoplast, cellular, and tissue levels in various plant species with varying degrees of success. About the same time, resistance for salts (sodium chloride) and heavy metals (aluminum) were also attempted through *in vitro* selection in herbaceous monocots and dicots.

The promise of genetic engineering in 1980s slowed *in vitro* selection research, as emphasis was shifted to the identification and cloning of useful genes that could confer resistance to biotic and abiotic stresses. As genetic engineering technology began to mature in the 1990s (see Chapter 2), research into *in vitro* selection effectively ceased. Our understanding of plant resistance to stress factors has advanced tremendously due to the development of molecular tools. These molecular tools can identify the changes that occur at DNA, RNA, or protein levels when plant cells are exposed to challenging environments. The use of molecular tools, such as randomly amplified polymorphic DNA (RAPD), restriction fragment length polymorphism (RFLP), DNA amplification fingerprinting (DAF), and amplified fragment length polymorphism (AFLP), allows for rapid and accurate study of the occurrence and nature of variation.

APPLICATIONS OF VARIATION DERIVED FROM TISSUE CULTURE

Cell culture systems offer plant breeders a well-defined environment where selection pressures can be imposed on thousands of genetically uniform single cells, each capable of growing into a whole plant. The effect of environmental variation is minimized, so that escapes or adaptations that can revert back to original genetic background are also reduced. This controlled growth atmosphere in a minimal space gives the plant breeder new options for introducing variation. In addition, a scientist can study a tropical species in a temperate region or vice versa because specialized environmental conditions can be provided anywhere. Although it is not studied much today, culture-derived variation has the potential to improve crops, especially in perennial species, which are hampered by a narrow germplasm base and long regeneration cycles. Some of the important applications for induced variation are discussed below, with one or two classical examples from the literature.

DEVELOPMENT OF DISEASE-RESISTANT PLANTS FROM TISSUE CULTURE

Hammerschlag (1992) points out the following as the most important among several criteria to be met before effective *in vitro* selection for disease resistance can proceed.

First, an effective selection agent that can be produced and utilized in an *in vitro* system must be identified. The identified selection agent should act at the cellular level and should be an important factor in the disease process.

Second, a reliable protocol must be developed for regenerating whole plants from single cells for the species in question. The protocol must allow the cells to withstand several cycles of selection in a stringent environment and still be able to regenerate whole plants.

In addition to these important factors, effective tools to determine if selected cells are truly resistant to the pathogen at the cellular level and whole plant level are necessary. Possible diagnostic tools include a bioassay that can be performed on selected cells and polymerase chain reaction (PCR) analysis to identify genetic changes. Greenhouse and field exposure to the pathogen are required tests.

During late 1970s and early 1980s, phytotoxins were employed as selection agents to impart disease resistance. A partial list of important work done in this area is summarized in Table 26.1. However, information regarding the heritability or stability (in perennial species) of such *in vitro* derived resistance is lacking. Often, enhanced resistance is accompanied with other undesirable characters, such as low yield. For instance, continued selection with a phytotoxin resulted in potato regenerants that were resistant, but the tubers were small and inferior in edible qualities. One possible reason is that these phytotoxins are produced by the pathogen in a very timely and specific manner and in very low quantities during the disease process. When plant cells are subjected to

TABLE 26.1

In Vitro **Selection for Disease Resistance against Phytotoxins**

Crop	Pathogen	Phytotoxin	Results

Alfalfa
  Medicago sativa
	Colletotrichum
	  gloeosporioides
		Culture filtrate	Enhanced disease
			  resistance

Asparagus
  Asparagus offcinalis L.
	Fusarium oxysporum f. sp.
	  asparagi
		Pathogen inoculation	Enhanced disease
			  resistance

Banana
  Musa spp.
	Fusarium oxysporum f. sp.
	  cubense race 4
		Fusaric acid	Disease-resistant clones

Citrus/lemon
  Citrus limon L.
	Phoma tracheiphila	"Mal secco" toxin	Resistant embryogenic
			  cultures

Coffee
  Coffea arabica L.
	Colletotrichum kahawae	Partially purified culture
		  filtrate
			Plants with increased
			  resistance

Maize
  Zea mays
	Helminthosporium maydis	T-toxin	Phytotoxin-resistant cells
			  and plants

Mango
  Mangifera indica
	Colletotrichum
	  gloeosporioides
		Colletotrichin and culture
		  filtrate
			Resistant embryogenic
			  cultures

Oat
  Avena sativa L.
	Helminthosporium
	  victoriae
		Victorin	Inheritable disease
			  resistance

Peach
  Prunus persica L.
	Xanthomonas campestris
	  pv. *pruni*
		Culture filtrate	Disease-resistant clones

Potato *Solanum tuberosum*	*Phytophthora infestans*	Culture filtrate	Phytotoxin-resistant plants
Rice *Oryza sativa*	*Xanthomonas oryzae*	Culture filtrate	Filtrate-resistant plants

Strawberry
  Fragaria sp.
	Fusarium oxysporum f. sp.
	  fragariae
		Fusaric acid	Resistant shoots

Sugarcane
  Saccharum officinarum L.
	Helminthosporium
	  sacchari
		Culture filtrate	Disease-resistant clones

Tobacco *Nicotiana tabacum*	*Pseudomonas tabaci*	Methionine sulfoximine	Phytotoxin-resistant plants

Tomato
  Lycopersicon esculentum
  Mill.
	Fusarium oxysporum	Fusaric acid	Plants with increased
			  resistance

higher doses of these toxins, they not only affect the ability of the cells to resist the phytotoxin, but also cause some unwarranted genetic damage to the cells. Another problem is that sometimes regenerants resistant to the phytotoxin were not resistant to the pathogen. These results exposed problems of using phytotoxins to select for disease resistance.

Another strategy involved the use of crude fungal or bacterial culture filtrate as a selection agent to impart disease resistance. In most cases, the level of resistance expressed in the regenerants was higher than the original parent, and these changes were also genetic, as the disease resistance was transmitted to the sexual offspring in a Mendelian fashion. However, some of the early studies, such as a study on the use of *Phytophthora citrophthora* culture filtrate to select citrus nucellar calli of orange, gave results that did not concur with what is known about the host's natural ability in the field. These studies showed, surprisingly, that field resistance and *in vitro* resistance were in reverse order (Vardi et al., 1986).

As there were more failures than successes in using crude culture filtrate for developing disease resistance, this approach was not preferred. The ineffectiveness of crude culture filtrate as selection agent can be attributed to several factors. The most important are as follows:

1. Involvement of secondary metabolites such as PGRs produced by the pathogen, which enhance the growth of plant cells during selection
2. Differential response of the host plant to the culture filtrate at the cellular level (*in vitro*) and at the whole plant level (*in vivo*)
3. Ineffective selection methods that allow more "escapes" (Daub, 1986)

However, another problem contributing to the failures of *in vitro* selection may have been the use of an inadequate culture environment. For example, most of the *in vitro* selection studies were carried out on a semisolid medium, supplemented with a known dose of the crude culture filtrate or phytotoxin. This likely allowed for the formation of a gradient in actual selection pressure within the explanted tissue such that those cells farthest from the medium surface were able to escape selection and/or become adapted to the culture environment. Often, these adapted cells were regenerated into plants in medium containing the culture filtrate and then reported as resistant genotypes. Resistance in such regenerants usually breaks down at the whole plant level, as it is not genetic resistance. A solution to this problem has been to use cell suspension cultures bathed in selection medium so that all cells are subjected to complete and rigorous selection. Escapes or epigenetic variations following *in vitro* selection using suspension cultures are very minimal compared to semisolid medium. Cells and plants regenerated from such cultures following selection in suspension culture show elevated levels of the resistance against which they were selected.

Our understanding of the plant–pathogen interaction during the disease process has tremendously increased in recent years. One of the main findings is that there are compounds other than phytotoxins produced by the pathogen that are involved in the disease process. For instance, "harpin" proteins produced by the pathogen have been shown to elevate plant resistance against a diverse group of pathogens. Using these compounds as a whole unit (as in a crude culture filtrate) in suspension cultures could be a better approach for bringing out the true genetic resistance of the plant.

INDUCTION OF SALT AND HEAVY METAL TOLERANCE THROUGH TISSUE CULTURE

A considerable amount of land is unfit for agriculture because the soil is high in salts such as sodium chloride or in heavy metals like aluminum. Development of crop plants suitable for these regions is still a high priority, as the availability of arable land is continuously shrinking. Tolerance to sodium chloride using *in vitro* selection has been achieved in several crop species, such as rice, potato, sugarcane, and tomato (Table 26.2). However, in most cases the resistance was epigenetic. In some instances, salt tolerance was also accompanied by other undesirable characters such as low yield or lack of fertility. Only a few studies have been carried out to assess the use of somaclonal variation for heavy metal tolerance. Since there were many more failures than successes, current research is more directed toward the understanding of molecular mechanisms involved acquiring such tolerance. The identification and cloning of genes that could elevate resistance to salt and heavy metals are major breakthroughs for plant geneticists.

IDENTIFICATION OF VARIATION IN TISSUE CULTURE

It is important to identify desirable variation in culture as early as possible so that the long waiting period before the regenerated plants can show the phenotypic variability can be avoided and the possible accumulation of additional undesirable traits due to stringent selection can be reduced. If the variation is directed against a particular trait of interest, a specific test or bioassay will be helpful. A classic example is the selection of putative transformants following a gene transfer experiment. Gene transfer experiments routinely utilize a selectable marker, such as antibiotic resistance, to identify the transformed cells. Explants are placed in a medium containing the particular antibiotic to which resistance was introduced. Only those cells that have acquired the

TABLE 26.2
In Vitro **Selection for Abiotic Stress**

Crop	Selection Factor	Result
Almond	Iron deficiency	Resistant shoots
Prunus amygdalus Batsch.		
Citrus/sweet orange	Sodium chloride	Resistant plants
Citrus sinensis (L.) Osbeck		
Flax	Sodium sulfate	Resistant plants; inheritable
Linum usitatissimum L.		
Grapes	Sodium chloride	Resistant embryos
Vitis rupestris Scheele		
Potato	Sodium chloride; seawater	Resistant callus
Solanum tuberosum		
Rice	Sodium chloride; seawater	Resistant plants; not inheritable
Oryza sativa		
Sugarcane	Sodium chloride	Resistant callus
Saccharum officinarum		
Tobacco	Sodium chloride	Tolerant plants; inheritable
Nicotiana tabacum		

new gene will show resistance to the antibiotic and will grow, whereas all other untransformed cells will succumb.

In vitro bioassays are very easy to perform and the results are seen rather quickly. For instance, if a perennial crop is subjected to *in vitro* selection for disease resistance, it may take several weeks or months before the regenerated plants are ready to be tested for the particular disease. However, a dual culture assay, where the pathogen is grown in the same culture plate along with the selected cells, might determine if the selected cells express resistance. When successful, such an assay shows the pathogen growth to be significantly reduced by the selected cells when compared to the unselected cells (Jayasankar and Litz, 1998). This assay has worked quite effectively against fungal pathogens in several fruit crops such as *Citrus* spp., mango (*Mangifera indica* L.) and grapevine. It will also be easy to test if the induced resistance is specific against the host alone or if it is broad spectrum resistance. However, there are certain limitations to this approach. For example, the pathogen may not grow well in the plant growth medium. So far, this has been demonstrated effectively only with fungal pathogens.

In the past decade, several molecular techniques that can distinguish minor changes at the nucleic acid level have been developed. Amplification-based scanning techniques such as RAPD, AFLP, DAF, and arbitrary signatures from amplification profiles (ASAP) are used to identify if there is a permanent genetic change in the regenerents. In contrast to RFLP and other hybridization-based techniques, amplification-based scanning or profiling techniques can distinguish individuals and closely related organisms. Hence, these techniques can identify variation induced in tissue culture more precisely and quickly. The major advantage of these techniques is that they need only a small amount of tissue, from which information-rich nucleic acids can be used to generate critical data for analysis. Further, these techniques require much less effort than performing a phenotypic or cytological analysis (Caetano-Anollés and Trigiano, 1997). For instance, RAPD has been used effectively to identify the genetic changes that were induced following *in vitro* selection of mango embryogenic cultures against *Colletotrichum gloeosporioides* phytotoxins (Jayasankar et al., 1998). Resistant cultures exhibited significant differences in the genetic markers generated after RAPD analyses, in comparison with the unselected controls as well as the original parent trees. The phytotoxins may also act as mild mutagens that cause permanent genetic changes in the host cells, and these changes can be specific, such as the activation of the promotor in resistance genes. RAPDs have also been used to identify somaclonal variation in asparagus, beet, Norway spruce, peach, sugarcane, and wheat. AFLPs are more recent inventions, and they can identify variations more rapidly and in greater numbers. Furthermore, AFLP markers can be located easily in the genome, since only a known fragment of the genomic DNA is amplified for analysis. AFLPs are relatively easy to perform and are more reproducible than RAPDs. Currently, AFLPs have been successfully used to identify somaclonal variation in pecan (*Carya illinoinensis* Wangenh.), lettuce (*Lactuca sativa* L.) and chrysanthemum species. DAF is, perhaps, the most useful tool in identifying variants in perennial species that are regenerated through tissue culture. Minor variants that are poorly characterized phenotypically can be precisely identified using DAF. This helps breeders to eliminate the unwanted rogues very early on.

FUTURE PROSPECTS

It is evident that, despite their potential, variations from tissue culture have not been fully exploited. Induced variation is still the best route in perennial crop improvement, although one can argue in favor of the currently untapped potential of genetic transformation. However, it must be noted that tissue-culture induced variation does not face the socio-ethical hurdles of genetically modified crops (see Chapter 20). In addition, there are no significant technology ownership issues such as have become problematic with genetic engineering. Further, gene transfer techniques, though successful in herbaceous species, still have not been commercialized in perennial and woody species. Molecular techniques have greatly aided in understanding the plant cell response to biotic and abiotic stresses at the subcellular level. The future lies in utilizing these techniques to induce the species' own resources, such as disease resistance, to our benefit.

ACKNOWLEDGMENTS

I thank Dr. A.K. Yadav, Fort Valley State University, for critically reading the manuscript and making suggestions. Florida Agricultural Experimental Station Journal series No. R-06862.

LITERATURE CITED

Caetano-Anollés, G. and R.N. Trigiano. 1997. Nucleic acid markers in agricultural biotechnology. *Ag. Biotech. News Inf.* 9:235–242.

Carlson, P.S. 1973. Methionine-sulfoximine resistant mutants of tobacco. *Science* 180:1366–1368.

Daub, M.E. 1986. Tissue culture and selection of resistance to pathogens. *Annu. Rev. Phytopath.* 24:159–186.

Hammerschlag, F.A. 1992. Somaclonal variation. In: *Biotechnology of Perennial Fruit Crops.* (Ed.) F.A. Hammerschlag and R.E. Litz. CAB International, Wallingford, UK. 35–56.

Jacobs, M. et al. 1987. Selection programmes for isolation and analysis of mutants in plant cell cultures. In: *Plant Tissue and Cell Culture.* (Ed.) C.E. Green et al. Alan R. Liss Inc., New York. 243–264.

Jayasankar, S. and R.E. Litz. 1998. Characterization of embryogenic mango cultures selected for resistance to *Colletotrichum gloeosporioides* culture filtrate and phytotoxin. *Theor. Appl. Genet.* 96:823–831.

Jayasankar, S. et al. 1998. Embryogenic mango cultures selected for resistance to *Colletotrichum gloeosporioides* culture filtrate show variation in random amplified polymorphic DNA (RAPD) markers. *In Vitro Cell. Dev. Biol. Plant* 34:112–116.

Larkin, P.J. and W.R. Scowcroft. 1981. Somaclonal variation — A novel source of variability from cell cultures for plant improvement. *Theor. Appl. Genet.* 60:197–214.

Lee, M. and R.L. Phillips. 1988. The chromosomal basis of somaclonal variation. *Annu. Rev. Plant Physiol. Plant Mol. Biol.* 39:413–437.

Novak, F.J. 1992. *Musa* (Bananas and Plantains). In: *Biotechnology of Perennial Fruit Crops*. (Ed.) F.A. Hammerschlag and R.E. Litz. CAB International, Wallingford, UK. 449–488.

Vardi, A., E. Epstein, and A. Breiman. 1986. Is the *Phytophthora citrophthora* culture filtrate a reliable tool for the *in vitro* selection of resistant *Citrus* variants? *Theor. Appl.. Genet.* 72:569–574.

27 Commercial Laboratory Production

Gayle R. L. Suttle

CHAPTER 27 CONCEPTS

- It is not enough to produce high-quality, healthy plant material; one must also produce it in a timely fashion, giving customers what they want, when they want it, at a price they can afford.

- Successful micropropagation on a commercial scale requires an understanding of two equally complex and dynamic factors — the plants and the marketplace.

- The primary method of increase in woody plant micropropagation is axillary shoot proliferation, although adventitious shoot proliferation may occur with some kinds of plants, such as species of the family Ericaceae.

- The most obvious use for micropropagation is to get a jump-start on growing the newest and hottest items quickly. Micropropagation can cut 3 to 10 years off the time it takes to bulk up new selections and get them to market.

INTRODUCTION

Over the past 20 years, the horticultural nursery trade has found ever-increasing ways to utilize micropropagation as a practical and cost-effective production tool. This chapter will describe some of the basic steps involved in plant micropropagation, discuss some of the practical aspects of the commercial micropropagation business, and give some examples of how and why micropropagation is being used by growers today.

Many "scientific" papers have been published describing methods, formulae, and techniques for the successful micropropagation of a multitude of plant species. In theory, once one has the right recipe, all that is needed is a well-equipped kitchen, some trucks to haul the goodies to market, and a good bank to hold all the money when it comes rolling back in. In reality, successful micropropagation on a commercial scale requires an understanding of two equally complex and dynamic factors — the plants and the marketplace. It is not enough to produce high-quality, healthy plant material (a major challenge in itself); one must also produce it in a timely fashion, giving customers what they want, when they want it, at a price they can afford.

THE FACILITY

Well-designed laboratories generally have a good workflow, allowing for the actual movement of supplies, people, cultures, and the finished product in an easy and logical pattern. Many small laboratories, unable to build a facility from scratch, use existing residential housing or mobile homes and remodel them to accommodate their needs. In order to keep airborne contaminants to

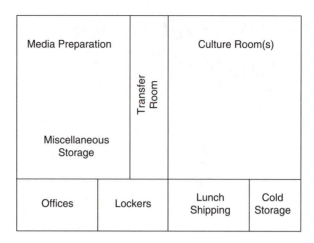

FIGURE 27.1 Design of a commercial tissue culture laboratory facility.

a minimum, the ideal facility has very few entry sites (doors) and often has large areas designated as "clean rooms" with purified air (high-efficiency particulate air [HEPA] filtered) flowing under positive pressure. The basic areas in any laboratory are listed in Figure 27.1.

MEDIA PREPARATION

This area is set up like a kitchen, with a lot of counter space and shelves for chemicals and stock solutions, a pH meter, scales, autoclaves (usually more than one), media dispensing equipment, and a dishwashing area. Also important here is adequate storage space for empty culture vessels. Autoclaves create a tremendous heat load and usually are placed on an exterior wall with good, clean ventilation.

MEDIA STORAGE

Once media are "cooked" or sterilized and dispensed into culture vessels, they must be cooled down (again with good, clean ventilation) and stored until needed. Often this area is combined with the media preparation room or situated in a section of the transfer room.

TRANSFER ROOM

This is the heart of any laboratory, filled with clean benches (laminar flow hoods) where all of the culture initiation and subsequent subculturing takes place. "Ripe" cultures (cultures with usuable shoots) are harvested, and cultures replanted (transferred or reinitiated) to clean, fresh medium. Up to 80% of the laboratory's labor force may be involved in this one room, so it must be a comfortable and pleasant working environment. Tools are sterilized at the hoods using glass bead sterilizers, bacticinerators, or alcohol and flame. Transfer rooms tend to be cramped; the heat load generated by people, hoods, and sterilizers must be compensated for with a good air conditioning system. People and plants suffer when it is hot and stuffy, and productivity declines.

CULTURE ROOM

Usually the largest area of any laboratory, this is where cultures are incubated on shelves under specific light and temperature regimes. Light quantity and quality can be key factors in culture success. Generally, regular or high-output cool white fluorescent lighting is used, with higher intensities achieved by using more lights per square foot or by bringing cultures closer to the lights. Once again, heat buildup can be a major problem. Many laboratories remove the ballasts from the fluorescent light fixtures and mount them in an external area for easier cooling. Good air movement is critical in equalizing temperatures throughout the culture room. Temperature requirements are often crop-specific, but generally range from 18 to 26°C. Different light or temperature regimes may require separate culture rooms for different crops. Usually air to the culture room is purified with HEPA filters.

Cold Storage

Cold storage is an absolutely essential component of any commercial micropropagation facility growing woody plants. Cold storage (2–4°C) is used for the following purposes:

1. To maintain stock blocks in culture for plants that are produced on a seasonal basis
2. To allow production to continue during the cold winter months by "banking" finished plant material ahead of time
3. To provide a chilling requirement for crops, such as apple (*Malus*) and pear (*Pyrus*) species, that respond in the greenhouse much more rapidly if given a "rest" or dormant period before out-planting

Shipping/Lunch Area

Ripe product (either microcuttings or rooted plantlets) must be harvested, boxed, packed, and otherwise prepared for shipping. This activity requires sinks to rinse the plants, and should be near the cold storage unit for holding until shipping. This area may serve double duty as a lunch or break area.

Locker Room

When first entering any commercial facility, employees usually must change their shoes, and may be required to wear additional clean clothing such as laboratory coats and hairnets in production areas. This locker area may also, as mentioned above, double as a lunchroom, and may include restrooms.

Offices

Perhaps the least important place in a facility from an operations standpoint, but essential for running the minor details such as sales, production planning, accounting and, of course, payroll, is the offices.

Miscellaneous Storage

No laboratory ever has enough general storage space. Supplies, chemicals, shipping containers, extra equipment, spare parts, and items, such as vessels, which are used on a seasonal basis, all require space. Logical locations for storage are near an external door (for receiving) or near the media preparation and shipping preparation areas, where supplies are most often used.

THE PROCESS

What to Grow

There are two different approaches in determining product focus. The first is growing what you think customers might want (perhaps a novel plant) based on what you hope is good market research and analysis, and then working hard to convince potential customers that they really do want it. While this approach is the very basis of our country's free enterprise system, it can be very risky if you happen to guess wrong. The most common reason for the failure of commercial laboratories over the past decade has been growing plants that the market either does not want or does not need. The second approach is to work with the customer on a more contractual basis, asking what their problems are, what they need and when they need it, and then going to work to try to solve the problems and fill the order. Arranging customers in advance is a much safer route to success, considering the costs involved. Regardless of approach, every plant must be evaluated for suitability for micropropagation by asking the following questions:

1. What is the minimum number of plants needed? Most laboratories have a minimum requirement (5000 to 10,000 or more), although smaller facilities may be willing to have lower minimums.
2. How many customers are interested in a specific product? The more the better, but sometimes exclusivity is important.
3. How long is the product needed? Is it just for mother block establishment, is it seasonal, or will it be needed on a continuous basis?
4. How difficult is the plant to culture? Is there existing literature or knowledge, or will the technology need to be developed?
5. Who "owns" the plant? Are there patent or customer restrictions on marketing the plants?
6. What is the estimated value of the plant once produced? Can you charge enough to recover costs and make a profit?

All of these factors must be considered before starting any project. Often the price of a product cannot be defined until considerable research has been done on protocols for the given plant. Guarantees of success are impossible to make. The up-front cost of protocol development may or may not be shared with the customer. If exclusivity is needed, then the customer usually funds the work. Once a plant is deemed worthy for an attempt at micropropagation, the process of producing it may begin.

STOCK PLANT PREPARATION

Healthy, well-labeled stock plants are essential. The most important job the customer must perform is to make absolutely sure that the stock wood sent to the laboratory is exactly the same material as that which will be wanted in large numbers when finished. This step sounds simple, but in reality, mistakes in labeling can and do happen far too frequently. This point cannot be overemphasized.

Ideally, 10 to 15 juvenile stock plants are placed under an intense fertilization and pesticide regime to provide rapid, healthy shoot growth. The plants are watered from the base to avoid getting moisture on the foliage, a breeding ground for contaminants. Sometimes a single, very old tree located thousands of miles away may be the only source, or a customer may have only a small amount of stock and may not want to risk sending it away. In these cases, wood must be collected periodically throughout the growing season and sent by overnight express.

STAGE I: CULTURE INITIATION

The process begins by carefully harvesting soft, active shoot growth from the stock plant. Usually, 5 to 15 cm cuttings are snipped early in the morning while the plants are fully turgid. The cuttings are immediately placed in a cooler until disinfestation can take place. The leaves, including most if not all of the petiole, are removed, and each stem is cut into one or two nodal pieces.

The disinfestation or cleanup process usually involves several cold soapy water washes, then soaking the shoots in a 10 to 20% commercial bleach (sodium hypochlorite) solution along with a surfactant such as Tween 20 for 10 to 20 minutes, followed by a series of sterile water rinses. Damaged tissue is trimmed away using aseptic techniques. The treated cuttings are then placed individually in test tubes filled with a nutrient gel (medium) composed of a "best guess" formula or range of formulae that may allow the buds to grow. The tubes are usually incubated in a culture room under 16 hours of light at 25°C.

Visual screening takes place every 3 to 5 days, and the dead or dirty (visibly contaminated) cultures are discarded. Visibly clean, viable buds are transferred individually to fresh medium as new growth appears, usually 2 to 6 weeks after disinfestation. A portion of the stem piece is placed in a rich liquid- or agar-based medium to detect bacterial contaminants that may be invisible to the naked eye. This contaminant screening process (also called indexing; see Chapter 28) allows us to discard any additional unseen dirty material from the pool of viable shoots. This procedure may be repeated periodically; unfortunately, however, even this procedure does not catch everything, and growing clean stock is a continuous struggle.

Culture initiation can be more of an art than a science. Often the concentration of bleach or the time of disinfestation must be adjusted based on the woodiness, hairiness, and source of the stock tissue. Several variations may be tried if there is an abundance of source material, and the process may be fine-tuned based on previous efforts and the time of year. If materials are limited, the best guess is tried and hopefully will succeed.

Timing

The best success in getting plants into culture is typically realized by taking the first flushes of vegetative growth in the early spring. A very successful second option is to force dormant buds from sticks in the winter after the full chilling requirement has been satisfied. These sticks are placed in a "forcing solution" of 30 g/l Floralife Floral Preservative (Floralife, Inc., 120 Tower Drive, Burr Ridge, IL 60521), incubated at approximately 20°C with a 10 to 12-hour photoperiod (typical office conditions). Floral preservatives generally have sugar, an antimicrobial agent, and an acidifier to prolong bloom life. The dormant buds of many plant species will begin to sprout and grow under such conditions. Flower buds are removed and discarded when they appear to allow better vegetative growth. When these soft shoots reach the length of 2 to 4 cm, they are easily removed from the stem and processed. They tend to be extremely clean, so that surface disinfestation and establishment in culture are usually easier to achieve.

One theory explaining why such early growth seems to go into culture so well is that endogenous plant growth regulators (PGRs) and growth factors may have primed the plants to spring into a rapid growth phase (inhibitory factors are at a low ebb). Fortunately, this new growth is often very "clean" — it has not had time to acquire a load of contaminating organisms. Whatever the reason, early spring growth is without question the best starting material. Starting early in the growing season also gives more time to generate numbers and bring more cuttings into culture if needed.

Late summer or outdoor field-grown stock pose particular problems from increased contamination and a lack of turgidity, due to heat stress, which causes plant tissues to be less tolerant of bleach disinfestation. Success rates for establishing material in Stage I can range from 0 to 95% due to all these factors and many more we do not yet comprehend. The ultimate goal of Stage I is to get the plant material clean and actively growing in culture. Once this is achieved, Stage II becomes the next challenge.

STAGE II: SHOOT PROLIFERATION

The primary method of increasing woody plant micropropagation is axillary shoot proliferation, although adventitious shoot proliferation may occur with some kinds of plants, such as species of the family Ericaceae. Axillary bud proliferation is preferred (Chapter 11) in order to avoid the potential of off-types (somaclonal variation; see Chapter 26) from occurring. Genetic mutation can occur by any vegetative propagation method, and if the characteristic is based on a relatively unstable chimera, as some variegations are, micropropagation can enhance reversion. Micropropagators must continuously monitor the quality of plants they produce to ensure trueness. In our 20 years of experience of micropropagating trees, genetic mutation has not been even a minor problem. Useful strategies we employ to help reduce the potential of off-types are limiting the number of subcultures or the length of time in culture, keeping PGR levels in the culture medium as low as possible, and always going back to the original stock source to start new lines.

Once a plant is clean and actively growing in culture, the goal of Stage II is to induce all (or at least some) of the axillary buds along the new, young, aseptically growing shoot to sprout and grow. In other words, we want the shoot to branch. Every 2 to 8 weeks (culture cycles vary depending upon the type of plant), we can subculture or harvest the new branches, "replant" or transfer them to fresh medium, and wait another 2 to 8 weeks until each of these new branches sprouts its own set of branches. There are rarely any roots involved during Stage II.

With woody plant micropropagation, there are two main methods of achieving multiplication of new axillary buds, depending on the type of plant material involved. The first involves promoting many branches on a single stem piece, resulting in a clump of new shoots. Species of apple and cherry (*Prunus*) would be two examples of plants best micropropagated by means of the clump method. During subculturing, the taller shoots are removed, and are either replanted in fresh multiplication medium to produce their own clumps, or planted in a rooting medium or shipped out as microcuttings to be rooted in the greenhouse. The basal clump mass is left intact and replanted to yield another crop of branches. The clumps get thicker and thicker with new branches and can be used as miniature stock blocks. Eventually the clumps decline in quality and quantity, usually after two to five culture cycles, and must be discarded and replaced.

The second form of axillary shoot proliferation can be described as the single-node method. Plants exhibiting strong apical dominance, such as lilac (*Syringa*) and maple (*Acer*) species, are best micropropagated by means of the single-node method. A single shoot grows up straight and tall, producing 2 to 10 sets of leaves. One or two node sections are separated and replanted, whereby each of these runs straight up again and is ready to be chopped into nodal sections again. During any subculture, the nice terminals or tips can be harvested and planted into rooting medium or shipped out as microcuttings to be rooted in the greenhouse.

MEDIA FORMULAE

The actual media formulae used during Stage I and II often are the same. There are three or four basic formulae that are adapted and modified as necessary, sometimes on a cultivar-by-cultivar basis. The proper ratios of the inorganic components in a formula actually are the main key to developing a successful protocol for a given plant. PGRs are also a key factor, but secondary to proper nutrition. It is not uncommon for a commercial laboratory to have more than 50 modifications of basic media on file for multiplication alone and another 40 or so Stage III modifications. When a new plant is brought in, one takes an educated guess as to what might work based on past experience and literature searches, but usually modifications must be made by a systematic trial and error method of marching through the various components. In general, Stage I and II media usually have higher levels of salts, sugars, and cytokinins. Stage III media generally have lower levels of salts and sometimes lower levels of sugars, and usually lack cytokinins, but include high levels of auxins.

STAGE III: ROOTING

When sufficient numbers of healthy shoots have been generated by one of the proliferation processes described above, healthy active terminal buds are selected either for *in vitro* rooting or for shipping directly to customers as microcuttings for rooting and acclimatization in the greenhouse. The decision whether or not to root *in vitro* is quite often made by the customer, based on several factors. *In vitro* rooting adds about 30% to the final cost of the product, and some crops, such as birch (*Betula*) and lilac species, root so well in the greenhouse as microcuttings that it simply does not make sense to root them *in vitro*. Other crops such as *Malus, Pyrus, Prunus*, and *Acer* are much easier to acclimatize as *in vitro* rooted plantlets. Perhaps the most important factor of all for the grower, other than survival, is that *in vitro* rooted plantlets require 2 to 5 weeks less time in the greenhouse to reach a size suitable for planting in the field. Many growers have limited greenhouse space and do not want to have to baby-sit tender microcuttings for such a long time. A grower may be able to process two, three, or four crops through his greenhouse in the same time it takes to raise one crop from microcuttings. Often, it pencils out as more economical to buy rooted plantlets.

STAGE IV: GREENHOUSE ACCLIMATIZATION

Fresh Stage II microcuttings (without roots) or Stage III *in vitro* rooted plantlets are both planted in any of several well-drained soilless potting mixes in the greenhouse under high humidity. Various types of fogs, tents, domes, and mist systems are used to wean the tender shoots into the real world by gradually reducing the humidity. The weaning process takes from 3 days to 6 weeks, depending on the plants, the weather, and if plants were rooted *in vitro*. Fertilization is not recommended until the plantlets become established and begin to grow. Some crops such as *Betula* can become stunted if over-fertilized.

PRODUCTION SCHEDULING

When preparing an actual production schedule for any given plant, many factors play a role. How many are ordered and in what form (microcuttings or rooted plantlets)? When are they wanted? Can the order be prepared ahead of time (does it require or tolerate cold storage)? How does it multiply (nodal or clump)? What are the multiplication rates during buildup? What is the number of rootable or harvestable microcuttings (yield), and does it change over time? What rooting percentage can be expected? Can smaller batches be produced and stored or must it be produced all at once? How much time is required between cycles (during buildup and during harvest cycles)? What are the estimated labor requirements for each cycle? Once these factors are determined, one works backward from the delivery date, calculating the number of cultures required at each of the various cycles along the way. Information on when each cycle must take place, what activities must be done at each step, and the estimated labor required are all plotted on a production planning calendar. This planning process must be accomplished for each cultivar on an individual basis, since multiplication rates, yields, number ordered, and dates ordered will vary from plant to plant. Critical labor peaks must also be smoothed out as well as possible ahead of time. Simply put, not everything can come out at once.

One of the most frustrating factors in planning a production schedule is the very dynamic and very unpredictable nature of the plant material itself. Plants are living systems and we simply do not control or even understand all the factors involved in *in vitro* propagation. Multiplication rates and harvestable yields, for example, can only be "guesstimated." Any plan has to be continuously updated and revised.

SEASONALITY

While a laboratory can produce material on a year-round basis, anyone working on temperate deciduous crops knows that there is a season for everything. Plants can be fooled into thinking it is summertime in culture by continuously being subjected to long (16 hour) days in the culture room, but as fall approaches, the rate of growth in the greenhouse slows considerably. To take maximum advantage of the growing season, customers want to get their product in the early spring through early summer. Use of cold storage to store crops ahead of time allows laboratory production of some plant material to take place in the winter, but much must come out fresh, making labor needs seasonal as well.

REASONS WHY MICROPROPAGATION IS USED

While micropropagated material represents only a tiny percentage of the total number of plants produced in the United States today, there is no doubt that many of the largest and smallest nurseries in the U.S. view micropropagation as an essential tool that helps them maintain their competitive edge by growing better plants more efficiently. The remainder of this chapter will be devoted to a discussion of some of the advantages that *in vitro* micropropagation offers over conventional production systems.

NEW INTRODUCTIONS

Perhaps the most obvious use for micropropagation is to get a jump-start on growing the newest and hottest items quickly. Micropropagation cuts 3 to 10 years off the time it takes to bulk up new selections and get them to market. For example, micropropagation was used by one grower to establish layer beds of a new apple understock. He sold more than one million rootstocks in the same amount of time his competition had bulked up to only a few thousands using conventional propagation methods. The market is always looking for something new and exciting. Nurseries on the forefront of introducing new plant material build their reputation as leaders in the industry, causing customers to come back year after year to find out what else is new.

In some cases, growers use micropropagated plant material to establish mother blocks and to fill in production shortages while mother blocks are too young to be productive. Conventional propagation methods such as hardwood or softwood cuttings, layer beds, or budding and grafting may take over long-term production needs once enough wood becomes available. In many other cases, micropropagation remains the method of choice for a variety of reasons.

RAPID RESPONSE TO MARKET DEMAND

Large mother blocks or scion orchards are time-consuming and expensive to establish and maintain. It is often difficult for growers to adjust quickly to the rise and fall in popularity of a given plant. With micropropagation, the stock block is maintained in a 10×10 foot cold storage unit. If a customer gets a call for an additional 10,000 liners of a particular blueberry, for example, they can simply call up and ask when is the earliest they can take delivery on the additional microcuttings. They then add the time they need for the greenhouse growing and call their customer back with a delivery date.

CLEAN PLANTS

Micropropagation is inherently a cleaner system for producing plants, compared to traditional production methods. Because the plants are grown in culture, diseases are not transmitted from the field into the greenhouse and on to subsequent generations. A single disease-free mother plant can theoretically produce unlimited disease-free daughter plants without the possibility of reinfection. Conversely, individual mother plants in a traditional virus-free cutting block must be tested again every year in order to maintain and ensure virus-free status (see Chapter 28). Testing fees add significantly to the expense of maintaining large mother blocks in the field.

EASE OF PROPAGATION

Bud incompatibility on budded or grafted stock and poor rooting percentages with softwood or hardwood cuttings make micropropagation the method of choice, or the only option, for many difficult-to-propagate plants such as *Syringa* and redbud (*Cercis*). Some red maple (*A. rubrum* L.) cultivars, such as "Karpick" and "Bowhall," are absolutely impossible to root from cuttings. Unreliable seedling availability and poor or unpredictable bud stands on *Betula*, *Tilia*, and *Morus* are problems growers are able to avoid by planting micropropagated material. Growing plants on their own roots offers major advantages for plants, such as contorted Filbert (*Corylus avellana* 'contorta'), where suckering of understocks can be a major problem.

Sometimes micropropagated material provides the grower with a nucleus of juvenile material from which additional cuttings can be rooted more easily. Each year or two, the customer starts over with a fresh batch of starter material from the laboratory.

SPEED

Research done at various universities indicates that it is possible to grow plants much more quickly to size than is traditionally seen in nurseries today, regardless of how the plants are propagated. Such rapid growth requires optimization of all growing conditions, including fertilizer, light, and temperature. However, amazing results can be achieved with even modest adjustments of growing practices. Several field growers are now producing well-formed, small-branched trees of *Prunus* "Kwanzan," *M. alba* "Chaparral" (mulberry) and others in one year instead of two. Blueberry plant production can be dramatically speeded up. Use of micropropagated cherry understock yields increased vigor and earlier fruitset (and earlier payback) for the orchardist.

BETTER BRANCHING

Because the internode length is greatly reduced on micropropagated plants, there is generally more opportunity to develop a more full head on the growing plant. Indeed, this is also one of the reasons why survival is often greater. If something happens to destroy the terminal bud (for example, damage caused by freezing, hungry rabbits, or poor pruning), there are other buds below to choose from. One grower accidentally sprayed a young block of *A. rubrum* liners twice with Surflan™, an herbicide. The stems of the young micropropagated liners were girdled right at soil level. All was not lost, for the grower dug the soil out from around the base of the plants and removed the damaged tops, and the buds below the girdling all pushed out again. The grower lost some height on his crop (about 1 to 2 feet), but shorter plants are better than no plants at all. Having more buds to choose from also allows a grower to cut back closer to the ground, producing straighter trees.

Better branching is a real advantage when it comes to growing well-formed shrubs. *Hydrangea quercifolia* "Snow Queen" (PP4458) produced from rooted cuttings tends not to throw many branches at an early age, whereas micropropagated plants are easily developed into a bushy habit with routine pinching. The increased branching of blueberry plants is seen as a great advantage by some growers, whereas others prefer the more traditional vase shape formed from rooted cuttings.

GREATER SURVIVAL AND UNIFORMITY

While cultural practices play an important role in the ultimate performance of any block of plant material, the two key attributes most often given in describing micropropagated material are superior survival rates and much greater uniformity. Better survival is explained, in part, by heavier root systems and more buds to choose from on the top. The fact that the root-to-shoot ratio is more balanced from the very beginning, and the fact that the small plantlets are highly uniform right from the start, help to explain why subsequent growth is more reliable and consistent.

CONCLUSION

As the use of micropropagation becomes increasingly varied and widespread, growers will find new and ingenious ways to take advantage of its power. Even so, growers must weigh the pros and cons of this relatively new technique against the old methods and determine for themselves, on a case-by-case basis, whether it is worth the trouble to change. It must help them solve problems, find new markets, or save time or money. The successful micropropagator will be one who fills these needs and many more.

28 Indexing for Plant Pathogens

Alan C. Cassells and Barbara M. Doyle

CHAPTER 28 CONCEPTS

- International standards have been established for the indexing of plants for certified stock production.

- Approved methods include the inoculation of indicator plant species and the enzyme-linked immunosorbent assay (ELISA).

- Methods not yet generally approved by certification authorities include PCR-based methods, such as RAPD-PCR and real-time PCR (using TaqMan, FRET, and molecular beacons).

- Classical methods, including PCR, focused mainly on the detection of individual pathogens (monoplex assays), whereas recent advances in PCR-based assays allow the detection of multiple pathogens in one assay (multiplex assay).

- Routine use of PRC-based methods in pathogen detection has yet to be established.

INTRODUCTION

Several factors account for the importance of producing plants of certified health status from tissue culture. First, while most pathogens are eliminated in seed propagation, they are transmitted at high frequency in vegetative propagation. This also applies to micropropagation, where the starting material is infected and where pathogens are not eliminated — e.g., by meristem culture at the establishment of the aseptic culture (Stage I; see Chapter 11) (George, 1993). Second, vegetative propagation is relatively slow and involves mature stock plants, so there is the likelihood that pathogen symptoms will be expressed. Pathogen symptoms, however, may not be expressed in tissues *in vitro*, and there is also the risk that uncultivable pathogens, especially fastidious bacteria, and inter- and intracellular endophytes may go undetected unless the tissues are specifically indexed for the latter. Third, there is the risk that the infected starting material may be clonally propagated and released locally or exported, where the contaminating microorganism may become pathogenic in the crop, establish a reservoir of a pathogen that will initiate infection in other crops, or, in an extreme case, introduce a new pathogen into a quarantine zone.

HEALTH CERTIFICATION OF PLANTS

Standards have been established internationally for the production of health certified planting material and for the granting of phytosanitary certificates for international shipments of plant material. Guidelines for the indexing of crops have been published by international organizations, such as the Food and Agriculture Organization of the United Nations (http://www.fao.org), and by regional organizations, such as the European and Mediterranean Plant Protection Organization (http://www.eppo.org) and the

North American Plant Protection Organization (http://www.nappo.org) for most major crops. In general, these guidelines can be extended to related minor crops or to the same pathogens in different crops.

The conventional strategy for the production of health-certified stock is based on the selection of symptomless individuals where available. These are grown in isolation greenhouses that are protected against pathogens and pathogen vectors. The individuals are indexed using approved methods (see later) and, if pathogen-free, may be used as stock plants for micropropagation. There is usually a limit on the number of subcultures or a time limit after which new stock must be used (Cassells, 1997). The plants resulting from micropropagation, that is, the established progeny, or those resulting from further cycles of vegetative propagation are again pathogen indexed and certified. Where pathogens are detected in the stock material, the plants may be subjected to *in vivo* pathogen elimination strategies such as thermotherapy and indexed again before being introduced into culture. Alternatively, meristem culture may be used to eliminate pathogens, but before these cultures are clonally propagated, samples should be grown in the greenhouse and the progeny tested in the same way as the parental stock (Cassells, 2000; Figure 28.1). Plants indexed in this way, using approved diagnostic techniques, may then be certified as free of the specific pathogen for which they were indexed. To satisfy the requirements of an importing country, such indexing may need to be carried out by an official laboratory.

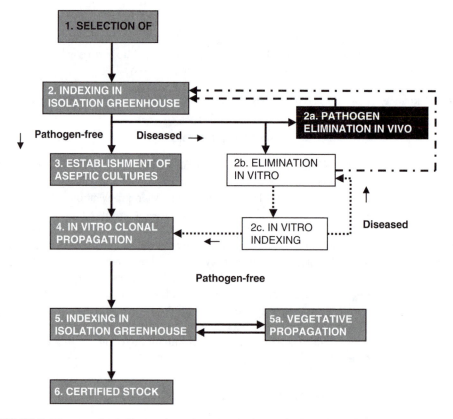

FIGURE 28.1 The grey-shaded boxes show the stages in the production of certified stock involving a tissue culture step (4). This scheme is based on the establishment of the cultures from a pathogen-free stock plant or plants. Where indexing shows the stock plant to be infected (2), there are the options of pathogen elimination *in vivo* (2a) or *in vitro* (2b). Following the use of *in vitro* methods to eliminate the pathogens, the progeny plants should be reestablished in the greenhouse and indexed there using approved methods. In many cases micropropagators index the *in vitro* cultures (2c) and, if the results are negative, use this *in vitro* stock for micropropagation. For a discussion of these strategies see Cassells (1997) and O'Herlihy and Cassells (2003).

PATHOGEN INDEXING *IN VIVO*

Certification authorities are conservative by nature and tend to rely on proven methods. These include symptomatology and biological testing — that is, mechanical, graft, and dodder or vector transmission to an indicator species, one that gives a characteristic response to the pathogen/pathogen strain. Microscopic examination is extensively used in fungal identification and for partial characterization of cultivable bacterial pathogens. The latter are also identified on selective growth media. Inoculation of indicator species, or a serological test, such as enzyme-linked immunosorbent assay (ELISA), is used mainly for viruses. There is also a range of more specialized techniques that are used in research and in commercial practice, but are not generally accepted by certification authorities. These include electron microscopy, the use of biochemical test kits, fatty acid profiling, matrix-assisted laser desorption ionization time-of-flight (MALDI-TOF) mass spectrometry, and the use of DNA probes and polymerase chain reaction (PCR), among others (Table 28.1). Historically, the different pathogen groups, fungi, bacteria and bacteria-like, and viral and virus-like organisms had indexing methods reflecting the characteristics of the different pathogen types. For classical methods of bacterial indexing, see Lelliott and Stead (1987); for viral indexing see Dijkstra and de Jager (1998); and for fungal disease diagnosis see Fox (1993). Nucleic-acid-based methods of indexing have been critically reviewed by Karp et al. (1998).

DIAGNOSTIC METHODS

Disease indexing has been revolutionized by DNA-based diagnostics (genomic probes), and increasingly proteomic analysis is being used. These methods are universal; they can be applied to all types of pathogens and offer the prospect of highly sensitive diagnostics that can be used to detect, in real time, many pathogens simultaneously (multiplex assays) in the same sample. As yet, genomic- and proteomic-based methods are not yet used extensively in official health certification schemes. In this chapter, both traditional and emerging techniques will be discussed.

Assays for pathogen detection should be "sensitive, specific, free of both false positive and false negative results, rapid and cost effective" (Candresse et al., 2000). The cost of any technique, especially expensive molecular techniques, has to be a main concern for any laboratory, and in particular for academic laboratories that may be working on limited budgets. According to the winner of the 2002 Nobel Prize for medicine, Sydney Brenner, the most important "-omics" in the biotechnology sector today are "economics".

With the development of PCR by Kary Mullis in 1984 (Schaad and Frederick, 2002), the area of plant pathogen detection was revolutionized. The sensitivity of this technique (which is reliant on the design of appropriate PCR primers), coupled with its amplification power and speed, makes it the method of choice in many cases of detection. Its ability to discriminate between two sequences that differ only by one nucleotide is possible when the correct primers are used, which is an important facet of pathogen identification as opposed to mere detection (Schaad and Frederick, 2002). Random amplified polymorphic DNA (RAPD) primers (10 to 12 nucleotides in length) are commercially available primers that can be used for the amplification, at random, of complementary DNA sequences in any genome. Here, no prior knowledge of the genome is needed. During a typical RAPD-PCR amplification process, a number of bands may be generated. Any unique bands can be removed from the gel, cloned, and sequenced, and new primers designed to match these unique sequences. The presence of bands of the same molecular weight in an electrophoresis gel (from RAPD analysis, for instance) is not indicative, however, of the presence of homologous sequences; hence, band homology needs to be confirmed using Southern analysis (Schaad and Frederick, 2002). This is one of the reasons why the technique is still not routine in many diagnostic laboratories.

Real-time quantitative PCR (Q-PCR), in which progress is being driven by the vast amounts of DNA sequence data being generated by genomic investigations for U.S., European, and Japanese

TABLE 28.1
Application of Indexing Methods to Pathogens*

Pathogen Type	Symptoms	Indicator Plants	Test Methods					
			Serological Diagnostics	Microscopy	Biochemical Test Kits	Selective Culture Media	Genomic Diagnostics	Proteomic Diagnostics
Virus	Widely used	Widely used to confirm pathogenicity and identify strains	ELISA is standard method	Limited use of electron microscopy	Not applicable	Not applicable	Detection of variable coat genes/conserved sequences	Detection of viral coat proteins
Viroid	Widely used	Widely used to confirm pathogenicity and identify strains	Not applicable	Not used	Not applicable	Not applicable	Standard method	Not applicable
Bacterium**	Widely used	Widely used to confirm pathogenicity	Not commonly used	Widely used for Gram and spore staining, etc.	Not commonly used for plant pathogens	Historically used in plant pathology	Detection of rDNA	Detection of bacterial biomarkers
Fungus	Widely used	Widely used to confirm pathogenicity	Not commonly used	Widely used for morphological identification	Not applicable	Widely used	Detection of rDNA	Detection of fungal biomarkers

* Genomics diagnostics are increasingly being used, while proteomic methods have been investigated more recently.

** Fatty acid profile analysis is used for the detection of bacterial plant pathogens.

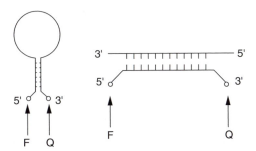

FIGURE 28.2 The structure of molecular beacons for the simultaneous detection of two plant viruses, fluorophore (F) and quencher (Q) (adapted from Klerks et al., 2001). At the 5′ end of the molecule a fluorophore is attached and at the 3′ end of the stem–loop structure a quencher molecule is attached. When the molecular beacon forms the stem–loop structure and the quencher is in close contact with the fluorophore, the fluorescence is reduced. During the amplification process, when the primer hybridizes to the target sequence, the quencher is separated from the fluorophore and the fluorescence increases, allowing for the simultaneous amplification and detection of the particular plant pathogens. Different color fluorophores can be combined in a multiplex assay.

databases (see the National Center for Biotechnology Information [http://www.ncbi.nlm.nih.gov/], the European Bioinformatics Institute [http://www.ebi.ac.uk/embl/], and the DNA Data Bank of Japan [http://www.ddbj.nig.ac.jp/], respectively), has many advantages over the more "classical" PCR methods. It can be used for the detection of bacterial pathogens where direct PCR can be carried out and where no DNA extraction is necessary prior to the amplification step, and for viral and fungal plant pathogens where RNA and DNA have to be extracted before PCR amplification (Schaad and Frederick, 2002). Real-time Q-PCR, unlike classical PCR, is based on the detection and quantification of a fluorescent reporter. An amplification plot can be produced whereby data from a fluorescence signal is plotted against the PCR cycle number, allowing for the quantification of products generated during a particular amplification cycle in the PCR process. It can be used for gene expression, pathogen detection and quantification, and the screening of mutants. Some PCR reaction inhibitors may interfere with the accumulation of product (some PCR reactions may produce more product than others), and therefore endpoint quantification of PCR product can be unreliable. With real-time Q-PCR, the particular cycle at which the reaction enters the log linear amplification phase is proportional to the amount of initial template; therefore, the concentration of an unknown sample can be calculated with reference to a known standard (Ball et al., 2003). Currently there are three fluorescence methods available for the detection of real-time Q-PCR amplification products: TaqMan™, fluorescent resonance energy transfer (FRET), and molecular beacons (for review, see Schaad and Frederick, 2002; Figure 28.2). With real-time Q-PCR, unlike classical PCR, there is no need to do Southern blotting to determine the nature of the amplification product, or amplicon; and the method can be used for the detection of multiple pathogens simultaneously (multiplexing).

MULTIPLEXING

Most molecular, serological, and biochemical detection systems for plant pathogens are capable of detecting only one pathogen per assay even though more than one pathogen may be present in the plant at the time of sampling. Multiplex PCR allows for the simultaneous detection of a number of pathogens in one assay, thus saving time and money. There are a number of published reverse transcriptase PCR (RT-PCR) protocols for the simultaneous detection of pathogens. One alternative to multiplex RT-PCR is isothermal multiplex AmpliDet™ RNA. An isothermal multiplex AmpliDet RNA detection system was developed for the simultaneous detection of PLRV and PVY in potato tubers (Klerks et al., 2001). The inclusion of molecular beacons into the nucleic-acid-sequence-based amplification (NASBA) system allowed for the simultaneous detection and amplification of RNA molecules. This was a closed system (a one-tube multiplex assay) and therefore had a reduced risk of carryover contamination. An advantage of the latter over the multiplex RT-PCR is that no gel electrophoretic analysis was necessary to detect the amplicons (Figure 23.2).

DETECTION VERSUS IDENTIFICATION

It is important to make the distinction between detection and the identification of pathogens. Detection of pathogens can be based on the presence of symptoms, the examination of specimens with a hand lens or light microscope to look for characteristic patterns, and knowledge of the crop. This approach can be extended to tissue culture, where cultivable contaminants such as bacteria and fungi can be seen growing in the cultures. Such cultures should be discarded on the premise that any contamination may comprise product quality. However, absence of visible symptoms per se is not proof that cultures are free of contaminants, and indexing of tissue explants should be carried out using selective media in support of visual examination of the cultures. As mentioned above, intracellular microorganisms may not express symptoms and may not be cultivable. They require specific tests (see below).

Pathogen identification is more complicated, in that many of the tests may not be capable of detecting the specific genes that determine pathogenicity, but may detect genes or gene products that are common to pathogenic and nonpathogenic isolates. With cultivable pathogens, the fallback position is to follow up the serological, biochemical, or molecular test with Koch's postulates, which involve obtaining a pure isolate of the pathogen, inoculating a susceptible host, establishing the disease, isolating the microorganism again, and confirming that the microorganism is the same as the one originally isolated (Lelliot and Stead, 1987). In the case of uncultivable microorganisms, direct inoculation of a susceptible host and observation of symptom development is carried out. DNA probes have the potential to detect both highly conserved sequences and unique pathovar-specific sequences. This is useful in quarantine applications, where the potential to detect uncharacterized pathogens is essential, and in certification application, where material must be free of specified pathovars (Cassells, 2000).

DETECTION AND IDENTIFICATION OF VIRUSES AND VIROIDS

A comprehensive and updated list of viruses, their symptoms, their modes of transmission, and suitable indicator plants is maintained by the Association of Applied Biologists (http://www.aab.org.uk). This list and disease compendia published by the American Phytopathological Society (http://www.apsnet.org) provide information on biological testing for plant viruses. Plant factors that are important include the tissue used, as virus distribution may be uneven in plant tissues *in vivo*. Also, there may be cyclical, seasonal, and developmental changes in titer (amount) (Hull, 2001). The second problem area is the transmissibility of the virus to the indicator plant. A third problem is that indicator plants have to be at the right stage of development in many cases to maintain sensitivity to inoculation. While some viruses are mechanically transmitted in plant sap to which a mild abrasive is added, in other cases it is necessary to prevent or eliminate interfering materials such as tannins. Many viruses cannot be transmitted mechanically, but require graft or dodder transmission. Finally, there is the problem of virus complexes (multiple infections of the same plant), which may involve the need to inoculate a range of host–nonhost combinations to separate the component viruses (Hull, 2001). In spite of these problems, indicator plants are still widely used, especially to confirm negative results from other tests. ELISA with confirmation of negative results by biological testing is the most widely used strategy in the virus indexing of officially certified crops (see http://www.eppo.org/).

ELISA is a widely used diagnostic method for viruses. There are a number of commercial companies that sell diagnostic kits for the major viral pathogens (e.g., Agdia [http://www.agdia.com], Bioreba [http://www.bioreba.com], and Leowe Biochemica [http://www.loewe-info.com]). Where these are not available, it is possible to raise antibodies to the pathogen and to construct ELISA kits (see Amersham Biosciences [http://www4.amershambiosciences.com]). The latter option does, however, require the purification or partial purification of the virus. This may be subcontracted to a university or to a state or commercial laboratory.

Electron microscopy, combined with the use of a virus or virus-strain-specific antibody, is a powerful tool in viral diagnosis. However, it is restricted in use by the capital costs of the equipment and by the requirement for highly trained technical support, especially where cell *in situ* detection is involved. Methods for virus detection are compared in Table 28.1.

As for viruses, symptomatology and transmission to indicator species was traditionally used for viroid indexing. Viroids lack a protein coat and so have been detected based on the structure of their double-stranded, nuclease-resistant genomes by means of polyacrylamide gel electrophoresis (Hull, 2001). PCR-based amplification of plant viruses using specific primer pairs has been described in numerous reports in the literature. The first report of the application of PCR to both virus and viroid detection in plants was in 1990. PCR-based tests have a much greater sensitivity (up to a few thousand times more sensitive) than the more traditional serological tests. Given that PCR inhibitors may be present in some plant samples, attention needs to be paid to pre-PCR sample preparation, which may include the use of dangerous and/or expensive chemicals. Immunomagnetic separation and immunocapture techniques use antibodies to separate the target organism from PCR inhibitors prior to the PCR amplification step. Multiplex RT-PCR assays have been described for the detection of a number of viruses. In economic terms and in terms of speed and efficiency, molecular-based assays may replace the currently used ELISA tests or the use of woody indicators for the detection of plant pathogens.

The detection of viruses in transgenic virus-resistant plants presents special problems. Since the first demonstration of transgenic virus resistance in plants in 1986, hundreds of plant cultivars have been transformed with viral sequences. The use of viral coat protein genes is the most common strategy, but a number of other viral genes have also been used; for example, replicase and protease genes and defective movement proteins (Aaziz and Tepfer, 1999). Nucleic acid hybridization and ELISA methods for the detection of viruses could be hampered by the presence of transgenic viral sequences. Also, what is not known to date is how these transgenic viral sequences and their protein products may interfere with the detection of other viruses present in the plant tissue which were acquired either prior to or after transformation. Interactions can occur between an incoming virus and the inserted viral transgene in three recognized ways: by hetero-encapsidation, by recombination, and by synergism (Aaziz and Tepfer, 1999).

Methods for proteomic analysis — e.g., using MALDI-TOF mass spectrometry — are also being evaluated for their potential to detect viral proteins in infected host tissue based on the detection of viral proteins in tissue extracts (Tan et al., 2000).

A problem common to all viral detection methods is the titer of virus in the target tissues, which may be below the threshold of detection for ELISA and other serological tests. This is theoretically less of a problem when using PCR or inoculation of indicator species, where virus replication and amplification can occur.

DETECTION AND IDENTIFICATION OF BACTERIAL PLANT PATHOGENS

The bacteria associated with plants may be classified as either epiphytes or endophytes, and can be either pathogenic or nonpathogenic. Classical methods for the detection of cultivable plant bacterial pathogens traditionally were based on separation into Gram positive and Gram negative groups, followed by keying out by means of culture responses and characteristics on elective and partially selective media (Lelliot and Stead, 1987). In recent years, the approach has been broadened by the use of biochemical detection kits (e.g., API kits; see bioMérieux [http://www.biomerieux.com]) following Gram staining and preliminary biochemical tests to identify the appropriate kit. A criticism of biochemical test kits is that identification relies on comparison with organisms in the kit database. If the organism is not in the database, the results may give little information as to its identity.

Identification of bacteria by fatty acid profile analysis, originally based on high-performance liquid chromatography (HPLC) analysis (Stead et al., 2000), has the advantage that, while it is also database-dependent, some fatty acids are taxonomic markers and thus allow at least partial identification of many plant-associated bacteria. Inability to cultivate some bacteria, particularly some mollicutes, has been a problem in the indexing of bacterial plant pathogens. In these cases, *in situ* staining with 4',6-diamidino-2-phenyl indole (DAPI; Clark, 1992) and examination of the stained sections with a fluorescent or confocal microscopy has been used. For cultivable bacteria, virulence testing was based on inoculation of indicator plants. In the case of fastidious and uncultivable bacteria, this was achieved by graft transmission to the indicator species. This biological indexing is still used in support of modern methods of detection.

According to Stead et al. (2000), the preferred DNA-based method for characterizing and identifying bacteria associated with plants would include the production of PCR profiles using repetitive sequences, whereas for detection of bacteria, novel PCR-based assays using, for example, fluorigenic probes would be the method of choice. Depending on the requirements of the test — e.g., if differentiation of the bacteria at the infraspecific level or strain level is required — fingerprinting methods are probably the best, and a number of tests for differentiating and identifying bacteria have been reviewed under the headings of cost, speed, and accuracy (Stead et al., 2000). The methods based on nucleic acids would include RNA- and DNA-based methods. Perhaps the most widely used RNA-based method is amplified ribosomal DNA restriction (ARDRA). This involves the amplification of the 16s rRNA using PCR, followed by restriction analysis (Stead et al., 2000).

For the DNA-based methods wherein a higher discriminatory power is needed (e.g., pathovars and biovars), RAPD-PCR has proven fairly valuable for comparing a small numbers of strains, but problems may still be encountered in terms of reproducibility (Stead et al., 2000). Repetitive sequence PCR profiles — for example, repetitive extragenic palindromic (REP), enterobacterial repetitive intergenic consensus (ERIC) sequence, and BOX-PCR — have proven effective, on the other hand, in discriminating between closely related taxa, with the latter being the most reproducible. Amplified fragment length polymorphism (AFLP), another fingerprinting technique, uses restriction enzymes and PCR and has been shown to be valuable for predicting the genus and species and also for the differentiation of bacterial strains. Bacterial identification relies on the isolation of pure cultures, whereas detection does not. The molecular-based tests for the detection of bacteria include hybridization techniques such as fluorescent *in situ* hybridization (FISH) and DNA amplification techniques such as real-time Q-PCR. Q-PCR has been used successfully in multiplex analyses to detect *Ralstonia solanacearum*, *Clavibacter michiganensis* subsp. *sepedonicus*, and *Agrobacterium* species. However, PCR has not replaced older methods such as plating or serological tests for the routine testing for bacteria associated with plants (Stead et al., 2000). There are a number of different strategies available for designing PCR primers for bacterial detection and identification, and primers that are complementary to pathogenicity genes may have the required level of specificity for pathovar detection.

Proteomic techniques, such as MALDI-TOF, have been widely used for the identification of human bacterial pathogens (Lay, 2001). These methods can be used with intact bacterial cells and can generate results in minutes. MALDI can also be applied to disrupted cells that generate a greater number of potential biomarkers. The power of resolution of the MALDI data can be increased with the use of pattern recognition software.

DETECTION AND IDENTIFICATION OF FUNGAL PLANT PATHOGENS

Traditionally, fungi have been identified on the basis of morphological attributes, such as the characteristics of the spore and the spore-bearing structure or fruiting body associated with the symptoms. In some cases, sporulation may have to be induced by incubation of diseased tissues (Fox, 1993).

Haugland et al. (2002) examined the use of real-time PCR for the quantitative detection of fungi. For information on the preparation of DNA as a PCR template from fungal conidia, see Haugland et al. (2002). Direct PCR (where DNA is not extracted prior to the reaction) does not work well for most fungi. PCR-based detection methods provide sensitive means for fungal detection but require further tests to identify the amplicon, such as Southern blot hybridization. An immunoenzymatic method like PCR-ELISA, which combines efficiency and sensitivity with specificity, eliminates the need for both gel electrophoresis and Southern blot hybridizations and has the potential to detect both routine and ambiguous pathogens reliably. In this method, the PCR products are labeled with digoxygenin-11-UTP (DIG) during the amplification process. This labeled DNA is then hybridized to a "capture probe" that has been biotin-labeled and immobilized in the wells of a microtiter plate (coated with streptavidin). After hybridization, the DIG can be detected in an ELISA-like reaction. Using PCR-ELISA, Bailey et al. (2002) were able to identify species of *Pythium* and *Phytophthora* using differences in the ribosomal internally transcribed spacer 1 region. Data from sequence alignment allowed for the preparation of a unique capture probe, permitting the identification of the two species. Proteomic methods as applied to bacterial identification also have potential application to fungal identification.

HEALTH CERTIFICATION OF PLANTS *IN VITRO*

The application of the above indexing methods to tissue cultures is not currently accepted by most certification authorities. The reason for this is that little is known about the distribution and amount of the respective pathogens in tissue *in vitro*. The Netherlands General Inspection Service for Floriculture and Arboriculture (NAKB) has published a scheme for the certification of *in vitro* material that involves testing samples of production in the greenhouse while multiplication continues (Van der Linde, 2000).

The client (i.e., the buyer) determines the extent of pathogen indexing that is carried out on *in vitro* cultures. Usually this involves indexing for specific pathogens that are problematic for the customer — e.g., *Xanthomonas campestris* pv. *pelargonii* in pelargonium. The indexing methods used are either specified by the client or left to the discretion of the micropropagator.

It is important to remember that pathogen-indexing procedures used in the certification of *in vivo* propagules have been validated with plant material of a defined growth stage and grown under natural environmental conditions. The implication is that only stock plants or tissue-culture progeny plants can be reliably certified. If the donor material is pathogen-infected, then methods of virus elimination, including meristem culture, may be used to eliminate the pathogen. However, for certification, progeny material from *in vitro* culture must be grown under environmental conditions to the same growth state at which *in vivo* testing is carried out. This may require one or more years of *post vitrum* plant culture and is clearly a major barrier to the use of micropropagation for the multiplication of pathogen-free material (Cassells, 1997).

In the multiplication of pathogen-free material in the environment, there is always the risk of reinfection with the pathogens. This is much less likely in *in vitro* mutiplication as there is less risk of pathogen or pathogen-vector contamination of cultures. However, the potential of maintaining pathogen-free material indefinitely *in vitro* has to be offset against the risks of genetic instability in long-term culture (see Chapter 26). For this latter reason most certification authorities require that new cultures be set up at intervals of 12 or 18 months, or after a defined number of culture cycles (Cassells, 1997).

Sometimes the term "axenic," meaning free of all pathogens and microbial contaminants, is used to describe the health status of tissue culture. In practice, it is very difficult to prove that tissues *in vitro* are axenic even for crops (e.g., potato) for which the health status of the parent plant, the explant donor, can be determined with good reliability using current diagnostic tests. At best, plants can be described as being indexed for characterized strains of known pathogens of the crop. Even in major crops (e.g., strawberry), the causal agents of some diseases have not been identified. The situation is even less clear with minor crops (e.g., ornamentals), which, due to their high propagule price, are propagated competitively by means of micropropagation.

CONCLUSIONS

For the control of diseases affecting plants globally, the timely and accurate identification of the causal organisms is of utmost importance (Levesque, 2001). Plant pathogen identification historically relied on symptomatology and biological indexing, which can be very difficult and time consuming. By contrast, modern molecular-based tests are capable of generating accurate results very quickly and relatively easily, once staff training is provided, from both symptomatic and asymptomatic plant material. There have been two major revolutions in the detection of plant pathogens since the 1970s. The first breakthrough came about in the mid-1970s with the use of antibody-based detection methods, and the second breakthrough followed soon after in the late 1980s with the advent of DNA-based technologies such as PCR. A further breakthrough is the development of multiplexing, where more than one pathogen can be detected per assay. Developments in the area of genomics have, in part, contributed to the speed with which this last revolution has erupted. The continued success of this latter revolution relies on the generation of a "critical mass of high quality sequence data" (Levesque, 2001). A limiting factor with respect to the DNA-based detection of plant pathogens such as fungi is the lack of sequence data. The vast majority of fungal species in the world remain undescribed. An examination of the fungal sequences deposited in GenBank (a public database maintained at the National Center for Biotechnology Information, Bethesda, MD) in 2001 showed that less than 1% of the putative one and a half million species of fungi were represented (Levesque, 2001).

Single base pair differences can now be detected between alleles (single nucleotide polymorphisms [SNPs]). This technique has been used to detect single base pair differences in fungal pathogens. According to Levesque (2001) and Schaad and Frederick (2002), DNA-based assays are still not routinely used for plant pathogen detection. The current standard method for the detection of *Bacillus anthracis* and other human pathogens is a serological test ("dip stick") followed by a cycle of PCR amplification (Schaad and Frederick, 2002). For the accurate detection of plant pathogens using DNA-based tests, there is a reliance on the development and use of highly specific primers. The public databases can be searched for sequences that can be used for primer design and subsequently for the detection of specific pathogens. Problems may arise in assay sensitivity when primers are designed for the detection of single copy sequences (e.g., the *argK-tox* gene in *Pseudomonas syringae* pv. *phaseolicola* and the *celA* gene of *Clavibacter michiganensis* subsp. *michiganensis*) as opposed to multiple-copy gene sequences (e.g., 16s RNA), but even for the latter, problems may arise in terms of primer specificity (Schaad and Frederick, 2002). Multiple-copy sequences in plasmids are also possible targets for DNA-based detection methods. A note of caution is necessary here because plasmids can be exchanged between bacteria (during horizontal transfer), and as a consequence noting their presence cannot be relied on as a definitive DNA-based detection method.

Commercially available random primers (i.e., RAPDs) have also been used successfully for the detection of plant pathogens. After amplification, any unique bands generated during the PCR cycles can be identified using gel electrophoresis, then excised and sequenced. Specific primers can then be designed for these polymorphic sequences. Due to the sequence variation found in the internal transcribed spacer region in the ribosomal genes in bacteria, between the 16s and the 23s ribosomal genes, specific primers designed for these regions hold much promise for the identification of different bacterial species (Schaad and Frederick, 2002).

The large-scale application of these PCR-based tests for the detection of viruses and viroids in "real life" situations has yet to occur. The routine use of real-time PCR for the detection of plant pathogens is limited by the availability of primers and fluorescent probes, but this will change very quickly as new sequence data become available in public databases. Additionally, the use of PCR-based detection methods can be limited by the presence of phenolic compounds that act as PCR inhibitors in the reaction mix. The future role for the diagnostician in the detection of plant pathogens may be as simple as choosing the correct dry beads and adding an aliquot of infected sample and water; however, pathogens mutate and recombine their genomes, so there will always be new disease problems requiring basic research. When molecular testing kits become a standard part of pathogen detection, particularly from diagnostics companies offering a service, it is likely that validation will be required by the International Organization for Standardization (ISO) (Levesque, 2001).

LITERATURE CITED

Aaziz, R. and M. Tepfer. 1999. Recombination in RNA viruses and in virus-resistant transgenic plants. *J. Gen. Virol.* 80:1339–1346.

Bailey, A.M. et al. 2002. Identification to the species level of the plant pathogens *Phytophthora* and *Pythium* by using unique sequences of the ITS1 region of ribosomal DNA as capture probes for PCR ELISA. *FEMS Microbiol. Lett.* 207:153–158.

Ball, T.B., F.A. Plummer, and K.T. HayGlass. 2003. Improved mRNA quantitation in lightcycler RT-PCR. *Intl. Arch. Allergy Immunol.* 130:82–86.

Candresse, T. et al. 2000. PCR-based techniques for the detection of plant viruses and viroids. In: *Proc. Intl. Symp. Meth. and Markers Qual. Assur. Micro.* (Ed.) A.C. Cassells, B.M. Doyle, and R.F. Curry. *Acta Hortic.* 530:61–68.

Cassells, A.C. 1997. *Pathogen and Microbial Contamination Management in Micropropagation.* Kluwer Academic Publishers, Dordrecht, the Netherlands.

Cassells, A.C. 2000. Contamination detection and elimination. In: *Encyclopedia of Plant Cell Biology.* (Ed.) R.E. Spier. John Wiley & Sons, Chichester, UK. 577–586.

Clark, M. F. 1992. Immunodiagnostic techniques for mycoplasma-like organisms. In: *Techniques for the Rapid Detection of Plant Pathogens.* (Ed.) J.M. Duncan and L. Torrance. Blackwell Scientific Publications, Oxford, UK. 34–45.

Dijkstra, J. and C.P. de Jager. 1998. *Practical Plant Virology: Protocols and Exercises.* Springer-Verlag, Berlin.

Fox, R.T.N. 1993. *Principles of Diagnostic Technique in Plant Pathology.* CAB International, Wallingford, UK.

George, E. F. 1993. *Plant Propagation by Tissue Culture.* Exegetics, Basingstoke, UK.

Haughland, R.A., N. Brinkman, and S.J. Vesper. 2002. Evaluation of rapid DNA extraction methods for the quantitative detection of fungi using real-time PCR analysis. *J. Microbiol. Meth.* 50:319–323.

Hull, R. 2001. *Matthew's Plant Virology.* Academic Press, New York.

Karp, A., P.G. Isaac, and D.S. Ingram. 1998. *Molecular Tools for Screening Biodiversity.* Chapman & Hall, London.

Klerks, M. et al. 2001. Development of a multiplex AmpliDet RNA for the simultaneous detection of *Potato leafroll virus* and *Potato virus Y* in potato tubers. *J. Virol. Meth.* 93:115–125.

Lay, J. O. 2001. MALDI-TOF mass spectrometry of bacteria. *Mass Spectrom. Rev.* 20:172–194.

Lelliott, R.A. and D.E. Stead. 1987. *Methods for the Diagnosis of Bacterial Diseases of Plants.* Blackwell Scientific Publications, Oxford, UK.

Levesque, C.A. 2001. Molecular methods for the detection of plant pathogens: What is the future? *Can. J. Plant Pathol.* 24:333–336.

O'Herlihy, E.A. and A.C. Cassells. 2003. Influence of *in vitro* factors on titre and elimination of model fruit tree viruses. *Plant Cell Tiss. Org. Cult.* 72:33–42.

Schaad, N.W. and R.D. Frederick. 2002. Real-time PCR and its application for rapid plant disease diagnostics. *Can. J. Plant Pathol.* 24:250–258.

Stead, D.E. et al. 2000. Modern methods for characterizing, identifying and detecting bacteria associated with plants In: *Proc. Intl. Symp. Meth. Markers Qual. Assur in Micro.* (Ed.) A.C. Cassells, B.M. Doyle, and R.F. Curry. *Acta Hortic.* 530:45–60.

Tan, S.W.-L., S.-K. Wong, and R.M. Kini. 2000. Rapid simultaneous detection of two orchid viruses using LC- and/or MALDI-mass spectrometry. *J. Virol. Meth.* 85:93–99.

Van der Linde, P.C.G. 2000. Certified plants from tissue culture. In: *Proc. Intl. Symp. Meth. Markers Qual. Assur. Micro.* (Ed.) A.C. Cassells, B.M. Doyle, and R.F. Curry. *Acta Hortic.* 530:93–101.

29 Entrepreneurship for Biotechnology Ventures: From Bench to Bag

David W. Altman

CHAPTER 29 CONCEPTS

- Secure intellectual property rights through a license or an asset purchase.

- Establish standard protocols for data storage and retention, conflict of interest, record-keeping, notification of inventions, disclosure, and financial reporting.

- Develop a business plan, preferably with comprehensive strategic planning.

- Complete business formation documents and patent filings.

- Identify and gain commitments from a management team with diverse fields of expertise.

- Raise adequate capital from an appropriate source after considering all options in addition to venture capital.

- Maintain focus during start-up by solidifying intellectual property, demonstrating proof-of-concept with the science, and operating within fiscally prudent guidelines.

INTRODUCTION

Now that you have come up with the greatest innovation the plant tissue culture and biotechnology universe has ever seen, what can you do to make sure that it gets into the hands of real people? As with many questions, the answer depends on your perspective. Of course the scientist thinks that business aspects are relatively mundane and should be utilitarian and subordinate to the technology. Businessmen and businesswomen often want to box up scientists and to keep them in the laboratory and out of trouble. Finally, investors who come up with capital to drive development are a mixed bag, but generally assume the science is great and are looking for a star team that works well together with both business and science elements.

This chapter will provide an outline in a how-to format for use by any stakeholder. The intention is to describe entrepreneurship conditions and processes for a typical U.S.-based plant tissue culture program in a stepwise fashion. Many of the principles and components could be adapted, however, for other life sciences or broader international situations. The first section will deal with protocols and systems that must be put in place in any modern plant science laboratory even before the thought of a commercialization route occurs. These requirements should be taught to all students so that they will not face major problems due to ignorance of standard protocols. Such standards should be considered in the same light as laboratory safety procedures and other typical operational items. The sections that follow will be specific to the formation and initiation of a business:

- Planning
- Personnel
- Legal filings
- Capital
- Start-up

In these five sections we will assume that rights have been secured for any inventions in the new venture from an employer or from other owners of discoveries to be used in the business.

Before proceeding, we must emphasize the importance of intellectual property considerations that are unique to plant tissue culture and to most life sciences research programs. From a business perspective, the focus in these types of research companies would be described as dependent on intangible assets. There are major entrepreneurship differences as a consequence of this distinction, which makes these businesses very different from a manufacturing operation making widgets or a retail shop selling knickknacks. My experience is that this central element often is ignored and limits the potential of many good concepts. The other salient point is that the plant tissue culture practitioner does not really need a bread-and-butter type of business expertise for key aspects of putting the enterprise together because plant tissue culture, or life sciences generally, requires certain unique business skills, such as intellectual property management and financial guidance for extensive periods without revenues. Chances of success are improved with special attention to enlisting business expertise that includes experience in a life science field; sometimes there is no other viable option.

STANDARD PROTOCOLS AND SYSTEMS FOR RESEARCH GROUPS

Except for the lone scientist working in his garage or in similar circumstances, nearly everyone working in plant tissue culture belongs to either profit or nonprofit organizations. Each organization has sets of rules that govern aspects of research touching on commercial endeavors, even if the researcher has no apparent interest in such endeavors at a particular point in time. At the beginning of employment or of a short-term engagement, each individual should determine what procedures and policies are in place for commercialization activities in his or her organization. In particular, every scientist should familiarize himself or herself with requirements for conflict of interest, recordkeeping, data storage and retention, notification of inventions, and financial reporting.

While most corporate organizations pay attention to these activities, public institutions often either do not take a proactive role in disseminating these policies or have outdated or inaccurate information. For any institution with government funding, the U.S. has certain requirements as a result of the Bayh-Dole Act of 1980, and guidelines from agencies such as the National Institutes of Health (NIH), the National Science Foundation (NSF), the United States Department of Agriculture (USDA), and others. Any competent group or research leader should have a reasonable understanding of these federal guidelines and be able to provide minimal explanation of requirements as well as knowing where to go for more in-depth information when the need arises. These policies prevent individuals from engaging in activities that could get them into trouble when venturing into entrepreneurship.

Most business organizations would insist on compliance with established guidelines in order to put any agreement in place, such as a research agreement, a licensing agreement, or other types of support (see Harrison [1999] for a review of industry requirements for compliance from public institutions). Sometimes, public institutions allow financial support from businesses as a gift, which can greatly simplify some of the minimal elements for commercialization, but this usually provides a stopgap that should not obviate the need for best practices.

The next important concern is to protect any discovery or invention that might occur. This principle applies even if you want to give the technology away. For this information, companies have legal departments that provide the necessary guidance, and public institutions usually have a technology transfer office. If these sources are not helpful, you can get information from the U.S. Patent and Trademark Office (http://www.uspto.gov) or refer to references about patents and other forms of intellectual property (IP). Although patents are the most commonly recognized commercial IP assets, other forms of IP include trade secrets, copyrights, trademarks, plant variety protection (PVP) certification, and a special type of U.S. patent protection for plants called plant patents. Plant tissue culture can derive particular IP protection from PVPs or plant patents that generally depend on whether the plant species are vegetatively propagated or sexually propagated, and this protection can be a very valuable asset that would increase the long-term viability of any entrepreneurial venture.

In order to properly protect any future IP asset, the laboratory needs to have standardized procedures for collection of data. In most instances, you might not be able to use the actual date of discovery from the research unless these procedures are appropriately followed. The recording of data is particularly critical. Each laboratory notebook should be bound, because loose-leaf notebooks or unbound files are more difficult to verify. Pencils are not acceptable for recording purposes, so pens or other indelible markers are required. Corrections in bound laboratory notebooks should not obliterate the original entry and should be accompanied by a brief notation. Pages in the bound notebooks should be initialed, dated and countersigned by another individual. The notebooks are best numbered sequentially and maintained in a separate location when completed and after being copied; only the copies should remain in the laboratory for routine reference. There are other procedures for the use of databooks, including procedures for inserting computer printouts, photographs, and other materials, so researchers need to have training in how to record all data appropriately for the purpose of protecting the IP.

Procedures also need to be established for data exchange, public dissemination, disclosure to business groups, and similar aspects of modern tissue culture research. No comprehensive description is offered here, but several highlights will be emphasized. Assuming that you have taken steps to protect discoveries, that you are following the guidelines in place at your place of employment, and that you are adequately collecting and recording results, the next criterion is to establish rules of engagement outside of the laboratory group. Most employers, either public or private, require assignment of rights to discoveries to the employer, and all research managers need to establish documentation for the employees they supervise in order to verify such assignment.

The next item is to decide who can be contacted externally and what procedures are necessary for making contact with other commercial groups, sending off publications, or conducting other similar activities. In the private sector, business development personnel take care of managing these aspects, but public researchers usually must take the initiative to comply with procedures established within their institutions. The first consideration for third-party contacts would be conflict of interest. In brief, you cannot fairly split the IP and commercialization rights so that you do the same venture, or in some cases overlapping ventures, with the same discoveries. You also cannot give unfair advantages that are outside of the legitimate assigned rights to special friends or to other groups with whom you wish to curry favor.

After passing the conflict-of-interest test, documentation procedures are needed to go forward with contacts that require disclosure of discoveries. Many employers require a discovery disclosure form or notification. In this manner, the appropriate manager can ascertain if any IP protection is necessary in advance of disclosure or if the researcher can proceed without IP protection. At this point, in the case of a publication, an abstract submission, or other types of similar disclosures that might impact future commercialization, the procedure is straightforward: namely, send it off.

However, if you are going to exchange materials (DNA, probes, proteins, new chemical entities, cell culture lines, and so on), then a material transfer agreement (MTA) is required in advance. If the researcher is going to enter into discussions, negotiations, or similar activities for possible entrepreneurship, then a confidential disclosure agreement (CDA), which might also be called a nondisclosure agreement (NDA), would be required. The format for MTA, CDA, and NDA documents is usually standard for most employers and can be obtained from the offices previously indicated.

PLANNING

Hopefully, all of the preceding has not dulled your enthusiasm for sailing off into the world of entrepreneurial ventures. The prerequisites are daunting, but following a prudent set of procedures ahead of time will get the intrepid novice or the seasoned veteran off to a good start. Before getting overconfident, do not forget the obvious need to do sufficient planning. Although planning might seem simple, the business world has particular parameters for typical new venture development, usually condensed down into the process of strategic planning and creation of a business plan. While not always necessary, the discipline of strategic planning and writing a formal business plan improves the chances of success, and many investors will insist on such materials before consideration of any funding.

Planning for an entrepreneurial effort depends considerably on an assessment of your goal with the technology. If the goal is to find a partner with business experience, then planning might simply be a matter of researching who are the primary players in that area of technology. The Internet is a good starting point for gathering information, and for those demanding detail, there are usually industry reports and other assessments by specialized consulting groups. Obviously, many veterans of entrepreneurial ventures tend to speak favorably of the virtues of launching a start-up over the more conservative track of finding an experienced partner. Timmons (1999) points out that the U.S. is unique in creating a remarkable environment that fosters entrepreneurship and labels this activity "the great equalizer and mobilizer of opportunity." That being understood, the exact formula for success has been elusive and the option to get a partner rather than take the plunge should not be dismissed flippantly.

If you have resources to complete a comprehensive strategic plan, there are several authorities on the process and the anticipated components of the resulting documentation (see Thompson and Strickland, 2001). Usually, a strategic plan would provide an assessment of the current status, an evaluation of internal and external factors, consideration of viable options for the business opportunity, and a recommended strategy, with benefits and costs, for executing the most advantageous option. The first item would be to develop the mission and vision for a new venture, as well as the objectives of the business. External factors include an analysis of the market, customers, competition, and the unique unmet need that your innovation will address. The internal factors can address management, marketing and sales, operations and product development, and distribution and other services. Finally, based on these other components, careful financial projections and an ownership plan would be essential to delineate.

PERSONNEL

Most scientists consider the most essential element of a new venture to be the technology. However, the investment community is going to assume the technology is viable, at least prior to initiating full diligence, and will focus instead on the management team. The necessity for cohesiveness and experience as an attribute of the management team cannot be overemphasized.

This aspect of entrepreneurship means that commitments are necessary from individuals with widely different backgrounds. Typically, science, business, regulatory affairs, law, and other disciplines are not used to cross-fertilization, but this convergence is the desired outcome for a new life science venture. The founder or founders for a new venture can come from any of these backgrounds, and the task for the entrepreneur is to gather other individuals to round out the management group. Usually, any delay in putting the team together will compromise the success of a new venture.

New ventures also have growth stages, similar in many respects to those of other new things. Just as a baby might have different needs compared with the toddler, the teenager, the young adult, and the middle-aged individual, a new venture does not have the same management requirements as a mature company. Individuals who desire stability and security would not typically be well suited for new ventures. Preferably, experience is an important attribute, and because many new ventures fail for a variety of reasons, the litmus test should not necessarily be experience with a successful new venture.

Another option would be to seek help from specialists in assembling a management team. Certain consulting firms can provide temporary management for start-up ventures in the formative period. This method of finding appropriate personnel has an added advantage of making a transition to new management easier as the company grows and has a need for management appropriate to another development stage. Venture capitalists who specialize in the very early stages of projects can often be another source of personnel, although this option can have other risks by potentially creating a conflict of interest with the new recruits. Specifically, the founder might be questioning if team members provided by venture capitalists serve company interests over those of the investment group. Another possibility would be to utilize an executive search firm or other "headhunters." This option can be very beneficial for those new ventures that have resources to hire such agencies. Some firms might defer fees, accept some equity, or make other flexible arrangements that render their use more feasible.

There are several pitfalls that inventors should avoid when putting together a management group. Even if some of these cautions seem obvious, they must be mentioned, and will hopefully provide guidance for readers without extensive human resources experience. Friends and relatives should not be the first option for completing a management team; such selections will be viewed skeptically by investors and other stakeholders. Particularly in making critical decisions about personnel, the prudent entrepreneur should not cut corners in recruitment. Some pointers include the following:

- Avoid relying too much on the interview, and be sure to standardize your interview questions as is typically recommended by human resource specialists
- Check references carefully before making a decision
- Involve all of the founders and other members of the management group
- Make sure that a penchant for entrepreneurship and risk-taking is an important trait for selection
- Choose individuals who mesh with other team members
- Seek diversity of skills within the team

LEGAL FILINGS

When beginning a new venture, certain legal filings are necessary for protecting the innovation in a reasonable fashion. All new ventures must be registered to conduct business, so certain filings are compulsory for any type of new venture. With tissue culture and other life sciences, the IP component has been mentioned and will be addressed again below.

Business registration usually begins by conceiving a legally acceptable name for the business. Typically, each state has a department that registers corporations and determines if a selected name is acceptable. Although every state can feasibly provide this service, several states are considered preferred jurisdictions for registration. Since a new venture is not required to have its principal place of business in the state of registration, Delaware is a frequent choice; there are several others, such as Nevada, that serve well for life science ventures.

After securing a name, the next decision is to decide what is the best type of company corporation to select. This function does not always require a lawyer, and there are Internet companies that can help the entrepreneur complete registration for a minimal expense. However, usually the entrepreneur can benefit from sound legal counsel and tax or accounting expertise when making this decision. One important factor is to limit liabilities by establishing a business incorporation. Many new ventures become established either as a limited liability partnership, a C-corporation, or an S-corporation, rather than a single proprietorship or a partnership. Because most new ventures must seek outside funding, the entrepreneur should carefully consider a business incorporation that will be conducive to receiving funding in exchange for equity in the venture.

An absolutely essential item for most life science ventures is filing for IP rights. Most individuals must seek outside legal advice for this component of the business. However, for the entrepreneur on an insufficient budget, assuming that an employer or institution has not already started the process, there are several stopgap measures. Contrary to popular belief, a lawyer is not required to file a patent. The U.S. Patent and Trademark Office allows registered patent agents or inventors to file patents as well. Patent agents typically have lower fees than patent lawyers, but the adage about getting what you pay for should be considered. A full patent application is also not necessary because protection for at least one year in the U.S. can be obtained by filing a provisional patent that has a lower submission fee. In any case, doing nothing is the worst option and will be punished in the marketplace. Thus, minimal filings, although not the preferred method, should be provisional patents submitted by the inventor or inventors.

CAPITAL

Capital is an essential item for launching an entrepreneurial venture. This requirement of any start-up business can be one of the most baffling and difficult for the plant tissue culture practitioner who normally does not think about the funding requirements of commercialization. Funding procedures for entrepreneurship are not similar at all to fundraising activities such as grant writing. Many references about new ventures give an introduction to methods for raising capital, so various options will only be briefly outlined here.

The first option would be to look at internal sources for getting past the first steps in launching a new business. Most individuals can access some funds by leveraging their own personal credit capabilities. Such an option would include home equity lines of credit, credit card debt, personal loans, and other similar sources. Many start-up businesses have only needed to tap into these types of reserves in order to get beyond the initial expenses and begin operations. The advantage of using these resources is that the entrepreneur does not need to convince other parties of the viability of the business model.

Another option could be a secured loan from a financial institution such as a bank. Unfortunately, most banks are very conservative, and the usual mode of operation is that these institutions offer loans only when the loans really are not necessary. However, sometimes the entrepreneur might be able to offer reasonable security to collateralize a commercial loan or line of credit, so this option should not be dismissed as completely impossible.

The most likely scenario for obtaining capital would be to approach venture capital firms that specialize in high-risk new ventures. However, these firms only deal with high risks because of anticipated high returns. Therefore, such firms will demand a substantial return on their investment, and the plant tissue culture practitioner must understand that this requirement is normal with this source of capital. In this situation, an experienced business person has the best chance of obtaining advantageous terms for funding.

In addition, specifics of the deal are extremely important, as is the opportunity to obtain the required funds. The most important element is determination of the "pre-money" valuation of the business opportunity, which is the value that both parties agree to place on the entire enterprise prior to investment of venture capital. Following many of the suggestions in this set of requirements should result in a higher pre-money value because the savvy entrepreneur would have protected the IP, assembled a credible management team, and put together a professional business plan after thoughtful strategic planning, besides having a great innovation from the beginning.

There are other important details, but everything depends on obtaining a satisfactory pre-money valuation. These details include the form of investment vehicle, favorable board representation, antidilution protection, reasonable commitment for assistance with future funding rounds, and other aspects. This should be the responsibility of the business people in the management team, but every entrepreneur should try to keep abreast of the issues and to stay informed about the status of negotiations with the venture capital groups. Without a term sheet or other written commitment, all of these aspects are negotiable and should be carefully considered before reaching a final decision.

Finally, even when you have agreed upon a term sheet, this does not necessarily mean that the fundraising task has been completed. Nothing is final until closing papers are signed by all parties in a funding transaction. In particular, most term sheet agreements are contingent upon due diligence by investors. This is when every aspect of the business is examined, sometimes in excruciating detail. If the entrepreneur has completed extensive planning and accumulated the necessary documentation, due diligence will proceed relatively smoothly.

There are occasionally some other options for obtaining capital. A large, established company in the industry might be willing to buy your new venture or other aspects of the business that have been developed. Usually, this option is better pursued once the venture has been able to operate for a period of time, so that milestones have been achieved and the innovation has been validated. However, many companies have divisions that act like regular venture capital firms. The advantages include the ability of larger companies to bring experienced business development groups into the venture with a proven capability to take discoveries to the marketplace. The founders might still maintain involvement through a consulting contract, or other advisory roles.

START-UP

After all this anticipation and work, the novice entrepreneur might think that the actual start-up of operations represents the achievement of the primary goal in launching the new venture. Actually, in reality, this start-up phase represents only the beginning of a new phase for business development. Start-up companies should never forget the sense of urgency that resulted in their successful launch. Frequently, start-ups do not have an effective implementation plan, and this can lead to pitfalls as well.

The most common error is to forget the need for positioning the new venture for the next round of financing. Usually, a single round of financing is not sufficient to bring the new company to the point of becoming a viable enterprise. The risk is that the science might prevent the young company from staying focused on what it takes to ensure long-term success. For life sciences companies, progress must usually be made in the start-up phase, enhancing IP and demonstrating proof-of-concept with those basic innovations that provided the reason to commercialize.

Furthermore, there is no restriction keeping your planning and basic strategy from undergoing a mid-course correction. Often, the ability to evaluate progress and adapt to changing dynamics is the most important attribute of a start-up venture. In addition, all entrepreneurs should carefully monitor cash expenditures, usually represented by an assessment of burn rate.

With all due respect to the science, entrepreneurship is governed by the dictum, "Cash is king." With life science and tissue culture ventures, cash can be the king, the queen, the prime minister, and anything else. A frequent problem is to fall into the trap of perennially burning cash and avoiding marketing and getting your first customers. The glory days when venture money fed high burn rates in companies with vague revenue models are probably a thing of the past. While scientific progress is important, be sure someone is monitoring cash, aggressively seeking customers and revenues, and holding down excessive expenditures. Sometimes this requirement can come into conflict with long-term goals, but the company needs to survive in order to reach them.

Another pitfall comes from the nature of the funding that usually provides capital to fuel new tissue culture ventures. Venture funding usually means that equity is diluted for founders and start-up employees who have put their hearts and souls into creating the business. Repeated financing rounds can excessively dilute founder and employee equity holdings, so these infusions should be limited to avoid excessive dilution. One frequently neglected aspect is the importance of debt financing as an alternative to equity financing. Although the business might not have been able to raise debt financing for the initial funding, once the start-up phase has begun, the venture should start an aggressive, sustained effort to open lines of credit, to utilize chattel loans, to be granted favorable payment terms with vendors, and to take advantage of every opportunity to establish a sound credit rating. This requires good financial leadership from the company's CFO or another manager with this responsibility.

Operations in general require specialized attention because mistakes can be costly to correct at a later stage and, in some cases, might even jeopardize the future of the venture. Tissue culture ventures will probably rely on IP rights of some sort, and these rights are intangible assets that must be appropriately entered on the company books. Many plain-vanilla accountants and book-keepers are oblivious to this aspect of a life science business. Similarly, the management of companies that typically have few initial revenues is a skill infrequently acquired by the average MBA. Be sure to insist that these aspects are addressed early and appropriately.

CONCLUDING COMMENTS

Entrepreneurship can be a stimulating and rewarding culmination of scientific achievements for plant tissue culture. Although scientific publications and other such end products have great value and contribute to our advancement of knowledge, without finding a commercial outlet the discoveries from the laboratory might not be easily accessible to the general public. Business is a democratic method for disseminating inventions, at least in the U.S. and most of the other nations around the world.

The tissue culture practitioner brings a vital element to the commercialization effort and is indispensable to the development of a product. However, the preceding material leads to the obvious conclusion that some other team members and components are necessary for a successful commercialization effort. Each person should do soul-searching at the start to decide if he or she has the temperament and stamina to try the new venture approach, and there is nothing wrong with choosing the path of finding an established business to buy the discovery and do the heavy lifting of completing the development part of research and development.

Finally, the ultimate key to success in a life science venture will always be the quality of the people involved in the enterprise. A science-based endeavor cannot afford to forget about this overriding consideration. People in an entrepreneurial business need to be motivated and to work together well as a team. Cohesiveness of the team will be critical when the new business faces those inevitable periodic crises that characterize entrepreneurship. Find the best people you can, reward them, and provide as many incentives as you can. Cover various business aspects as well as needed scientific expertise when you assemble the team. Also, because the entrepreneur cannot be an expert in all fields required for any business, do not be reticent in seeking outside expertise or in finding other resources within reasonable budget parameters. In the end, these guidelines can improve your chances of success, and hopefully the element of serendipity will be more likely to find its place in your new commercial enterprise.

LITERATURE CITED

Harrison, C.H. 1999. Industry-sponsored academic research in the health sciences: Regulatory, policy and practical issues in contract negotiations. *J. Biolaw Bus.* 2:9–25.

Thompson, A.A. Jr. and A.J. Strickland. 2001. *Crafting and Executing Strategy.* Irwin/McGraw-Hill, New York.

Timmons, J.A. 1999. *New Venture Creation: Entrepreneurship for the 21st Century.* Irwin/McGraw-Hill, New York.

Index

A

ABA, *see* Abscisic acid
Abies normanniana, 90
Abscisic acid (ABA), 11, 27, 88, 90
    regeneration processes and, 94
    somatic embryo maturation and, 93
Acacia, 89
ACCELL technology, 247
Acclimatization, 151, 155
Acetocarmine stain, 47, 48, 230
Activated charcoal, 25
Activation tagging, 124
    mutagenesis scheme, 178
    system, 205
Adenine derivatives, 91
Adenine sulfate, 11
Adhesive agents, 41, 42
ADMs, *see* Adventitious shoot meristems
Adobe Photoshop™, 50
Adventitious rooting, 25, 167
Adventitious shoot
    formation, 173
    meristems (ADMs), 180
    organogenesis, 181
Aequoria victoria, 274
AFLP, *see* Amplified fragment length polymorphism
Agar, 27, 43
Agarose, 215
Agglutination, 220
AGPs, *see* Arabino-galactan proteins
Agricultural Biotechnology Stewardship Technical
    Committee, 256
Agriculture, no-till, 258
Agrobacterium
    -mediated transformation, 242, 265
    T-DNA, 123
    tumefaciens, 120, 222, 242
AHP, see *Arabidopsis* histidine phosphotransfer protein
AHPs, 173
Ajuga
    pyramidalis, 287
    reptans, 287, 288
Albinism, 227, 303
Albizzia, 93
Alcohol, 36
Aldehydes, 40
Alginate gel, 283
Allergenicity
    IgE-mediated, 260
    testing, 258
Allium cepa, 225
Alternaria

alternata, 298
solani, 219
Amino acid(s), 25
    analogs, 221
    embryo rescue and, 236
    novel conjugates of, 242
4-Amino-3,5,6-trichloro-2-pyridinecarboxylic acid, 26
Ammonium nitrate, 218
Amplified fragment length polymorphism (AFLP), 130,
    221, 304, 308
Amplified ribosomal DNA restriction (ARDRA), 328
Analysis of variance (ANOVA), 61, 62
    differences among treatments and, 66
    procedure, contrast statements and, 68
    summary tables, 63
Anatomy textbooks, 73
Ancymidol, 93
Androgenesis, 226
Angiosperms, organs of, 73
ANOVA, *see* Analysis of variance
Anther culture media, 228
Anthocyanins, 287
Antibiotic
    resistance, 13, 221, 273
    sprays, 149
Antigibberellins, 93
Antioxidants, 150
Antirrhinum, 176
Apical dominance, 147, 316
Apical meristems, 146
Arabidopsis, 104, 169, 174, 201
    ethyl methane sulfonate-mutagenized, 178
    histidine phosphotransfer protein (AHP), 178
    thaliana, 120
Arabino-galactan proteins (AGPs), 203
Arbitrary signatures from amplification profiles (ASAP),
    307
ARDRA, *see* Amplified ribosomal DNA restriction
ARRs, 173
Artificial chromosomes, 249
Artificial seed technology, 199
ASAP, *see* Arbitrary signatures from amplification profiles
Ascorbic acid, 150
Asparagus offcinalis, 295
Auxins, 26, 88, 150
    chlorinated, 90
    rooting compound, 89
    use of, 203
Avena sativa, 295
Avogadro's number of molecules, 28
Axillary shoot proliferation, 145, 152, 315

343